CAMBRIDGE LIBRARY COLLECTION
Books of enduring scholarly value

Polar Exploration

This series includes accounts, by eye-witnesses and contemporaries, of early expeditions to the Arctic and the Antarctic. Huge resources were invested in such endeavours, particularly the search for the North-West Passage, which, if successful, promised enormous strategic and commercial rewards. Cartographers and scientists travelled with many of the expeditions, and their work made important contributions to earth sciences, climatology, botany and zoology. They also brought back anthropological information about the indigenous peoples of the Arctic region and the southern fringes of the American continent. The series further includes dramatic and poignant accounts of the harsh realities of working in extreme conditions and utter isolation in bygone centuries.

The Atlantic

In 1871 the British government agreed to support an expedition to collect physical and chemical data and biological specimens from the world's oceans. Led by Charles Wyville Thomson (1830–82), the expedition used H.M.S. *Challenger*, refitted with laboratories. They sailed nearly 70,000 nautical miles around the world, took soundings and water samples at hundreds of stops along the way, and discovered more than 4,000 new marine species. Noted for the discovery of the Mid-Atlantic Ridge and the Pacific's deepest trench, the expedition laid the foundations for modern oceanography. This acclaimed two-volume account, first published in 1877, summarises the major discoveries for the Atlantic legs of this pioneering voyage. Illustrated with plates and woodcuts, Volume 1 describes the laboratories and equipment, the observations from Portsmouth via Tenerife to the Caribbean, and the detailed studies on the Gulf Stream.

Cambridge University Press has long been a pioneer in the reissuing of out-of-print titles from its own backlist, producing digital reprints of books that are still sought after by scholars and students but could not be reprinted economically using traditional technology. The Cambridge Library Collection extends this activity to a wider range of books which are still of importance to researchers and professionals, either for the source material they contain, or as landmarks in the history of their academic discipline.

Drawing from the world-renowned collections in the Cambridge University Library and other partner libraries, and guided by the advice of experts in each subject area, Cambridge University Press is using state-of-the-art scanning machines in its own Printing House to capture the content of each book selected for inclusion. The files are processed to give a consistently clear, crisp image, and the books finished to the high quality standard for which the Press is recognised around the world. The latest print-on-demand technology ensures that the books will remain available indefinitely, and that orders for single or multiple copies can quickly be supplied.

The Cambridge Library Collection brings back to life books of enduring scholarly value (including out-of-copyright works originally issued by other publishers) across a wide range of disciplines in the humanities and social sciences and in science and technology.

The Atlantic

A Preliminary Account of the General Results of the Exploring Voyage of H.M.S. Challenger During the Year 1873 and the Early Part of the Year 1876

VOLUME 1

CHARLES WYVILLE THOMSON

CAMBRIDGE
UNIVERSITY PRESS

University Printing House, Cambridge, CB2 8BS, United Kingdom

Cambridge University Press is part of the University of Cambridge.
It furthers the University's mission by disseminating knowledge in the pursuit of
education, learning and research at the highest international levels of excellence.

www.cambridge.org
Information on this title: www.cambridge.org/9781108074742

© in this compilation Cambridge University Press 2014

This edition first published 1877
This digitally printed version 2014

ISBN 978-1-108-07474-2 Paperback

This book reproduces the text of the original edition. The content and language reflect
the beliefs, practices and terminology of their time, and have not been updated.

Cambridge University Press wishes to make clear that the book, unless originally published
by Cambridge, is not being republished by, in association or collaboration with,
or with the endorsement or approval of, the original publisher or its successors in title.

The Voyage of the 'Challenger.'

THE ATLANTIC.

C Wyville Thomson

Engraved by C. H. Jeens, from a Photograph

London Published by Macmillan & C?

The Voyage of the 'Challenger.'

THE ATLANTIC

A PRELIMINARY ACCOUNT OF THE GENERAL RESULTS

OF

THE EXPLORING VOYAGE OF H.M.S. 'CHALLENGER'

DURING THE YEAR 1873

AND THE EARLY PART OF THE YEAR 1876

BY

SIR C. WYVILLE THOMSON

KNT., LL.D., D.Sc., F.R.SS.L. & E., F.L.S., F.G.S., ETC.

Regius Professor of Natural History in the University of Edinburgh,
And Director of the Civilian Scientific Staff of the 'Challenger' Exploring Expedition.

IN TWO VOLUMES.
VOLUME I.

Published by Authority of the Lords Commissioners of the Admiralty.

London:
MACMILLAN AND CO.
1877.

TO THE

RIGHT HONOURABLE G. J. GOSCHEN, M.P.,

UNDER WHOSE ADMINISTRATION,

AS FIRST LORD OF THE ADMIRALTY,

THE 'CHALLENGER' EXPEDITION WAS ORGANIZED,

𝕿𝖍𝖊𝖘𝖊 𝖁𝖔𝖑𝖚𝖒𝖊𝖘

ARE RESPECTFULLY DEDICATED,

AS A MARK OF WARM APPRECIATION OF THE EFFECTIVE INTEREST

WHICH HE HAS TAKEN IN THE ADVANCEMENT OF SCIENCE,

AND OF SCIENTIFIC INSTRUCTION IN THE NAVY.

EDINBURGH,
January 2nd, 1877.

PREFACE.

THESE two volumes consist chiefly of an abstract of the less technical portions of my own journal for the first year of the voyage of the 'Challenger' (1873), and the early part of the fourth year (1876), when she again traversed the Atlantic on her way home. With these I have incorporated some of the more novel results of the observations of my colleagues, and I have added, usually in the form of Appendices, such Meteorological Tables, Tables of depths and of sea-temperatures, and Tables of specific-gravity determinations, as shall be sufficient to place in the hands of our fellow-workers in the field of Physical Geography at all events an abstract of the data on which our conclusions and speculations are founded, and enable them to judge for themselves.

I had much hesitation in attempting to prepare these volumes for publication, at sea; for the last four years the duties of each day have nearly fully occupied our time: our minds have been continually distracted by new impressions and by the necessity of devoting undivided attention to new objects of interest; ship-life is generally unfavourable to steady work, and during a great part of the time the motion of the ship makes it impossible to have even the limited space at one's command in his cabin, littered with papers and journals and memoranda in the orderly confusion which is inseparable from comfortable literary work. Although there was a reference library on board the 'Challenger,' excellent under the circumstances, it was of course by no means complete enough to enable us to go with safety into matters of detail. Still, notwithstanding these difficulties, it seemed to me so desirable that a sketch of our proceedings and of the more important results of the Expedition should be made public at as early a date as possible, that I determined to make the attempt.

At first I hoped that it might have been possible to make the publication of the journal almost keep pace with the voyage, or that at all events the rough results of one year's work might have been published during the next, and accordingly a great part of the first volume was written and put into type during the year 1874. I found, however, that this was impracticable; the uncertainty of communication became greater as the voyage proceeded; and inevitable delays having occurred, further observations constantly made it necessary to modify previously formed opinions, or desirable to strengthen them by additional evidence.

We arrived in England on the 24th of May, 1876,

PREFACE. xi

with the rough journal of the cruise in a sense complete, but not in a form for publication. From the amount of. work thrown upon my hands in getting the collections into order, and having arrangements made for the working out of the scientific results, the difficulties in the way were even greater than they had been on ship-board; and now, nearly a year after our return, the first two volumes only are completed.

In a former volume[1] I have given a general history of deep-sea investigations. I have described with some minuteness the appliances and methods employed during the preliminary cruises of the 'Lightning' and 'Porcupine' in ascertaining the exact depth, in taking bottom and intermediate temperatures, and in determining the distribution of the deep-sea fauna by means of the dredge; and I have stated the general conclusions on various questions to which we were at that time led by the study and generalization of our results. Although the additional experience which we have now gained has caused me to alter my views on one or two important points, I adhere in the main to the opinions and statements contained in that book, and to save unnecessary repetition I shall regard 'The Depths of the Sea' as a general introduction to the series of volumes giving an outline of the 'Challenger' Circumnavigating Expedition.

This last undertaking was on a very different scale from the trial cruises of the gun-boats, and

[1] 'The Depths of the Sea.' Second Edition. Macmillan and Co.

the first chapter is occupied with the description of the much more complete arrangements for scientific work on board the 'Challenger.'

After the termination of the cruise of the 'Porcupine' I had a very strong conviction that the foraminifera of the genera *Globigerina*, *Orbulina*, and *Pulvinulina*, which are chiefly concerned in forming the modern chalk, lived on the bottom; our later observations have, however, satisfied me that they *never* live on the bottom; but that they inhabit the surface, and the water to a limited depth beneath it.

My general views with regard to ocean-circulation remain unaltered; I think, however, that we have now good reason to believe that the indraught of water at a low temperature into the Atlantic and Pacific—gulfs as we may almost call them,—from the Southern Sea, is to a great extent due to an excess of precipitation over evaporation in the 'water-hemisphere' and a corresponding excess of evaporation over precipitation in the 'land-hemisphere;' that in fact a part of the circuit of general ocean circulation passes through the atmosphere.

The Zoological descriptions and determinations both in 'The Depths of the Sea' and in this preliminary sketch of the proceedings of the 'Challenger' must be regarded as to a certain extent provisional; some of the more remarkable and novel forms have been selected for illustration to give a general idea of the *facies* of deep-sea life, but the amount of new material is very large, and it will take some years of the patient work of many hands,

PREFACE. xiii

with full access to libraries and collections, to bring it into order.

As in 'The Depths of the Sea,' the metrical system of measurement and the centigrade thermometer scale are employed in these volumes. In case the centigrade notation should not be familiar to all my readers, I introduce a comparative scale, embodying those of Fahrenheit, Celsius (Centigrade), and Réaumur, for reference.*

With few exceptions the illustrations are from drawings by Mr. J. J. Wild; and as in the former volume I have to give my best thanks to Mr. Cooper for the excellent manner in which Mr. Wild's beautiful drawings have been rendered on the wood.

After the first chapters of these volumes were written many changes took place in our staff. Captain Nares, F.R.S., was recalled at the end of the

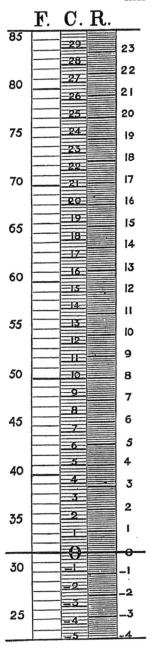

* 1° Fahrenheit = 0·55° C = 0·44° R.
  1° Centigrade = 0·80° R. = 1·80° F.
  1° Reaumur = 1·25° C. = 2·25° F.

second year, to take command of the Arctic Expedition. This was a heavy blow to the 'Challenger.' Captain Nares, from his experience as a surveying officer, was eminently fitted to direct the executive in such an undertaking, and the deep interest which he took in every branch of our investigations, and the intelligent knowledge which he possessed of their scope and objects, were the best possible guarantee for the various operations being thoroughly and conscientiously carried out. We were indeed most fortunate in the choice of a successor to Captain Nares, for Captain Frank Thomson, although identified with another branch of the service, and lying under the further disadvantage of wanting the experience in the new field of research which we had laboriously gained during the two previous years, showed himself in every way anxious to carry out the system initiated by Captain Nares, and to adopt any reasonable suggestion for the more complete performance of our task.

I think nearly all of us, naval and civilian, wildly volunteered to follow our old Captain—to the Pole or anywhere else. The services of one officer only, Lieutenant Pelham Aldrich, were accepted; and another most unwelcome blank was made in our circle. Lieutenant Bromley succeeded Aldrich as First Lieutenant, and we were again remarkably fortunate in Lieutenant Alfred Carpenter joining us from the 'Iron Duke' to fill the vacancy.

During the third year of the cruise the Civilian Staff lost one of its most valuable members by death.

Dr. Rudolf von Willemöes-Suhm, to whom the study of the annulosa had been more especially relegated, had an attack of erysipelas in the face on the passage from Hawaii to Tahiti; his health had previously suffered somewhat, probably from his never having got thoroughly used to a sea-life; the disease rapidly took a virulent form, and he sank into a state of coma and died on the 13th of September, 1875.

This sad event naturally threw a gloom over our little party. From the commencement of the voyage Dr. v. Willemöes-Suhm had devoted himself with unremitting industry and zeal to the objects of the expedition. He had already published in the Transactions of the Linnæan Society and elsewhere, a number of valuable papers in connection with our work. He leaves a fine series of drawings, with full descriptions, chiefly illustrating the development of surface Crustacea; for example, species of the genera *Euphausia*, *Sergestes*, and *Amphion* are traced through all their stages; he leaves also an ample official journal. The loss of his valuable aid in working up the final results of the expedition must, I fear, seriously affect their completeness. I regarded Rudolf v. Willemöes-Suhm as a young man of the highest promise, certain, had he lived, to have achieved a distinguished place in his profession, and I look upon his untimely death as a serious loss, not only to this expedition, in which he took so important a part, but also to the younger generation of scientific men, among whom he was steadily preparing himself to become a leader.

It would have been selfish to regret the departure of two of our younger messmates, Lieutenant Lord George Campbell and Lieutenant Andrew Balfour, who got their promotion at Valparaiso, and went home across the Andes.

Writing now after the commission has come to a close, I think I am justified in saying that the objects of the expedition have been fully and faithfully carried out. The instructions of the Lords Commissioners of the Admiralty, founded upon the recommendations of a Committee of the Royal Society, were followed so far as circumstances would permit. We always kept in view that to explore the conditions of the deep sea was the primary object of our mission, and throughout the voyage we took every possible opportunity of making a deep-sea observation. We dredged from time to time in shallow water in the most remote regions, and we have in this way acquired many undescribed animal forms; and collections of land animals and plants were likewise made on every available occasion; but I rather discouraged such work, which in our case could only be done imperfectly, if it seemed likely to divert our attention from our special object.

Between our departure from Sheerness on the 7th of December, 1872, and our arrival at Spithead on the 24th of May, 1876, we traversed a distance of 68,890 nautical miles, and at intervals as nearly uniform as possible, we established 362 observing stations.

At each of these stations the following observations were made, so far as circumstances would permit. The position of the station having been ascertained—

1. The exact depth was determined.
2. A sample of the bottom averaging from 1 oz. to 1 lb. in weight was recovered by means of the sounding instrument, which was provided with a tube and disengaging weights.
3. A sample of the bottom water was procured for physical and chemical examination.
4. The bottom temperature was determined by a registering thermometer.
5. At most stations a fair sample of the bottom fauna was procured by means of the dredge or trawl.
6. At most stations the fauna of the surface and of intermediate depths was examined by the use of the tow-net variously adjusted.
7. At most stations a series of temperature observations were made at different depths from the surface to the bottom.
8. At many stations samples of sea-water were obtained from different depths.
9. In all cases atmospheric and other meteorological conditions were carefully observed and noted.
10. The direction and rate of the surface current was determined.
11. At a few stations an attempt was made to ascertain the direction and rate of movement of water at different depths.

The somewhat critical experiment of associating a party of civilians, holding to a certain extent an independent position, with the naval staff of a man-of-war, has for once been successful. Captain Nares and Captain Thomson both fully recognized that the expedition was intended for scientific purposes, and I do not think that in one single case the operations of the combined scientific staff were hampered in the least by avoidable service routine. All the naval officers, without exception, assisted the civilian staff in every way in their power and in the most friendly spirit; if I wished anything done I had only to consider who was the man, naval or civilian, who was likely to do it best; and the consequence has been that with the entire sanction of Captain Nares and Captain Thomson, the parties sent to camp out or detailed for any special service have always been mixed, to the great advantage I believe of all concerned.

My thanks are perhaps more specially due to Commander Maclear and the First Lieutenant for the wonderful temper with which they tolerated all the irregularities, some of them very trying to the ministers of cleanliness and order, which were inseparable from the peculiar nature of our *imperium in imperio;* to Staff-Commander Tizard, the chief of the naval scientific staff, for the friendly readiness with which he has always assisted us, and placed his valuable observations and data at our disposal; and to Lieutenant Aldrich, Lieutenant Bromley, Lieutenant Bethell, and Lieutenant Carpenter, for the patience and care with which they superintended the dredging and trawling operations, and the determinations of ocean temperature, all of which fell into their immediate province. And here I must not omit to record my debt of gratitude to my friends the blue-jackets, who, greatly to their credit, treated us civilians throughout with as much respect and consideration as they did their own officers.

The members of the Civilian Scientific Staff under my direction have already in various ways given evidence of their industry; I think, however, I should not be doing my duty if I did not take this opportunity of recording my personal obligation to Mr. John Murray, who, while he has been bringing out highly important results by his own investigations, has undertaken the task of cataloguing and seeing to the security of the vast collections which have accumulated during the voyage.

The friendly hospitality and the ready assistance which we everywhere received, not only from the representatives of England and America, but also from those of foreign nations, added much to the pleasure of the cruise. Some few of these many benefits will be acknowledged in their places in the journal, but the names alone of the kind friends who have come forward to welcome and aid us would fill a goodly volume.

I cannot, however, close without recording, on my own part and on that of the Civilian Scientific Staff, our deep sense of the courteous consideration which we uniformly received from the Hydrographer to the Navy, the head of the department with which we were associated; and our hearty thanks to Mr.

Blakeney and the other officers of the department for the accuracy with which they kept us supplied with the latest scientific periodicals, and with such instruments and books of reference as we required.

This account may perhaps appear unnaturally *couleur de rose*, still, looking over it again, I can detect no exaggeration. There is of course another side to the picture. The work was done with the regulation expenditure of tissue; the strain both mental and physical was long and severe, and it has told a good deal upon all of us.

C. WYVILLE THOMSON.

EDINBURGH,
January 2nd, 1877.

CONTENTS.

CHAPTER I.

THE EQUIPMENT OF THE SHIP.

The Causes which led to the despatch of the 'Challenger' Expedition.—The Staff of Officers, Naval and Civilian.—The Special Arrangements for Scientific Work.—The Natural History Work-room.—The Chemical Laboratory.—The Apparatus for boiling out the Gases from Sea-water; for determining the Carbonic Acid; for Gas Analysis.—The 'Slip' Water-bottle.—'Buchanan's' Stop-cock Water-bottle.—The Hydraulic Pressure-gauge. — The 'Baillie' Sounding-machine. — The 'Valve' Sounding-machine.—Improvements in the Dredge and in the mode of handling it.—The Steam-pinnace ... ... *Page* 1

APPENDIX A.—Official Correspondence with reference to the 'Challenger' Expedition extracted from the Minutes of Council of the Royal Society ... ... *Page* 61

APPENDIX B.—List of the Stations in the Atlantic at which Observations were taken during the year 1873 ... ... *Page* 93

CHAPTER II.

FROM PORTSMOUTH TO TENERIFFE.

Departure from England.—Rough weather in the Channel.—Lisbon—Trawling in deep water.—Deep-water Fishes.—Surface animals.—Gibraltar. — *Cystosoma neptuni.* — Venus' Flower-basket. — *Naresia cyathus.* — The 'Clustered Sea-polype.'—Madeira.—Temperature Observations.—Meteorological Observations.—Teneriffe ... ... *Page* 107

xxii   CONTENTS.

APPENDIX A.—Particulars of Depth, Temperature, and Position at the Sounding Stations between Portsmouth and Teneriffe ; the Temperatures corrected for pressure... ... ... *Page* 162

APPENDIX B.—Comparative Table of the Indications of 'Stevenson's Mean Thermometers,' and the ordinary Maximum and Minimum Thermometers in Air, for the six months from the 1st May to the 31st October, 1873 ... ... *Page* 164

CHAPTER III.

TENERIFFE TO SOMBRERO.

First Deep-sea Section.—*Leiosoma limicolum.*— A Grove of Deep-sea Coral.—*Poliopogon amadou.* — Red Clay.— Phosphorescence.— Surface Fauna.– Blind Crustaceans.—Fishes' Nests.—The Paucity of the Higher Forms of Life.—Deep-sea Annelids.—The Structure and Mode of Formation of Globigerina Ooze.—The Habits of the Living *Globigerina.*—*Orbulina universa.*—*Pulvinulina.*—' Coccoliths ' and ' Rhabdoliths.'—The Origin and Extension of the 'Red Clay.'—Radiolarian Ooze.—The Use of the Tow-net —The Vertical Distribution of Temperature throughout the Section.—Specific Gravities ... ... *Page* 169

APPENDIX A.—Table of Temperatures observed between Teneriffe and Sombrero Island ... ... *Page* 248

APPENDIX B.— Table of Specific Gravities observed between Teneriffe and Sombrero Island... .. .. *Page* 251

CHAPTER IV.

ST. THOMAS TO BERMUDAS.

Dredging in moderate depths in the West Indian Seas.—New Blind Crustaceans.— Deep-sea Corals.—*Hyalonema toxeres.*—An accident.—A Deep sounding.—The Miller-Casella Thermometers.—Temperatures.—Arrival at Bermudas.—History of the Islands.—Their general Appearance.—'Red' and 'blue' Birds.—The Corals which form the Reefs.—The Geology of Bermudas.—General Nelson's description.—Æolian Rocks.—Calcareous concretions simulating Fossils.—The Topography of the Islands.—Their Products.—Their Climate.—Their Vegetation. *Page* 253

APPENDIX A.—Report from Professor Abel, F.R.S., to H. E. General Lefroy, C.B., F.R.S., on the Character and Composition of Samples of Soil from Bermudas ... ... *Page* 348

APPENDIX B.—Abstract of Temperature Observations taken at Bermudas from the Year 1855 to the Year 1873 ... ... *Page* 354

CHAPTER V.

THE GULF STREAM.

Departure from Bermudas.—Sounding and Dredging near the Islands.—*Madracis asperula.*—The determination of Surface and Deep Currents.—Difficulty and uncertainty of our present method of Observation.—The Current-drag.—Sounding in the Gulf Stream in rough weather.—The Temperature of the Stream.—*Aceste bellidifera.*—*Porcellanaster ceruleus.*—*Aërope rostrata.*—Dredged a huge Syenite Boulder.—Le Have Bank.—Mirage.— Halifax.— Ice-markings.— Recross the Gulf Stream.— General considerations.— Comparison between the Gulf Stream and the Japan Current.—*Calymne relicta.*—*Ophioglypha bullata.*—*Lefroyella decora.*—Return to Bermudas ... ... *Page* 358

APPENDIX A.—Table of Serial Temperature-Soundings taken between St. Thomas, Bermudas, and Halifax ... ... *Page* 405

APPENDIX B.—Table of the Bottom Temperatures taken between St. Thomas, Bermudas, and Halifax ... ... *Page* 408

APPENDIX C.—Specific Gravity Observations taken between St. Thomas, Bermudas, and Halifax ... ... *Page* 409

APPENDIX D.—Table of Meteorological Observations made in crossing and recrossing the Gulf Stream ... ... *Page* 410

LIST OF ILLUSTRATIONS.

WOODCUTS.

FIG.		PAGE
1.	THE NATURAL HISTORY WORK-ROOM	12
2.	THE CHEMICAL LABORATORY	19
3.	APPARATUS FOR COLLECTING THE ATMOSPHERIC GASES FROM SEA-WATER	23
4.	THE CARBONIC-ACID APPARATUS	25
5.	THE GAS-ANALYSIS APPARATUS	28
6. 7.	ARRANGEMENT OF THE CAPILLARY PORTION OF THE GAS-ANALYSIS APPARATUS	29
8.	GAS-GENERATING APPARATUS	32
9.	SEA-GOING SAND-BATH	33
10.	THE 'SLIP WATER-BOTTLE'	35
11.	INSTRUMENT FOR SLIPPING THE CYLINDER AT INTERMEDIATE DEPTHS	36
12.	BUCHANAN'S 'STOPCOCK WATER-BOTTLE'	38
13.	THE HYDRAULIC PUMP	45
14.	THE BAILLIE SOUNDING-MACHINE	47
15.	THE 'VALVE' SOUNDING-LEAD (IN SECTION)	49
16.	THE DREDGE	50
17.	THE DREDGING AND SOUNDING ARRANGEMENTS ON BOARD THE 'CHALLENGER'	53
18.	THE DEEP-SEA TRAWL	56
19.	BELEM CASTLE, LISBON	109
20.	THE PORCH OF SANTA MARIA, BELEM	110
21.	QUADRANGLE OF THE MONASTERY OF SANTA MARIA, BELEM	112
22.	CLOISTERS OF THE MONASTERY OF SANTA MARIA, BELEM	113
23.	CORYPHÆNOIDES SERRATUS, Lowe. Half the natural size (No. 4)	118

LIST OF ILLUSTRATIONS.

FIG.		PAGE
24.	CARINARIA ATLANTICA. Natural size. Surface. (No. 4)	121
25.	CLIO PYRAMIDATA, Browne. Slightly enlarged. Surface. (No. 4)	124
26.	TRIPTERA COLUMELLA. Twice the natural size. (No. 4)	125
27.	CYSTOSOMA NEPTUNI, Guerin-Méneville. Slightly reduced. (No. 107)	130
28.	EUPLECTELLA ASPERGILLUM, Owen. Half the natural size.	134
29.	EUPLECTELLA SUBEREA (sp. n.). Reduced one-third. (No. 5)	139
30.	NARESIA CYATHUS (sp. n.). Slightly enlarged. (No. 6)	143
31.	SALENIA VARISPINA, A. Ag. Four times the natural size. (No. 6)	145
32.	SALENIA VARISPINA, A. Ag. Showing the structure of the apical disk	146
33.	PHORMOSOMA URANUS (sp. n.). Natural size. (No. 6)	146
34.	PHORMOSOMA URANUS (sp. n.). Apical surface. Natural size. (No. 6)	147
35.	PHORMOSOMA HOPLACANTHA (sp. n.). Southern Sea between Australia and New Zealand. Portion of the ventral surface of the test. Reduced one-third	148
36.	UMBELLULARIA GRŒNLANDICA, L. Natural size. (No. 7)	150
37.	DIAGRAM SHOWING THE DIRECTION AND FORCE OF THE WIND (DIRECTION SOUTH-WEST BY COMPASS; FORCE=7 BY BEAUFORT'S SCALE) AT MIDNIGHT, JANUARY 1st, 1873	160
38.	POLIOPOGON AMADOU (sp. n.). One-third the natural size. (No. 3)	175
39.	ARCA (sp.) (No. 5)	179
40.	LIMOPSIS (sp.) (No. 5)	179
41.	LEDA (sp.) (No. 5)	180
42.	WILLEMŒSIA LEPTODACTYLA, v. Willemœs-Suhm. Natural size. (No. 13)	189
43.	SOLARIUM (sp.) Greatly enlarged. (No. 13)	191
44.	ANTENNARIUS MARMORATUS. Natural size. From the surface	195
45.	AVICULA (sp.) Greatly enlarged. (No. 16)	195
46.	GLOBIGERINA BULLOIDES, from the surface	211
47.	ORBULINA UNIVERSA, D'Orbigny. From the surface. Fifty times the natural size	214
48.	PULVINULINA MENARDII, D'Orbigny, a, the upper, b, the under surface. Thirty times the natural size. Dead shells from the bottom, at a depth of 1,900 fathoms	218
49.	A 'RHABDOSPHERE.' From the surface. Five hundred times the natural size	221
50.	A 'RHABDOSPHERE.' From the surface. Two thousand times the natural size	222

LIST OF ILLUSTRATIONS.

FIG.		PAGE
51.	CALCAROMMA CALCAREA (sp. n.). With the pseudopodia contracted. From the surface. Two hundred times the natural size ...	233
52.	DICTYOPODIUM (sp. n.). From the surface. Two hundred times the natural size ...	234
53.	XIPHACANTHA (sp. n.). From the surface. One hundred times the natural size. The skeleton only...	235
54.	HALIOMMA (sp. n.). From the surface. Two hundred times the natural size ...	236
55.	CURVES CONSTRUCTED FROM SERIAL SOUNDINGS BETWEEN TENERIFFE AND SOMBRERO ...	238
56.	CURVES CONSTRUCTED FROM SERIAL AND BOTTOM SOUNDINGS. A, in the Bay of Biscay. B, off the coast of Portugal; and C and D, between Teneriffe and Sombrero. ...	241
57.	DIAGRAM CONSTRUCTED FROM SERIAL SOUNDING No. 10...	243
58.	DIAGRAM CONSTRUCTED FROM SERIAL SOUNDING No. 1 ...	243
59.	WILLEMŒSIA CRUCIFER, v. W-S. × 2. (No. 23)...	257
60.	ASTACUS ZALEUCUS, v. W-S. (No. 23)...	260
61.	CHÆTODERMA NITIDULUM, Lovén. (No. 24) ...	264
62.	TROCHOCYATHUS CORONATUS, Pourtales, × 2. (No. 24) ...	266
63.	DELTOCYATHUS AGASSIZII, Pourtales, × 4. (No. 24)...	269
64.	DELTOCYATHUS AGASSIZII, Pourtales. Stellate variety. From a depth of 200 fathoms, near Bermudas ...	271
65.	CRYPTOHELIA PUDICA, Milne-Edwards. Twice the natural size. (No. 24) ...	272
66.	HYALONEMA TOXERES (sp. n.). Upper surface, natural size. (No. 24) ...	274
67.	HYALONEMA TOXERES (sp. n.). Lower surface of the sponge, natural size. (No. 24) ...	275
68.	HYALONEMA TOXERES (sp. n.). Part of the membrane from the upper surface, × 40. (No. 24) ...	277
69.	HYALONEMA TOXERES (sp. n.). A young specimen, × 2. (No. 24)	278
70.	THERMOMETER TUBES BROKEN BY PRESSURE AT STATION 25. A, Thermometer No. 39; B, No. 42 ...	282
71.	CURVES CONSTRUCTED FROM SERIAL TEMPERATURE SOUNDINGS BETWEEN ST. THOMAS AND BERMUDAS...	286
72.	GROUP OF GRU-GRU PALMS ON THE CROQUET-LAWN, MOUNT LANGTON ...	301
73.	STRATIFIED 'ÆOLIAN' ROCKS, BERMUDAS ...	309
74.	'SAND-GLACIER' OVERWHELMING A GARDEN, ELBOW BAY, BERMUDAS. (From a Photograph.)...	310
75.	CHIMNEY OF A COTTAGE WHICH HAS BEEN BURIED BY A SAND-GLACIER, ELBOW BAY. (From a Photograph.) ...	312

xxviii LIST OF ILLUSTRATIONS.

FIG.		PAGE
76.	'ÆOLIAN' LIMESTONE BEDS IN PROCESS OF FORMATION, SHOWING STRATIFICATION, AND THE REMAINS OF A GROVE OF CEDARS WHICH HAS BEEN OVERWHELMED, ELBOW BAY, BERMUDAS. (From a Photograph)	313
77.	SECTION EXPOSED IN EXCAVATING THE BED FOR THE FLOATING DOCK, IRELAND ISLAND, BERMUDAS	319
78.	ENTRANCE TO THE 'CONVOLVULUS CAVE,' WALSINGHAM, BERMUDAS. (From a Photograph)	325
79.	CALCAREOUS CONCRETION SIMULATING A FOSSIL PALM-STEM, BOAZ ISLAND, BERMUDAS	330
80.	CALCAREOUS CONCRETION IN ÆOLIAN LIMESTONE, BERMUDAS	331
81.	CALCAREOUS CONCRETION IN ÆOLIAN LIMESTONE, BERMUDAS	332
82.	CALCAREOUS CONCRETION, BERMUDAS	332
83.	CALCAREOUS CONCRETIONS IN ÆOLIAN ROCK, BERMUDAS	333
84.	CONCRETIONS IN ÆOLIAN ROCK, BERMUDAS	334
85.	CEDAR AVENUE, HAMILTON, BERMUDAS. (From a Photograph)	338
86.	NATURAL SWAMP-VEGETATION, BERMUDAS. (From a Photograph)	341
87.	PAPAW-TREES (*Carica papaya*), IN THE GARDEN AT CLARENCE HILL. (From a Photograph)	344
88.	MADRACIS ASPERULA	360
89.	THE 'CURRENT-DRAG' AND 'WATCH-BUOY'	363
90.	RESULT OF CURRENT OBSERVATIONS MADE ON THE 24TH OF APRIL REDUCED TO A DIAGRAMMATIC FORM	365
91.	DIAGRAM SHOWING THE RELATION BETWEEN DEPTH AND TEMPERATURE AT STATIONS 41 AND 42	372
92.	DIAGRAM SHOWING THE RELATION BETWEEN DEPTH AND TEMPERATURE AT STATION 43	372
93.	CURVES CONSTRUCTED FROM SERIAL TEMPERATURE-SOUNDINGS BETWEEN BERMUDAS AND SANDY HOOK	374
94.	DIAGRAM SHOWING THE RELATION BETWEEN DEPTH AND TEMPERATURE AT STATION 44	375
95.	ACESTE BELLIDIFERA (sp. n.). A, Upper surface; B, Under surface. Twice the natural size. (No. 44)	376
96.	ACESTE BELLIDIFERA (sp. n.). Inner surface of the test. Twice the natural size. (No. 44)	377
97.	PORCELLANASTER CERULEUS (sp. n.). Oral surface. Natural size. (No. 45)	379
98.	PORCELLANASTER CERULEUS (sp. n.). Dorsal surface. Natural size. (No. 45)	380
99.	AEROPE ROSTRATA (sp. n.). Four times the natural size. (No. 45)	381
100.	CURVES CONSTRUCTED FROM TEMPERATURE-SOUNDINGS IN THE NORTH ATLANTIC	392

LIST OF ILLUSTRATIONS.

FIG.		PAGE
101.	CURVES CONSTRUCTED FROM TEMPERATURE-SOUNDINGS IN THE NORTH PACIFIC	394
102.	CALYMNE RELICTA (sp. n.). A, oral ; B, apical aspect, slightly enlarged. (No. 54)	397
103.	CALYMNE RELICTA (sp. n.). Oral and apical aspects of the denuded test	398
104.	OPHIOGLYPHA BULLATA (sp. n.). Dorsal aspect. Three times the natural size. (No. 54)	400
105.	OPHIOGLYPHA BULLATA (sp. n.). Oral aspect. Five times the natural size. (No. 54.)	402
106.	LEFROYELLA DECORA (sp. n.). Natural size. (No. 56)	403

VIGNETTES.

H.M.S. 'CHALLENGER'	60
GIBRALTAR, FROM THE SEA	161
ÆOLIAN ROCKS, BERMUDAS	347
SAN VICENTE	404

EXPLANATION OF THE PLATES.

PLATE	To face page
I.—THE TRACK OF THE SHIP FROM PORTSMOUTH TO TENERIFFE	116
II.—METEOROLOGICAL OBSERVATIONS FOR THE MONTH OF DECEMBER, 1872	129
III.—METEOROLOGICAL OBSERVATIONS FOR THE MONTH OF JANUARY, 1873	136
IV.—THE TRACK OF THE SHIP FROM TENERIFFE TO SOMBRERO, SHOWING THE DATE, THE POSITION, THE DEPTH, AND THE NATURE OF THE BOTTOM AT EACH OBSERVING STATION..	170
V.—DIAGRAM OF THE VERTICAL DISTRIBUTION OF TEMPERATURE BETWEEN TENERIFFE AND SOMBRERO	176
VI.—METEOROLOGICAL OBSERVATIONS FOR THE MONTH OF FEBRUARY, 1873	180
VII.—METEOROLOGICAL OBSERVATIONS FOR THE MONTH OF MARCH, 1873	200
VIII.—THE TRACK OF THE SHIP BETWEEN ST. THOMAS, BERMUDAS, AND HALIFAX	254
IX.—DIAGRAM OF THE VERTICAL DISTRIBUTION OF TEMPERATURE BETWEEN ST. THOMAS AND HALIFAX	272
X.—CHART OF BERMUDAS	290
XI.—DIAGRAM OF THE VERTICAL DISTRIBUTION OF TEMPERATURE BETWEEN BERMUDAS AND SANDY HOOK	360
XII.—DIAGRAM SHOWING THE SURFACE TEMPERATURES OBSERVED IN CROSSING AND RECROSSING THE GULF STREAM	365
XIII.—METEOROLOGICAL OBSERVATIONS FOR THE MONTH OF APRIL, 1873	366
XIV.—METEOROLOGICAL OBSERVATIONS FOR THE MONTH OF MAY, 1873	370

THE ATLANTIC.

THE ATLANTIC.

CHAPTER I.

THE EQUIPMENT OF THE SHIP.

The Causes which led to the despatch of the 'Challenger' Expedition.—The Staff of Officers, Naval and Civilian.—The Special Arrangements for Scientific Work.—The Natural History Work-room.—The Chemical Laboratory.—The Apparatus for boiling out the Gases from Sea-water; for determining the Carbonic Acid; for Gas Analysis.—The 'Slip' Water-bottle.—'Buchanan's' Stopcock Water-bottle.—The Hydraulic Pressure-gauge.—The 'Baillie' Sounding-machine.—The 'Valve' Sounding-machine.—Improvements in the Dredge and in the mode of handling it.—The Steam-pinnace.

APPENDIX A.—Official Correspondence with reference to the 'Challenger' Expedition, extracted from the Minutes of Council of the Royal Society.

APPENDIX B.—List of the Stations in the Atlantic at which Observations were taken during the year 1873.

It may perhaps be well, before going into the story of our own experiences, to sketch in a few words the train of circumstances which led to the despatch of H.M.S. 'Challenger' on a voyage of scientific research and discovery "round the world." The wonderful project of establishing a telegraphic communication between the old world and the new directed the attention of practical men to a region, about which up to that time but little had been known with certainty, and about which there had been a great deal of hazy misconception—the bottom of the deep sea. To procure information sufficient to enable them to prepare for the laying of a telegraph cable, sounding expeditions were organized, by both of the Governments specially interested, across the Atlantic Ocean. Ingenious contrivances were suggested and applied, not merely for determining the exact depth, but for bringing up samples of the bottom sufficient to test the composition and character of the deposits in process of formation on the sea-bed.

In the meantime another class of students, working for the increase of knowledge, though perhaps with less immediate bearing upon the progress of the human race or the advance of their own interests, had been investigating the forms and natures of living things, the external conditions upon which their frail lives depend, and the laws under which they are localised or distributed upon the surface of the earth; and, judging from the scanty data laboriously accumulated with the imperfect appliances at that time at their disposal, had come to the conclusion that life at the bottom of the sea was confined to a narrow border round the land; that at a depth of 100 fathoms plants almost entirely disappeared and animals were scarce, and represented those animal groups only which are among the most simple in their organization; while at 300 fathoms the sea-bottom became a desolate waste, the physical conditions being such as to preclude the possibility of the existence of living beings.

Samples of the sea-bottom, procured with great difficulty and in small quantities from the first deep soundings of the Atlantic, chiefly by the use of Brooke's sounding-machine, an instrument which by a neat contrivance disengaged its weights when it reached the bottom, and thus allowed a tube, so arranged as to get filled with a sample of the bottom, to be recovered by the sounding-line, were eagerly examined by microscopists; and the singular fact was at length established that these samples consisted, over a large part of the bed of the Atlantic, of the entire or broken shells of certain foraminifera, and the bulk of the evidence seemed to be in favour of the animals which inhabited these shells having lived in the situation in which they were found, and not, as was at first supposed, having lived in the sunshine on the surface of the sea, and having gradually sunk into the abysses after death. Dr. Wallich, the naturalist to the 'Bulldog' Sounding Expedition under Sir Leopold M'Clintock, reported that star-fishes with their stomachs full of the deep-sea foraminifera, had come up from a depth of 1,200 fathoms on a sounding-line, and doubts began to be entertained whether the bottom of the sea was in truth the desert which we had hitherto supposed it to be, or whether it might not prove a new zoological region open to investigation and discovery, and peopled by peculiar faunæ suited to its most peculiar conditions.

This new view however progressed but slowly, for it was almost as difficult to believe that creatures comparable with those of which we have experience in the upper world could live at the bottom of the deep sea, as that they could live in a vacuum, or in the fire. Of many of the conditions at great depths we as yet knew nothing, but some of them were as easily determined by calculation as by direct experiment, and we knew that an animal at a depth of 1,000 fathoms must bear a weight of a ton on the square inch, and one at a depth of 3,000 the almost inconceivable weight of three tons; and we had every reason to believe that the sun's light is almost entirely cut off at a depth of 50 fathoms, and that therefore the existence of plants upon which animals primarily depend for their food is impossible at great depths. These considerations alone seemed almost sufficient to place this question beyond the region of reasonable inquiry, and it was not until a considerable amount of evidence had been brought forward, that what was called the 'antibiotic' prejudice was in any degree overcome.

About this time, another class of facts which gave the whole subject a singular interest, were forcing themselves upon the attention of naturalists. Some dredgers, particularly our indefatigable brother-naturalists of Scandinavia, pushed their dredging operations to the utmost limit practicable in the northern seas by ordinary means, to depths of from 300 to 400 fathoms, and they found, contrary to the general impression of the British school, that at these depths there was no lack of animal life, and that further, many of the animal forms were new and unfamiliar, while many showed a much closer relation to the inhabitants of the seas of former geological periods than to the marine fauna of the present day.

In the year 1868, when the question was thus undecided, Dr. Carpenter and I, looking at the matter chiefly as one of scientific interest, but at the same time fully recognising the practical importance of many of the results of such an investigation, induced the Council of the Royal Society to apply to the Admiralty to place means at our disposal to go into the whole question of the physical and biological conditions of the sea-bottom, in the neighbourhood of the British Islands but beyond the range of ordinary boat work. The Admiralty assented, and, in the autumn of 1868, through about two months of wretched weather, we knocked about in the 'Lightning,' a somewhat precarious little gun-boat, between Scotland and Fäeroe.

Nine tolerable days fortunately checkered the uniformity of the heavy weather, and on these we registered some remarkable results.

We found that there was abundance of animal life at the bottom of the sea to a depth of 600 fathoms at least, and that the life there was not confined to the more simply organized animals, but extended very irrespectively through all the invertebrate classes, and even included some true bony fishes. We found that the general character of the fauna at these depths was not such as to indicate a mere gradual disappearance of the known fauna of the British ocean, but was in many respects peculiar, and presented many characters in common with the faunæ of older times. We found that, instead of having a constant and universal temperature of 4° C.

beyond a certain depth, as had previously been very generally supposed to be the case, the sea was warm or cold at all depths, according to the source from which each particular layer or current of water was derived, and that in accordance with this arrangement we might have two regions separated from one another by an invisible and impalpable boundary of liquid contact, differing widely in climatal conditions, and showing all the consequent wide differences in faunæ; we found that from the surface to the bottom the water of the sea contained organic matter in solution, or in suspension, and that therefore the Protozoa, which appear to pave the floor of most parts of the sea in a continuous sheet, derived by surface absorption the soft jelly of their bodies with the same ease and from the same source as they derive the carbonate of lime and the silica of their outer casings.

These results and many others were attained or suggested by our first season's very imperfect work, and they were regarded as so interesting and suggestive that with even greater willingness than before the Admiralty placed a gun-boat at the disposal of a committee, consisting of Dr. Carpenter, F.R.S., Mr. Gwyn Jeffreys, F.R.S., and myself, for the two succeeding summers, during which time one or more of us prosecuted the same line of inquiry, confirmed the result of the previous years, and added many new facts. The 'Porcupine,' which we used in 1869 and 1870, was much better suited in every way to our purposes than the 'Lightning.' The weather was more favourable, and we succeeded in dredging to the depth of 2,435 fathoms, and establishing the fact that even at that depth the invertebrate subkingdoms are still fairly represented.

Another great advance was made at this time. The registering thermometers which we used in the 'Lightning' gave uncertain indications, and on submitting them to experiment in a hydraulic press, it was found that their error depended upon their bulbs being irregularly compressed by the enormous pressure to which they were subjected, the fluid being thus forced up mechanically in the tube and giving an indication higher than that due to heat; the amount of excess depending upon the thickness and uniformity of the glass of the bulb. Prior to our second expedition the late Professor W. A. Miller had adopted a plan of defending the bulb from external pressure by inclosing it in an outer shell of glass, with alcohol and a bell of vapour between. With this construction the outer wall bears the whole of the pressure, compressing in turn the included bell of vapour and relieving entirely the inclosed thermometer bulb. By the use of this 'Miller-Casella' modification of Six's registering thermometer our deep-sea temperature determinations have been rendered fairly trustworthy.

Public interest was now fairly aroused in the new field of research. The rapid development of ocean telegraphic communication made all these results which affected telegraphy in any way—the precise depth, the nature and composition of the bottom, the presence or absence of animals capable of making inroads into hemp or gutta-percha, the temperature of the water through which telegraph cables might have to pass—of the highest practical value; while the novelty and peculiarity of many of the observations awakened a wide-spread curiosity and interest in even the purely scientific bearings of the inquiry.

Our cousins across the Atlantic had been working along with us *pari passu*, and ere long several of the European States sent out deep-sea expeditions more or less effective. None of these were attended with any great amount of success, and it seemed evident that England must give, at all events the first broad outline of the physical conditions of the bed of the ocean. How this was best to be done was a matter of the most serious consideration and frequent consultation among those to whom the earlier stages of the inquiry had been intrusted; and finally Dr. Carpenter addressed a letter to the First Lord of the Admiralty, urging the despatch of a circumnavigating Expedition, thoroughly equipped and with a competent scientific staff, to traverse the great ocean basins and prepare sections showing their physical and biological conditions, along certain lines. Dr. Carpenter's letter was referred in due course to the Hydrographer to the Navy, who at once threw himself cordially into the project and prepared a report, which resulted in the Lords of the Admiralty agreeing to the despatch of such an Expedition if the Royal Society recommended it, and provided them with a feasible scheme. A committee was appointed by the Royal Society, and a comprehensive scheme was drawn up.

The sagacious minister who at that time held the purse-strings regarded this as an important matter beyond the reach of private enterprise, and it was with the cordial assent of the House of Commons that Mr.

Lowe agreed to defray from the public purse what additional funds might be required to equip a Surveying ship in commission with all the necessary appliances for scientific research, and to associate with her complement of scientific officers a civilian staff of specialists in departments which do not come within the scope of the ordinary work of naval surveyors.

The Committee of the Royal Society, with Admiral Richards as one of its most influential members, met from time to time and offered practical suggestions. The 'Challenger,' a spar-decked corvette of 2,306 tons with auxiliary steam to 1,234 horse-power, and usually mounting eighteen 68-pounders, was chosen for the service; and Captain Nares, a surveying officer of great experience and singularly well suited in every way for such a post, was selected to take command. When it was suggested to me at the commencement of the negotiations to join the Expedition as Director of the civilian scientific staff the sacrifice appeared in every way too great; but as the various arrangements progressed, so many friendly plans were proposed on all hands to smooth away every difficulty, that I finally accepted a post which to a younger naturalist without the ties of a family and a responsible home appointment would be perhaps among the most delightful the world could offer.

A staff of officers, many of them surveyors and already distinguished by their knowledge of various branches of practical science, were carefully selected from a large number of volunteers, and a civilian scientific staff, consisting of a director, his secretary, three naturalists, and a chemist, were appointed on the recommendation of the Committee of the Royal Society. I extract the list of officers from the Navy List for April, 1873 :—

Captain	George S. Nares.
Commander	J. F. L. P. Maclear.
Lieutenant	Pelham Aldrich.
,,	Arthur C. B. Bromley.
,,	George R. Bethell.
Navigating Lieutenant	Thomas H. Tizard.
Paymaster	Richard R. A. Richards.
Surgeon	Alexander Crosbie.
Assistant-Paymaster	John Hynes.
Chief Engineer	James H. Fergusson.
Sub-Lieutenant	Henry C. Sloggett.
,, ,,	Lord George G. Campbell.
,, ,,	Andrew F. Balfour.
,, ,,	Arthur Channer.
Navigating Sub-Lieutenant	Arthur Havergall.
,, ,,	Herbert Swire.
Assistant-Surgeon	George Maclean, M.A., M.B.
Engineer	William J. J. Spry.
,,	Alfred J. Allen.
Boatswain, Second Class	Richard Cox.
Carpenter, Second Class	Frederick W. Westford.
Assistant Engineer, Second Class	William A. Howlett.
,, ,,	William J. Abbott.

CIVILIAN SCIENTIFIC STAFF.

Professor C. Wyville Thomson, F.R.S.
J. Y. Buchanan, M.A.
H. N. Moseley, M.A.
John Murray, Esq.
Dr. von Willemoës-Suhm.
J. J. Wild, Esq.

Only one change has taken place during the first year; Sub-Lieut. Sloggett has returned to England

and has been replaced by Sub-Lieut. Henry C. E. Harston.

The particular build of the 'Challenger' gives her an immense advantage for her present purpose, as she has all the accommodation of a frigate with the handiness and draught of water of a corvette. Sixteen of the eighteen large guns have been removed, and the main-deck is almost entirely set aside for the scientific work. The after-cabin is divided into two by a bulkhead, and the two very comfortable little rooms which are thus formed are occupied by Captain Nares and myself. The fore-cabin, a handsome room 30 feet long by about 12 wide into which these private cabins open, runs athwart-ships, and is lighted by a large port at either end and two cupola sky-lights. The Captain and I use this as a sitting-room, the port end with writing-table and work-table and book-cases packed with old home favourites, being appropriated to my use and that of my secretary Mr. Wild, to whose facile pencil we are indebted for beautiful illustrations of our novelties, and who sits with me gathering in the various threads, which we combine into a symmetrical web as best we may.

Two sets of cabins have been specially built on the after part of the main-deck for the different departments of the scientific work. The chart-room, the head-quarters of the naval scientific staff, is a commodious apartment on the starboard side with ranges of shelves stocked with charts and hydrographic, magnetic, and meteorological instruments. All the work in these departments, as well as the whole of the practical operations in dredging, sounding, and taking bottom and serial temperatures, is

12 THE ATLANTIC. [CHAP. I.

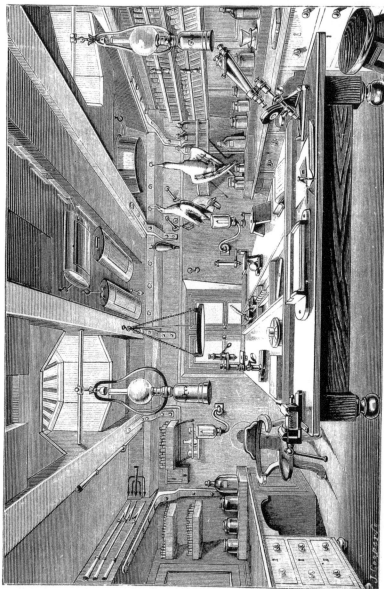

Fig. 1.—The Natural History Work-room.

conducted by the naval officers. The natural history work-room corresponds with the chart-room on the port side, and as this is a novel addition to the equipment of a surveying ship, I will describe it somewhat in detail.

The room (Fig. 1) is 12 feet wide by 20 long, the height between decks 6 feet 10 inches. It is lighted by a large square port, a small scuttle, and two cupola sky-lights, and the side towards the main-deck is closed by moveable glazed sashes. At either end are fitted broad mahogany dressers, with knee-holes and spacious cupboards and tiers of drawers beneath, and book-shelves and cupboards against the bulkheads above. At the back of the dressers all round are racks with holes to fit the fish-globes and the bottles of various sizes which are in constant use, and similar racks are fitted wherever there is available space against the ship's side. For convenience of working at sea it is impossible to have too many such racks where bottles may be instantly put in safety in case of the vessel suddenly rolling. Racks for test-tubes, which are simply thick slabs of mahogany drilled with deep holes to fit the tubes set as closely as possible, are fitted against the walls. Similar slabs of smaller sizes are also used for standing on the table while tubes are being filled with specimens.

Some of the drawers in the dressers are fitted with racks for smaller bottles, for specimens under examination or for reagents, and others which contain forceps, tools, corks, and all the innumerable small things of a rougher description required for all our complicated operations, are cut up by vertical partitions into small compartments to prevent their contents being shaken together. The instrument cases have each its own compartment in the drawers and cupboards in which they are secured by battens. A fresh-water tank and sink occupy a space against the side bulkhead, and spirit of wine is laid on to a locked-up tap from a cistern in the nettings above. Long shelves with ledges, running parallel with the beams overhead, give a great deal of stowage room, and various implements such as harpoons, botanical vasculums, an injecting copper, &c., are conveniently suspended from the beams and deck by hooks. A long table is placed across the centre of the work-room running right up to the port, so that two persons sitting opposite one another at the end of the table close to the port have a good side-light for their microscopes. The most convenient height for the table, using principally Hartnach's microscopes, was found to be 2 feet 9 inches. The microscopes are secured to the table by brass holdfasts like those in common use on carpenters' benches. The holdfast when brought to bear upon the back of the footpiece holds the instrument rigidly firm; two holes are bored in the table for the holdfast, one for holding the microscope in position when in use and the other for securing it when set aside.

The centre of the table is divided by low fixed battens into oblong compartments for micro-reagents, canada-balsam, glycerine jelly and the paraphernalia used in examining objects with the microscope and mounting microscopic preparations, ink-stands, and drawing materials. Two large moderator lamps swing below the cupolas, moveable branches for candles are screwed to the bulkheads, and for exa-

THE EQUIPMENT OF THE SHIP.

mining minute surface animals at night when they are frequently in greatest abundance the Bockett microscope-lamp made by Collins is found most useful.

Three of Hartnach's small model microscopes with objectives 2, 4, 7, 8, and 10 are in constant use in the work-room, but one of Smith and Beck's binoculars is found more convenient for observing objects such as the large foraminifera, by reflected light. There are also several other microscopes by Ross, Zeiss, and other makers available, and a number of the ordinary dissecting microscopes.

The heat of the tropics affects unfavourably many of the substances in common use in mounting microscopic preparations; thus glycerine jelly will scarcely set at all but remains nearly fluid, and the different varnishes and lacs remain soft and sticky. It is unsafe to put preparations on edge, and we find small pine-wood cases supplied by Baker, Holborn, containing each twelve horizontal trays with accommodation for six dozen slides, most suitable for storing. It is almost inconceivable how difficult it is to keep instruments, particularly those which are necessarily made of steel, in working order on board a ship; or how rapidly even with the greatest care they become destroyed or lost. For this reason it is necessary to have an almost unlimited supply of those in most frequent use, such as scissors, forceps, and scalpels of all sizes; reserves being rubbed over with mercurial ointment and stowed away where they can be looked at from time to time.

The operations carried on in the work-room on ship-board are of course very much the same as the ordinary routine work of a museum work-room and a physiological laboratory, and the processes are much the same, only modified by the special nature of our work. We are provided with all the necessary apparatus and arrangements for skinning, mounting, and preparing specimens in all ways, and for dissecting and injecting. By far the greater number of the animals obtained are preserved in spirit, and to the stowage of spirit and of spirit preparations the entire fore-magazine is devoted. The spirit is stored in cylindrical iron vessels, containing each four gallons and closed by screw-taps; they are stowed in racks in the magazine and taken up as required and emptied into the tank in the nettings. Stowed also in racks in the magazine are a series of cases of wide-mouthed specimen bottles. The cases are numbered and arranged in the racks in order, so that it is only necessary to give the number to the gunner's mate who has charge of the magazine, and any case required is at once brought up into the work-room for inspection. An exact list is kept of the contents of each case and of each bottle, so that it is never necessary for any member of the scientific staff to go down into the magazine. The bottles in which the greater part of the specimens are preserved are those known in the trade as 'drop-bottles' manufactured for holding sweetmeats of various kinds. They are of pale-green glass, very transparent, and are closed by glass stoppers with cork rims. Three sizes are in use, the diameters of the bottles being 6 inches, $4\frac{1}{2}$ inches, and $3\frac{3}{4}$ inches, with mouths $3\frac{3}{4}$, $2\frac{3}{4}$, and $2\frac{3}{4}$ inches respectively. The bottles are all of the same height, 9 inches, and they pack conveniently upright in cases with wooden partitions and hinged lids, and padded at the bottom with cork. These jars are extremely convenient and wonderfully cheap; 200 cases complete containing 2,300 jars were supplied by E. Breffit and Co., Upper Thames Street, at a cost of 70*l*. Besides these large store-bottles there are many thousands of smaller stoppered bottles and corked test-tubes of different sizes and forms. Larger animals are packed in cylinders of zinc, which are made on board by a tinsmith as required.

For preserving salpæ, heteropods, and other surface-animals containing much water a solution of picric acid in water has been found very useful. A saturated solution at the ordinary temperature of England (say 5° C.) answers well, but picric acid becomes rapidly more soluble as the temperature rises, and in the tropics a saturated solution is much too strong, and shrivels up delicate tissues. We have now on board a *Carinaria* which, having lost its shell, was put into the picric acid solution as an experiment, and after ten months it is still wonderfully perfect, retaining the form and the transparency of the thick gelatinous mantle unimpaired. *Pterotrachea* and *Firola* have been similarly prepared with success, and a portion of a huge *Pyrosoma* 5 feet long which was brought up in the trawl thus treated is in excellent condition. Soft and pulpy animals steeped for a few hours in a weak solution of chromic acid before being put in alcohol have their tissues hardened and retain their form. This process answers well for oceanic cephalopods and holothurians, which should be put in living and allowed to die in the acid. A very weak solution of osmic acid is of great

c

value for killing and hardening small gelatinous animals for microscopic preparations. A drop of a solution, one-tenth to one-fortieth per cent., may be added to a watch-glass of sea-water in which the creatures are; but they must not be allowed to remain more than a few minutes in contact with the acid, as they rapidly turn black. They may then be mounted in glycerine jelly, or in a solution of tartrate of potash.

Artificial heat is absolutely necessary for drying plants in quantity at sea, particularly in hot regions where the air is constantly saturated with watery vapour; and an excellent drying-room has been discovered in a space in the funnel-casings. The plants are placed between single sheets of botanical drying-paper and piled, with numerous wire ventilators interspersed among them, into bundles, which are drawn tight together with cords which stand the heat better than straps. The bundles are placed in the drying-room, care being taken to tighten the cords from time to time, and in two days the plants are quite dry without the trouble of changing the papers. A somewhat elaborate botanical press which occupies a corner of the work-room has been entirely abandoned for this method.

Two shelves fitted along the fore-bulkhead of the work-room sustain a detachment of the very valuable little library of books of reference with which we are provided; the remainder of the books find room where they can in the fore- and after-cabins, in the ward-room and elsewhere.

The chemical laboratory (Fig. 2) is on the starboard side, nearly amidships. The following brief

FIG. 2.—The Chemical Laboratory

account of the laboratory and of some of the principal apparatus and processes bearing upon our special line of research is abstracted from a careful description prepared by Mr. Buchanan, which will appear in full elsewhere. The laboratory is 10 ft. 4 in. long by 5 ft. 3 in. wide and 6 ft. high. It is lighted by a large square port and by glass sashes in the door and bulkhead towards the main-deck. The port is closed by a French window in two sashes opening inwards. The fittings consist of a working bench, a locker-seat, a blow-pipe table, a writing-table, and drawers. The working bench fills up the space between the port and the after bulkhead; it is 4 ft. long, 2 ft. wide, and 3 ft. 10 in. high. It is built of teak, the top in two slabs $1\frac{1}{2}$ in. thick, below which are arranged a number of drawers and some shelves for the reagents and apparatus in constant use. The reagents are contained in bottles of four sizes, large and small for liquids and large and small for solids, with flat stoppers. The large hold about 350 cc. and the small 50 cc. The large bottles occupy three drawers divided into 18 compartments each, and the small two drawers each with 60 compartments. A number of small drawers are fitted to receive the everyday laboratory wants,—filtering paper, blow-pipe apparatus, corks, india-rubber, &c.; and one is specially set apart for nails, screws, and hooks, things not without their uses in a laboratory on shore and absolutely indispensable at sea, where every article, even the smallest, must not only have its place but must be secured in it.

The top of the bench is fitted with shifting battens to keep things from falling off, and at one corner a leaden sink is let into it, communicating with the sea by a pipe which passes through a scupper.

The locker-seat stretches across the forward end of the laboratory. It is 5 ft. 9 in. long, 2 ft. 6 in. wide, and 2 ft. high, and is divided into three compartments for the storage of apparatus not in constant use. The top of the locker is cushioned, and serves for a lounge; and above it are two book-shelves stocked with books of reference in chemistry, physics, and geology.

The blow-pipe table is 2 ft. 9 in. high and 17 in. square, and carries a folding leaf. The bellows are circular, 8 in. in diameter, and the table is so fixed near the inner bulkhead that they can be conveniently worked from the locker-seat.

The writing-table can be raised for use or folded down out of the way at pleasure. It is close to the window, 2 ft. 6 in. long, by 21 in. wide. Every available wall space is occupied by shelves, and when standing at the working bench one has the glass apparatus in ordinary use such as beakers, flasks, test-tubes, &c., conveniently arranged on shelves to the right hand and behind him. As in the natural history work-room deep shelves are run along the beams, and these serve for the stowage of glass tubing, note-books, portfolios, and miscellaneous articles. On wall-spaces not adapted for shelves small articles are fixed by hooks or nails, or in whatever way is most suitable, to be ready at hand. A convenient way of stowing glass tubes, small pipettes, parts of larger apparatus, &c., as well as pens and pencils, is to slit up a piece of india-rubber tubing an inch or so long of suitable bore. and fix it to the bulkhead by a tack. A light india-rubber clamp is thus formed sufficiently strong to grasp and retain anything light. If the tube be long each end may be supported in this way.

An ingenious modification of Bunsen's apparatus, by Dr. Jacobsen of Kiel, is used for boiling the atmospheric gases out of the sea-water. It consists of three principal parts,—the flask, the bulbed tube, and the receiver for the gases. The flask is spherical, with a strong lip; the one at present in use contains 940 cc. The peculiarity of the apparatus consists in the arrangement of the bulbed tube. The bulb a (Fig. 3) in which the water is boiled to expel the air from the apparatus is of the pear-shape represented in the figure, in order to have the exit-tube as nearly as possible at its highest point, so as to prevent the accumulation of any air in the upper part of the bulb. Its capacity is about 60 cc. The lower end of the tube is closed, but about half-an-inch from the end it has a very small hole c in the side. The perforated india-rubber cork d fits the neck of the flask accurately, and through the perforation the tube passes air-tight and with some friction. The receiver b holds from 50 to 60 cc , and has the entry- and exit-tubes contracted in the way shown in the figure. It is joined to the bulbed tube by an air-tight india-rubber connection, and carries at its exit another piece of tubing for a purpose to be mentioned presently. The upper part of the apparatus is supported by the clamp m, and by the bent rod f, which clamps firmly on to the lower part of the bulbed tube. The flask is supported in the water-bath g by the clamp h attached to the retort-

CHAP. I.] THE EQUIPMENT OF THE SHIP. 23

Fig. 3.—Apparatus for collecting the atmospheric Gases from Sea-water.

stand k, which in its turn is lashed to the blow-pipe table.

When the apparatus is to be used, a sufficient quantity of boiled distilled water is introduced into the bulb, and the cork d pushed over the opening c. The sea-water to be examined is run directly into the flask from the deep-sea water-bottle by means of a tube with a narrow opening reaching to the bottom of the flask, the tube being gradually withdrawn until the flask is overflowing. The opening c in the tube is then brought just below the lower surface of the cork, which is pressed tightly into the neck of the flask. A certain amount of water displaced by the cork rises into the bulb, and the tube is carefully drawn upwards till the opening is well within the cork and therefore closed. A small vacuum is thus produced, causing the immediate appearance of air-bells in the water. The receiver b is now attached, and the water brought to boiling by a hand spirit-lamp, and kept so until the whole of the air has been expelled, which takes from six to eight minutes. While the water is still boiling the india-rubber tube on the exit tube of the receiver is closed with a glass stopper so tapered that at the point it slips easily into the tube, and being pressed in closes it tightly. The receiver is now hermetically sealed at the upper contraction, and connection made between the bulb and the flask by pushing down the tube until the hole c is below the cork. A lively disengagement of gas commences which is kept up by heating the water in the water-bath, the water being brought slowly to the boiling point, at which temperature it is retained for some time. When it is judged that the gas has been wholly expelled the receiver is sealed up at the lower contraction and the operation ended.

The arrangement employed for boiling the carbonic acid out of sea-water is represented in Fig. 4.

The flask a has a capacity of about 500 cc., and receives the sea-water to be operated upon usually

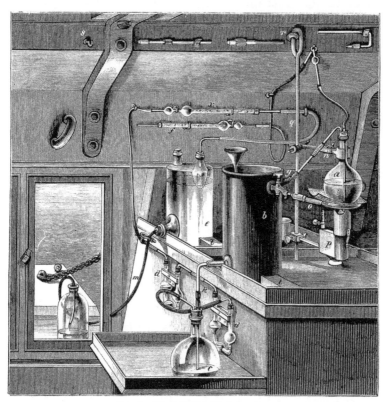

Fig. 4.—The Carbonic-acid Apparatus.

to the amount of 200 to 250 cc. It is closed by an india-rubber cork, through which pass two tubes; one reaching to the bottom communicates with the atmosphere by means of the soda-lime tube f, to to which it is attached by a flexible tube; the other, opening but little below the cork, communicates with the condenser b, a cylindrical copper vessel $5\frac{1}{2}$ in. in diameter with a block-tin worm. The lower end of the worm is attached to the receiver c by a bent-glass tube with a flexible joint k, from which a glass tube leads to the bottom of the receiver. The flexibility thus obtained is of the greatest use in practice, enabling one by shaking to expose constantly fresh surfaces of the baryta-water to the passing gases. The receiver c is connected by india-rubber tube with the two bulbed U-tubes d, d; the aspirator e enables a stream of air to be drawn through the apparatus, and between it and the U-tubes there is a soda-lime safety tube x, the water running from the aspirator out at the port by the tube m which passes through a hole in the sash.

The flask a is supported on a ring by the clamps n and o; both of these, along with the spirit-lamp p, are fixed in the usual way to the iron rod q, which is attached to the projecting beam of the ship's side by the eye-bolt r, in which it has a play of rather more than an inch. When not in use the rod is pushed up out of the hole in the working-table in which it is inserted and laid along the roof, its lower end being supported by the hook s.

The carbonic acid is determined by boiling it out of sea-water and receiving it in baryta-water of known strength; the amount of baryta neutralized is then ascertained by titration. The sea-water is boiled in the flask a, and the baryta-water is distributed between the receiver c and the U-tubes d, d. When a sample of water is to be examined, the apparatus is put together and a stream of air freed from carbonic acid passed through it; the corks in the receiver and tubes are then eased, and 15 to 20 cc. of baryta-water, usually about tenth-normal strength, run into them. The water to be examined is introduced into the flask a, and 10 cc. of a concentrated chloride of barium solution added to it to precipitate the sulphates. The apparatus is then put together and heat applied to the flask a. The boiling is continued until less than 50 cc. remain in the flask, a constant stream of air being drawn through all the time. The great bulk of the carbonic acid comes over in the first 30 to 40 cc., and during this time and occasionally throughout the whole operation it is well to shake the receiver gently, so that fresh baryta-water shall always moisten the walls. When this is attended to only a slight turbidity appears in the first U-tube and the second remains perfectly clear. When sufficient water has been distilled over, air is allowed to pass for some time, and the contents of the U-tubes and the receiver are collected in the latter, and the alkalinity determined by means of tenth-normal hydrochloric acid, the point of neutralization being indicated by rosolic acid.

A very important subject for investigation in the chemistry of the ocean is the nature and quantity of the atmospheric gases dissolved in the water. These are extracted by boiling *in vacuo* in the apparatus already described, and at the end of the operation are obtained hermetically sealed in glass tubes, in which they may be preserved for an indefinite time. Of course it is, as a rule, more convenient to retain such specimens until they can be carefully analysed on shore; but in a long cruise it may sometimes be of importance to make an approximate analysis, as the composition of the dissolved gases, particularly in volcanic regions, may bear upon other questions.

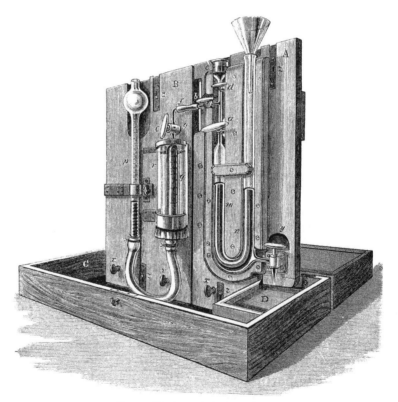

Fig. 5.—The Gas-analysis Apparatus.

The apparatus represented in Fig. 5 was designed by Mr. Buchanan to fulfil this purpose. It consists essentially of two U-tubes. The one, A, which according to precedent we may call the 'Laboratory tube,' is wholly of glass; the other, the eudiometer, B, has the legs of glass united by an india-rubber tube of suitable length. These two parts are connected by a capillary portion ($g\ f\ b\ d\ a$), of which the part ($b\ d\ a$) belonging to A is shown separate in section in Figs. 6 and 7. The stopcock a is pierced with two tubes, the one (Fig. 7) affording direct communication upwards between the two portions of the

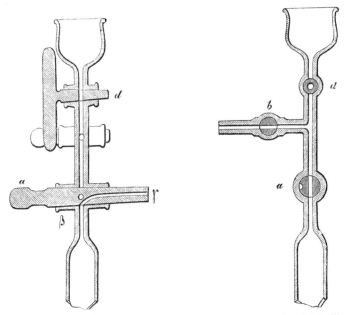

FIG. 6.—Arrangement of the Capillary portion of the Gas-analysis Apparatus.

FIG. 7.—Arrangement of the Capillary portion of the Gas-analysis Apparatus.

capillary, the other (Fig. 6) communicating with the atmosphere through the prolongation of the stopcock. b, c and d are ordinary stopcocks pierced to the bore of the capillary. A general idea of the use of the different parts of the apparatus will be obtained from the description of the analysis of a sample of air extracted from sea-water. Suppose the instrument to be set up, and with the gaseous mixture in the eudiometer where its volume has been ascertained. The mixture consists, we shall say, of oxygen, nitrogen, and carbonic acid. The last of these is determined first by absorption by caustic potash. For this purpose mercury is run out of n by the stopcock y, p is raised and the stopcocks $c\ b$ and a opened. The air is thus driven over out of q into m, mercury being allowed to fill the capillary, when the stopcocks are again shut. The cup e is now filled with strong solution of caustic potash, and, the level in n being still kept low, the stopcock a is opened full and d very carefully, thus allowing caustic potash to run down through a into m, where it meets the gas in the most advantageous way for quick absorption. When enough caustic potash has been allowed to enter, d is closed, some mercury poured into e, and d again opened, when the mercury drives all the solution out of the capillary tube into m and occupies its place. When the absorption is finished the level of the mercury in n is again raised and the stopcocks b and c opened. The stop-cock a is now very carefully opened, the flow of the gas being further regulated by raising or depressing p, and the gas allowed to pass over into q until the potash solution just touches the lower surface of the stopcock a, which, being open, has the position shown in Fig. 7. The position of a is now changed to that shown in Fig. 6, which enables the potash to be eliminated from the apparatus. The position of a is then brought back to that of Fig. 7, and the gas remaining in the capillary swept out by mercury and measured as before.

CHAP. I.] THE EQUIPMENT OF THE SHIP. 31

If the oxygen is to be determined by absorption, the manipulations are exactly the same as in the case of carbonic acid, alkaline pyrogallic acid being used instead of caustic potash. If the oxygen is to be determined by eudiometry, then, after the carbonic acid has been absorbed, the gas remains in the eudiometer, the stopcocks c and b being closed. The stopcock d is now opened and a turned the reverse way to that shown in Fig. 6, that is to say with the side communication $\beta \gamma$ communicating with the capillary *above* a. The capillary is thus emptied of mercury, and the hydrogen evolving apparatus is connected with γ by an india-rubber tube and the hydrogen allowed to stream through $a\, d\, e$ until all the air is swept out; the stopcock d is then closed and a brought back to the position in Fig. 6, when the gas enters m. When enough hydrogen has passed in, a is brought to its position in Fig. 7, and the hydrogen apparatus is dispensed with. Mercury is now poured into e, and d opened, and the hydrogen in the capillary driven into m, its place being taken by mercury. The hydrogen is now passed over into the measuring tube and measured, and the explosion is made.

When the analysis is finished the mercury is emptied out of the tubes, the parts A and B are separated from each other at f by relieving the india-rubber connection, and from the box by taking out the screws $x\, x\, x$. The mercury receptacle D is removed and A and B deposited each in its own wing of the box, to which it fits, and is fixed by the bolts $z\, z$. The whole gas-analysis apparatus is now contained in a box which measures when closed, over the outside, 19 inches by 9 inches square.

Fig. 8 represents an apparatus for preparing oxygen, hydrogen, or knallgas, for use in gas analysis. It is of the well-known form of the lecture-apparatus known as 'Hofmann's tubes.' It consists of two tubes, A and B, united at their lower extremities by a short tube C, which connects them at the same time by means of the tube D with the reservoir E. At their upper extremities A and B terminate in capillary tubes provided with stopcocks F F. Communication with the reservoir can be made or interrupted by the stopcock G. One of the tubes A is provided with two platinum electrodes, the other, B, has but one. Bent delivery-tubes not shown in the drawing fit on the tubes above the stopcocks F F. When about to be used, all the stopcocks are opened and dilute sulphuric acid poured in through the reservoir until it has eliminated all air and is running out at the delivery-tubes. The stopcock G is now shut, and the battery connected as circumstances may require, either with the electrodes in A or with one in A and one in B. Gas is allowed to escape freely until one can be sure that all dissolved air is removed; the stopcocks F F are now closed

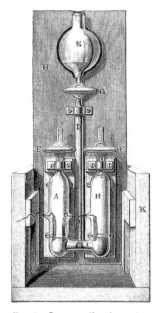

FIG. 8.—Gas-generating Apparatus.

and G opened, and the liquid in A and B allowed to sink until it just covers the electrodes. G is then closed and F F opened, when the gases may be introduced into the eudiometer in the ordinary way. The whole apparatus is attached to a mahogany slab H which fits into the box K—shown cut through the middle in the drawing—either when in use, as represented, or when not in use, as a lid,

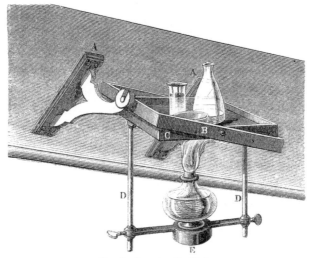

Fig. 9.—Sea-going Sand-bath.

with the apparatus within the box. The box thus answers the double purpose of a convenient stand and a safe packing-case.

For evaporating or heating in flasks or beakers a small sand-bath suspended on gimbals has been found very useful... A A (Fig. 9) are iron brackets screwed to the ship's side; B is the outer frame, made of cast-iron and moving on an axis parallel to the ship's length; C is the inner frame, also of cast-iron, and moving on an axis at right angles to that direction. The size of the iron frame was arranged so as to receive one of Bunsen's thermostats in ordinary use in laboratories, and was furnished with a cast-iron plate when used as a sand-bath, with a piece of strong copper-wire gauze stretched over a frame for boiling purposes, and with two cast-iron plates with large holes to receive water-baths. The half-inch iron rods, D D, are fixed to the lower side of the inner frame, and the leaden counterpoise E is moveably attached to them by screws.

For collecting water from the bottom we use a water-bottle, originally, I believe, the invention of a Swede, but which was first suggested for use in the 'Challenger' by the visit of the German North-Sea Expedition to Leith, a visit which we have to thank for numerous other most useful hints.

It consists essentially of a brass cylinder, A (Fig. 10. I, II), which slides up and down a metal shank, B, of at least twice its length. When the water-bottle is sent down, the cylinder is fixed in the upper part of the shank, as in Fig. 10. I; and when it arrives at the bottom it is released and falls down to the lower part, as in II, where it rests on two accurately ground valves, C and D, which fit into two conical surfaces on the inside of its upper and under edges. Thus the water which surrounds the shank at the moment of slipping is securely enclosed. The proper working of the instrument is dependent on the shank remaining straight; any bend in it would cause the valves to leak. In the instrument used in the German Expedition this was sufficiently well provided for, for shallow soundings with light

CHAP. I.]   THE EQUIPMENT OF THE SHIP.   35

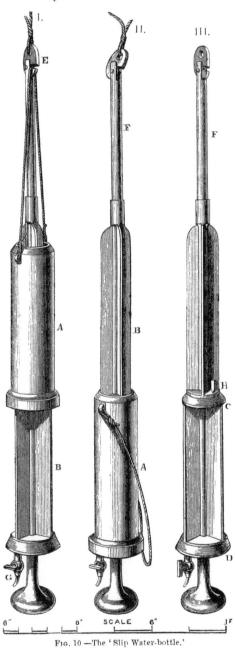

Fig. 10.—The 'Slip Water-bottle.'

weights, by the two valves being connected by a short iron rod, and the upper valve with the slipping arrangement by means of four slighter ones; but for deep soundings, where it is attached to a line along with a weight of three and often four hundredweight, greater strength is necessary to enable it to withstand the knocks which, even with the greatest care, it is exposed to, in being got over the ship's side in a sea-way. Mr. Milne of Edinburgh, into whose hands the construction of the instrument was put, has secured this end in a way which adds equally to the elegance and to the strength of the instrument. The shank and valves are one solid brass casting of the shape shown in the figure, the cylinder is another, and the slipping arrangement E, fixed to the end of a rod F of suitable length and great stoutness, is screwed into the top of the shank, the screw being secured by a rivet. The water enclosed is removed by means of a tap G, passing through the lower valve, air being at the same time admitted at the top by the removal of a plug H, from a hole in the upper valve. The lower valve and stopcock are protected from damage when striking against the ground by the casting extending about six inches below the valve. The arrangement and dimensions of the parts are

FIG 11.—Instrument for slipping the cylinder at intermediate depths.

sufficiently apparent from the plate to make further description unnecessary. The slipping arrangement is in principle the same as that used on Brooke's sounding-rod.

In order to adapt this water-bottle to collecting water at intermediate depths, it is fitted with a slipping plate F, Fig. 11, furnished with a metal flap Q, which depresses it when the motion of the instrument is reversed. It is inserted into a slot S, immediately below the usual slipping plate to which the sounding-line is attached, and differs from the latter in having a deeper notch R, and having a slot instead of a hole for the reception of the pin T, round which it turns. The object of this slot is, that after the string has been cast free, the flap may fall down close alongside the rod and afford as little resistance as possible in pulling up. In using the instrument, it must be let go before the flap enters the water, and not checked until the depth desired has been reached. For collecting water at any given depth below the surface and retaining the gases dissolved in it, Mr. Buchanan has devised the very ingenious instrument (Fig. 12), which, in careful hands, gives satisfactory results. It consists of a brass tube A, two inches and a half in diameter and of a length suitable to the capacity desired, closed at both ends by stopcocks B B, with $\frac{3}{4}$-inch clear passage, attached by flanges, screwed down upon washers. The stopcocks are connected by a straight brass beam C, and, when fully open, the levers D D which work them stand up at an angle of 45° to the axis of the instrument, and when fully shut

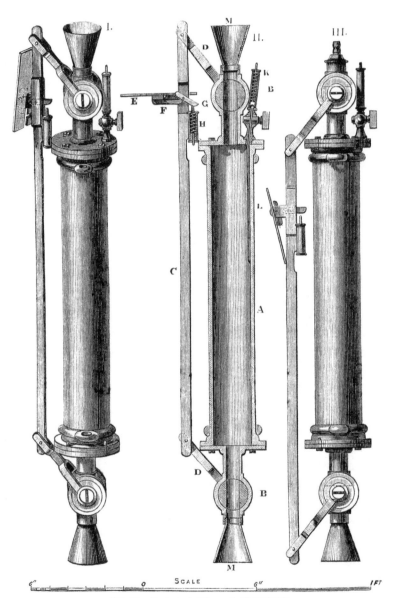

Fig. 12.—Buchanan's 'Stopcock Water-bottle.'

they point downwards at about an equal angle. The beam connecting them carries a metal plate E of about thirty square inches area, secured by a pin F, round which it moves freely through two right angles, when pushed upwards; but when pulled downwards it is arrested in a position at right angles to the beam by a tongue G, resting upon a spring H, the strength of which is such that, before giving way, the rush of the water past the plate will shut the stopcocks. The plate then passes the spring, and any reversal of the motion of the apparatus will only reset the plate, and, on heaving up, shut the stopcocks closer. As the water is thus hermetically enclosed it is necessary to provide for its expansion on coming to the surface. This is secured by means of a safety-valve K, the tube L of which penetrates well into the interior; so that, supposing the water to be overcharged with gas, it would, on coming to the surface, only suffer *water* to escape, the gas remaining at the top of the instrument. Brass funnels M M are fitted at top and bottom, so as to give a greater draw through. They unship, and can be replaced by nozzles N, screwing down upon washers air-tight, to the top of one of which can be affixed a gas-collecting apparatus, such as Bunsen used for boiling the gases out of water. The brass water-bottle thus replaces the flask ordinarily used, and renders transvasing unnecessary. In order that the instrument may answer this purpose the parts must be fitted with the accuracy of an air-pump. Unfortunately those in use on board the 'Challenger' had to be finished so hastily that this condition could be only imperfectly complied with.

The instrument arranged for sinking is represented in Fig. 12. I; in Fig. 12. II in section when the line is checked on commencing to haul in, and in Fig. 12. III in perspective after it has been brought up, and with one of the funnels replaced by a nozzle. Care is necessary in using the instrument to see that the stopcocks work easily, but not so easily as to make them shut by the weight of the beam. They can be adjusted to any degree of stiffness by means of the screws in the keys of the stopcocks and those attaching the levers to the beam. When the instrument has been set and fixed to the sounding-line, it should be lowered as nearly as possible to the surface of the water and then let go, care being taken that the line is not checked till the desired depth is attained. If the vessel from which the soundings are being taken is not very high out of the water, it is better, as soon as the water-bottle is fixed, to let it go without previous lowering, avoiding thereby the danger of the stopcocks being shut by the line being lowered by jerks.

That the instrument really collects the water at the depth to which it is sunk, or, in other words, that the water really passes through it without sticking, was proved by some experiments in Linlithgow Loch. The bottle was filled with water containing ferrocyanide of potassium in solution and sunk to various depths, the water brought up being tested with perchloride of iron. Water, brought up from any depth over $1\frac{1}{2}$ fathoms, showed no trace of prussian blue.

For determining the specific gravity of the sea in different parts of the ocean Mr. Buchanan uses a hydrometer of the following construction and dimensions:—It is a glass instrument of the ordinary shape, loaded with mercury and carrying an arbitrary scale. It is made to float at the lowest mark on the stem at about 15° C. To the top of the stem is fitted a small table of thin sheet brass, of such a weight that it depresses the instrument, in distilled water of 15° C., to the highest division on the stem. This table is destined to carry such weights as shall sink the instrument within the limits of the scale in the liquid whose specific gravity is to be determined. The instrument ceases to be useful when the weight becomes so great as to render it top-heavy. By use of a series of six weights, specific gravities between 1 and 1·034 can be determined with one hydrometer, and the results be accurate to 5 in the fifth decimal place.

The method of construction is as follows :—A piece of glass tube of suitable and uniform diameter is chosen for the stem, and the cylindrical body attached to it. A paper millimetre scale is inserted into the stem, and mercury poured in until it sinks to the lowermost division in distilled water of 15° C. It is now necessary to calibrate the stem. This is done by immersing the instrument in distilled water of 15° C. and loading the stem by placing known weights on the top of it and reading the consequent depression. In order that this may be done conveniently the stem is not sealed up at the top, but slightly widened out to a funnel-shaped opening, on the edges of which decigramme weights rest securely.

The value of the scale having been thus determined, the stem is sealed up, and the now finished instrument is very carefully weighed, and its weight *in vacuo* calculated. By immersing it in distilled water of different temperatures its volume for different temperatures is obtained. The instrument is now finished so far as determining specific gravities very little above that of distilled water is concerned. In order to extend the range of the instrument a series of weights is made, such that it shall always be possible to sink the instrument within the limits of this scale in solutions whose specific gravities lie between 1 and 1·034. The lightest of these weights is a small table, capable of being placed securely on the top of the stem, and destined, when greater weight is required, to carry any of the others. These are made of about the calculated weight, after which their true weight is accurately determined. A cheap and ready way of making these weights was found to be, to cut the lightest out of sheet brass of suitable thickness, then, as it is quite sufficient to make the weight of the others as nearly as possible simple multiples of the first, to cut out a number of pieces of brass of the same size as the first, for the second weight solder two of them, for the third three of them, together, and so on. They are then trimmed with the file and accurately weighed.

For use a curve is laid down giving the volume of the stem for every division of the scale, measured from the lowest one, and another curve, giving the volume of the body of the instrument, upon the lowest division of the scale in the stem, for different temperatures. When, therefore, in calculating the specific gravity of a liquid, we require to find the volume of the instrument immersed, from the one curve we get the volume of the stem, and from the other that of the body of the instrument immersed. The sum of the two gives the volume of the liquid displaced. The weight *in vacuo* of the hydrometer, together with plate and weight divided by the volume so found gives the specific gravity of the liquid for the temperature during observation, that of water at 4° C. being unity.

The instrument at present in use has the following elements :—

Weight of hydrometer *in vacuo*	. . . .	160·2128	grammes.
Weight of table	. . . . . .	0·8360	„
Weight of weight I.	. . . . . .	0·8560	„
„ „ II.	. . . . . .	1·6010	„
„ „ III.	. . . . . .	2·4225	„
„ „ IV.	. . . . . .	3·2145	„
„ „ V.	. . . . . .	4·0710	„
„ „ VI.	. . . . . .	4·8245	„

The diameter of the stem is as nearly as possible 3 millimetres, the volume of 100 millimetres (the length of the scale), being 0·86 cc. Had the volume of the stem been determined by weighing the instrument in water at 4° C. instead of 15° C., the volume would have been found to be 0·8607 cc., instead of 0·86 cc. The difference between the two is wholly inappreciable.

The volume of the body of the instrument at 0° is 160·3 cc., and at 22°·2 160·4, the coefficient of expansion having been found by immersion in distilled water of different temperatures to be 0·000027.

The bulk and weight of the body, taken along with the slimness of the stem, make great care in handling it necessary, and, in order to get good results, the instrument must be kept perfectly clean, being always wiped dry with a clean cloth after use. Repeated experiments with the same liquid give results always agreeing within one scale division. In taking specific gravities at sea the operation is performed on a swinging table; the motion of the ship gives the hydrometer just so much oscillation as to overcome all effect of sticking, and to make the reading as much more satisfactory than on shore as that of a balance is, when it is allowed to oscillate, in preference to being brought to rest with the tongue on the zero. As it is very difficult to place the cylinder perfectly vertically on the swinging table, the hydrometer generally has a certain list to one side or another, which also favours its freedom of motion.

The space on the port side of the main-deck between the chart-room and the laboratory is occupied by such of our gear as would not pack into the work-rooms, including the apparatus devised by Mr. Siemens, F.R.S. for telegraphing the temperature from all depths; and his photometric apparatus for determining, by the exposure of sensitive paper for a certain length of time, the depth to which the chemical rays of the sun penetrate into the water of the sea. Fig. 13 represents a hydraulic pump for reproducing the pressure to which thermometers and other instruments are subjected at great depths, and thus affording us a means of determining their error under certain measured pressures before sending them down.

The pump A is of the ordinary construction, only with a very narrow cylinder, the diameter of the cylinder and piston being ¼ inch. The water is pumped into the reservoir B, a cast-iron tube of 3 inches internal and 9 inches external diameter, closed above by the plug C, which is held in its place

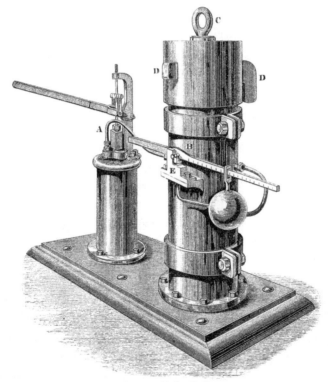

Fig. 13.—The Hydraulic Pump.

by the bolt D. The instruments to be tested are placed in B; the plug C is inserted and made fast by the bolt, and water is pumped in until the desired pressure has been obtained. This is indicated by water issuing from the safety-valve E, which is of the ordinary construction. The machine, which was made by Messrs. Milne of Edinburgh, works up to a pressure of 4 tons on the square inch.

On the port side opposite the laboratory two cabins have been fitted up as a light-room and a dark-room for photographic work, and a corporal of Royal Engineers, C. Newbold, a very skilful photographer, has been established there. The management of a photographic studio during a long sea-cruise is a matter of great difficulty. All the circumstances—the motion, the dampness of the air, and its vitiation by vapours of various kinds, and the extremes of climate which affect the different reagents and materials, all tell against the photographer; yet, in spite of these disadvantages, Newbold has already produced a large number of very satisfactory pictures.

Hundreds of miles of line, of strength and material suited to different purposes, are reeled and coiled in every available spot on the forepart of the main-deck and elsewhere. When we left England we were provided with 25,000 fathoms of bolt-rope for the dredge, 10,000 fathoms 3-inch rope, 10,000 fathoms 2½-inch, and 5,000 2-inch, and the supply has since been renewed. The methods of sounding and dredging, and the appliances used, are very much the same as those which we employed in the 'Porcupine,' and which I have described fully in a former volume.[1] For deep sounding we now very generally use a

---

[1] The Depths of the Sea. An Account of the general Results of the Dredging Cruise of H.M.SS. 'Porcupine' and 'Lightning' during the summers of 1868, 1869, and 1870, under the scientific direction of Dr. Carpenter, F.R.S., J. Gwyn Jeffreys, F.R.S., and Dr. Wyville Thomson, F.R.S. London: Macmillan and Co., 1873.

neat modification of the 'hydra' machine, devised by Navigating-Lieut. C. W. Baillie. The 'Baillie' sounding instrument is represented in perspective in

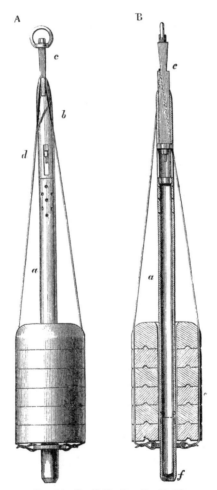

Fig. 14.—The Baillie Sounding-Machine.

the position in which it is let go in Fig. 14 A, and in section in the same position at B. The tube a is about 5 ft. 6 in. in length by $2\frac{1}{2}$ inches in diameter.

The bore is 2 inches, so that the wall is $\frac{1}{4}$-inch thick. The principal part of the tube is of iron. It is bored near its upper end with a number of holes to let out the water; it unscrews into two at *e*, and at its lower end *f*, there is a pair of butterfly-valves working inwards. A strong brass cylinder *b*, with a diameter equal to that of the tube, is firmly attached to the upper end; a heavy piece of iron *c*, works in the brass cylinder like a piston to the extent of the length of the slots *d*, in the sides of the cylinder, in which it is retained by a strong square bolt. The piston-iron is flattened, and it is provided at *c* with a projecting shoulder, which, when the piston is drawn out—the bolt being a tthe top of the slot as in the figure—is well above the top of the cylinder; but when the piston is down and the bolt at the bottom of the slot, the shoulder is just within the upper part of the cylinder. The wall of the upper part of the cylinder is bevelled away to a long rounded slope. When to be used the instrument is hung by the ring to the sounding-line, and a sufficient number of weights are suspended on an iron-wire sling, as in the 'hydra' machine, the tube passing through the middle of them, and the sling hooking upon the shoulder of the piston-iron. When the tube and the weights touch the bottom the brass cylinder is pushed upwards the length of the slots, and the sling is slipped off the shoulder of the piston-iron by the upper rim of the cylinder, and allowed to slide down over its bevelled upper end. This is a very simple plan, and the doing away with the steel spring of the 'hydra' is an advantage. The larger tube also brings up a better and fuller sample of the bottom.

For less depths, where it is possible to recover the weights, a modification of the old cup-lead has been found very serviceable (Fig. 15). A 140 lb. or other deep-sea lead is cast round an iron rod which terminates above in an eye and below in an iron disk the diameter of the lead with a short wide-threaded screw. On this is screwed a stout hollow cylinder of iron up to 13 inches or so in length, ending in a pair of butterfly-valves opening inwards. 'Valve-leads' on this plan are made of different sizes. The larger, which work easily down to 1,000 fathoms with the No. 1 sounding-line, bring up a most satisfactory sample of the bottom. We still use ' Ball's dredge,' and only some slight modifications have been the result of further experience. Fig. 16 represents the form of dredge which we find most suitable for work at all but the very greatest depths, when one of a smaller size is used. The dredge-frame of hammered iron is 4 ft. 6 in. long, and 1 ft. 3 in. broad. The scrapers are 3 inches wide, and are connected at the ends by bars of $1\frac{1}{4}$ in.-round iron. The arms are of inch-round iron and slightly curved; they are bolted together to a stout iron bar which ends above in a swivel and ring. Two bars of square iron of some strength are attached by eyes to the round cross-bars at the ends of the dredge-frame, and have the other ends lashed to the iron bar which bears the hempen tangles. These rods

Fig. 15.—The 'Valve' Sounding-lead (in section.)

keep the dredge-bag at its full length, and prevent it

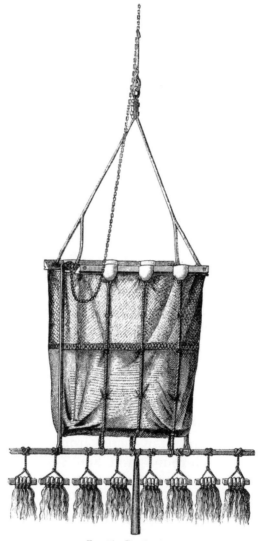

Fig. 16.—The Dredge.

or the tangles from folding over the mouth of the dredge. The dredge-bag is 4 ft. 6 in. in length; the lower half is of twine netting, so close as to retain everything except the finest mud, which indeed only partially washes through, and the upper half is of twine netting with the meshes an inch to the side. The bag is guarded by three loops of bolt-rope attached to the frame of the dredge, to the bottom of the bag, and finally to the tangle-bar. The canvas pads represented in the figure on the dredge-frame are only to protect the seizings of the loops. The dredge is suspended from an inch-iron chain, which forms the first few fathoms of the dredging-line. The chain is not, however, directly fastened to the ring at the end of the arms, but is made fast to one of the end-bars of the dredge-frame, and is stopped to the ring by a single strand of bolt-rope. If the dredge get caught on a rock, a rare occurrence fortunately in 'deep water, the stop carries away, the direction of the strain on the dredge is altered, and it probably relieves itself and comes up end upwards. In deep water a 28 lb. deep-sea lead is usually hung from the centre of the tangle-bar with four tangles on each side.

It is altogether a new experiment to dredge and to take deep-sea observations from so large a ship, and it seems to present some special difficulties, or at all events to require great management. The weight of the ship is so great that there can be no 'give and take' between her and the dredge, such as we have in smaller vessels. If there be any way on, the impulse to the dredge is irresistible; and it seems to tend to jerk it off the ground. The roll of the ship, her height above the water, her want of flexibility of movement compared with

E 2

the vessels which had been previously employed for the purpose, raised new questions as to the method of working.

Dredging and sounding are carried on in the 'Challenger' from the main yard-arm. A strong pendant is attached by a hook to the cap of the main-mast, and by a tackle to the yard-arm (Fig. 17). A compound arrangement of fifty-five to seventy of Hodge's patent accumulators is hung to the pendant, and beneath it a block, through which the dredge-rope passes.

The donkey-engines for hoisting the dredging and sounding gear are placed at the foot of the main-mast on the port side. They consist of a pair of direct-acting, high-pressure, horizontal engines, collectively of eighteen horse-power nominal. Instead of a connecting rod to each, a guide is fixed to the end of the piston-rod with a brass block working up and down the slot of the guide. The crank-axles run through the centre of the blocks, and the moveable block, obtaining a backward and forward motion from the piston-rod, acts on the crank as a connecting-rod would do. This style of engine is commonly used for pumping, the pump-rods being attached to the guide on the opposite side from the piston-rod. At one end of the crank a small toothed wheel is fixed, which drives one thrice the multiple on a horizontal shaft, extending nearly across the deck and about three feet six inches above it. At each end of this shaft a large and small drum are fixed, the larger having three sheaves cast upon it of different sizes, the lesser being a common barrel only. It is to these drums the line is led, two

CHAP. I.] THE EQUIPMENT OF THE SHIP. 53

Fig. 17.—The Dredging and Sounding Arrangements on board the 'Challenger.'

or three turns being taken round the drums selected. In hauling in, the dredge-rope is taken to a gin-block secured to a span on the forecastle, then aft to the drum of the donkey-engines on the port side, then to a leading block on the port side of the quarter-deck and across the deck to a block on the starboard side, then to the drum of the donkey-engines on the starboard side corresponding in diameter with the drum used on the port side, and from this it is finally taken and coiled. The strain is of course greatest at the yard-arm and the first leading-block, and by this arrangement it is gradually diminished as the wire passes round the series.

One slight change has been made in the handling of the dredge which has certain advantages. Instead of attaching the weights directly to the dredge-rope and sending them down with the dredge, as was our former practice, a 'toggle,' a small spindle-shaped piece of hard wood, is attached transversely to the rope at the required distance, 200 to 300 fathoms, in advance of the dredge. A 'messenger,' consisting of a figure of eight of rope with two large thimbles in the loops, has one of the thimbles slipped over the chain before the dredge is hung, and the other thimble made fast to a lizard. When the dredge is well down and has taken its direction from the drift of the ship, the weights, usually six twenty eight pound deep-sea leads, in three canvas covers, are attached to the other thimble of the 'traveller,' which is then cut adrift from the lizard and allowed to spin down the line until it is brought up by the toggle. By this plan the dredge takes a somewhat longer time to go down; but since we adopted it we have not had a single case of the fouling of the dredge in the dredge-rope, a misadventure which occurred more than once before, and which we were inclined to attribute to the weights getting a-head of the dredge in going down, and pulling it down upon them entangled in the double part of the line.

For the first two or three hauls in deep water off the coast of Portugal the dredge came up filled with the usual 'Atlantic ooze,' very tenacious and uniform throughout, and the work of hours in sifting gave the smallest possible result.

We were extremely anxious to get some idea of the general character of the fauna, and particularly of the distribution of the higher groups; and, after various suggestions for modifying the dredge, it was proposed to try the ordinary trawl. We had a compact well-balanced trawl with a fifteen-feet beam on board, and we sent it down off Cape St. Vincent to a depth of 600 fathoms. The experiment looked hazardous, but to our great satisfaction the trawl came up all right, and contained, along with many of the larger invertebrata, several fishes. The plan seemed to answer so well that we tried it again a little farther south in 1,090 fathoms, and again it was perfectly successful. Since that time we have used the trawl frequently, and particularly in very deep water where there is a certainty of finding a smooth bottom free from rocks, and where the large area covered by the trawl greatly increases the chance of bringing up some record of the scanty

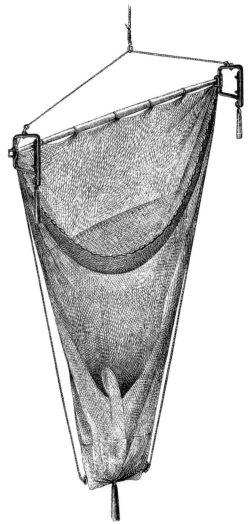

Fig. 18.— The Deep-sea Trawl.

and sparsely-scattered fauna. The deepest haul taken

with the trawl was on the voyage from Halifax to Bermudas at a depth of 2,650 fathoms.

Fig. 18 represents the deep-sea trawl at present in use. A conical bag 30 feet in length is suspended by one side to a beam of hard wood by half-a-dozen stops; the other side of the mouth of the net hangs loose, and is weighted with close-set rolls of thick sheet-lead to drag along the bottom. Two iron runners like the runners of a sledge are fixed one at either end of the beam, and the runners are so weighted themselves, and so weight the beam, that they tend, if fairly launched, to keep beneath with the beam above them. A second bag or 'pocket' open at the bottom hangs in the outer net reaching about three-fourths of its length, and acts as a valve, preventing the washing out of its contents, and about a yard of the narrow extremity of the net is lined with 'bread-bag' to give a chance of bringing up small things and a small sample of the bottom. The trawl is suspended by a bridle of rope, which is made fast to the runner at each end of the beam and then continued down on each side and fastened to the end of the trawl-bag. The trawl is usually sent down with the additional weight of three fourteen-pound hand-leads, and as in the case of the dredge, weights to about a hundredweight and a half are slipped to a 'toggle' 400 to 500 fathoms before it on the line.

The tow-net has been constantly worked during the cruise, and Mr. Murray, who has taken charge of this department, has latterly got most remarkable results from using the tow-net just as we use a dredge, pulled along by the drift of the ship with a weight a few fathoms in advance, at different depths down to 100 fathoms. Mr. Murray finds that frequently, when scarcely anything is to be found on the surface, all the pelagic animals which congregate there under favourable circumstances are to be met with at different depths below, and on more than one occasion he has taken animal forms with the deep tow-net which he has never seen on the surface. The tow-net used is of the ordinary form: a stout iron ring a foot or eighteen inches in diameter and a conical bag three to four feet deep of muslin, of buntine, or sometimes of 'bread-bag.'

The 'Challenger' is provided with a steam-pinnace, which is an invaluable adjunct for dredging and sounding in shallow water and in a smooth sea. She is a life-boat 36 feet in length with two pairs of engines, one pair for propelling her and another for heaving in the dredge-line. The propelling engines are a pair of high-pressure direct-acting vertical engines of six horse-power (nominal), with a horizontal tubular boiler and a disconnecting shaft and screw. At full speed the engines travel at about 240 revolutions a minute; and, on the trial for speed over the measured mile the boat averaged 8 knots an hour. The dredging engines are fitted to the top of the boiler; they are a direct-acting horizontal pair, the cylinders at the after end of the boiler and the crank-shaft forward. The shaft extends beyond the boiler on both sides, and at each end a drum is fixed. The drums are constructed with two sheaves, and it is to the greater or the lesser of these that the dredging line is led.

THE EQUIPMENT OF THE SHIP.

It is impossible in the limited space at present at my disposal to do more than sketch the more prominent of our equipments and methods; but even now, after nearly a year's experience, I feel justified in expressing my opinion that the arrangements for scientific work on board leave very little to be desired.

The 'Challenger' was ready for sea early in December 1872, and before she left Sheerness some of the Lords of the Admiralty and the committee of the Royal Society visited the ship.

A party of sixty sat down in the handsome wardroom where we now have our general mess, and good wishes and hopeful anticipations were warmly exchanged. We shall not soon forget the hearty British cheer of encouragement which rang out from a chorus of the voices which most influence the destinies of their country and their time, as our illustrious guests bade farewell to the 'Challenger' from the deck of the steamer which was to take them to the shore.

Thus, with every possible advantage, and in the highest hope of being able to fulfil her difficult mission, the 'Challenger' cast off from the jetty at Portsmouth at 11.30 A.M., on Saturday, the 21st December 1872. Leaving England with the drum up in the middle of an unusually trying winter, it was not to be expected that we should escape without roughing it a little. This we certainly did for a week or so in the mouth of the Channel and the Bay of Biscay, and with this good result, that we ourselves and all our manifold apparatus and machinery experienced at the beginning about as rough usage as we are likely to meet with during the voyage. This at once brought all our weak points to light, and it is well that we got into quiet water again with increased confidence in the stability of our arrangements.

H.M.S. CHALLENGER.

APPENDIX A.

Official Correspondence with reference to the 'Challenger' Expedition. Extracted from the Minutes of Council of the Royal Society.

June 29th, 1871.

Read the following Letter from Dr. Carpenter:—

"University of London, Burlington Gardens, W.
"June 15, 1871.

"Dear Prof. Stokes,—The information we have lately received as to the activity with which other nations are now entering upon the Physical and Biological Exploration of the Deep Sea, makes it appear to my colleagues and myself that the time is now come for bringing before our own Government the importance of initiating a more complete and systematic course of research than we have yet had the means of prosecuting.

"The accompanying slip from last week's 'Nature' will make known to the Council what is going on elsewhere, and the feeling entertained on the subjects alike in the scientific world and (as I have good reason to believe) by the public generally.

"For adequately carrying out any extensive plan of research, it would be requisite that special provision should be made; and as the Estimates for next year will have to be framed before the end of the present year, no time ought now to lost, if the matter is to be taken up at all.

"In order that the various departments of Science to which these researches are related should be adequately represented,—so that any Application made to Government should be on the broadest basis possible,—I should suggest that the Council of the Royal Society, as the promoters of all that has been already done in the matter, should take the initiative; and should appoint a Committee to consider a Scheme, in conjunction with the

President of the British Association, and the Presidents of the Chemical, Geographical, Geological, Linnean, and Zoological Societies. Such a Committee might meet before the Recess, and decide upon some general plan; and this would be then considered as to its details by the Members representing different departments of Scientific Enquiry, so that they might be able to report to the Council, and enable it to lay that Scheme (if approved) before the Government by the end of November.

"Believe me, dear Prof. Stokes,
" Yours faithfully,
" WILLIAM B. CARPENTER."

"*Prof. Stokes.*"

Resolved,—That the subject of Dr. Carpenter's Letter be taken into consideration at an early Meeting of the Council after the Recess.

October 26th, 1871.

In reference to the subject of Dr. Carpenter's Letter of the 15th June, read at the last Meeting of Council, the Secretary stated that he had received a subsequent Letter from him, dated Malta, 29th Sept., which was now read. In this Letter Dr. Carpenter urges the expediency of making arrangements for the proposed circumnavigating Expedition without delay, and communicates a correspondence with the First Lord of the Admiralty, from which it appears that H.M. Government will be prepared to give the requisite aid in furtherance of such an Expedition on receipt of a formal Application from the Royal Society; and in consequence of this information, Dr. Carpenter now suggests a modification in the composition of the Committee to which, in his former Letter, he had proposed that the matter should be referred.

Resolved,—That a Committee be appointed to consider the plan of operations it would be advisable to follow in the proposed Expedition, the staff of scientific superintendents and assistants to be employed, and the different provisions and arrangements to be made, with an estimate of the probable expense, and to submit to the Council for approval a scheme which might be laid before H.M. Government, if the Council see fit, at as early a period as may be convenient. The Committee to consist of the President and Officers of the Royal Society, Dr. Carpenter, Dr. Frankland, Dr. Hooker, Professor Huxley, the Hydrographer of the Admiralty, Mr. Gwyn Jeffreys, Mr. Siemens, Sir William Thomson, Dr. Wyville Thomson, and Dr. Williamson, with power to add to their number.

November 30th, 1871.

The following Report of the Committee on the proposed voyage of circumnavigation was read:—

"*Report of the Committee appointed at the Meeting of the Council held October 26th, to consider the Scheme of a Scientific Circumnavigation Expedition.*

"The Committee, having before them the correspondence which has already taken place between the First Lord of the Admiralty and Dr. Carpenter, are of opinion that it is advisable that the Council should make immediate Application to Her Majesty's Government for the means of carrying out the objects therein referred to; but that it would not be expedient that such Application should include more than a *general specification* of those objects,—which may be stated as follows:—

"1. To investigate the *Physical Conditions* of the *Deep Sea*, in the great Ocean-basins,—the North and South Atlantic, the North and South Pacific, and the Southern Ocean (as far as the neighbourhood of the great ice-barrier); in regard to Depth, Temperature, Circulation, Specific Gravity, and Penetration of Light; the observations and experiments upon all these points being made at various ranges of depth from the surface to the bottom.

"2. To determine the *Chemical Composition* of *Sea Water*, not merely at the surface and bottom, but at various intermediate depths; such determinations to include the Saline Constituents, the Gases, and the Organic Matter in *solution*, and the nature of any particles found in *suspension*.

"3. To ascertain the *Physical* and *Chemical* characters of the *Deposits* everywhere in progress on the Sea-bottom; and to trace, so far as may be possible, the sources of those deposits.

"4. To examine the Distribution of *Organic Life* throughout the areas traversed, especially in the *deep* Ocean-bottoms and at different depths; with especial reference to the Physical and Chemical conditions already referred to, and to the connection of the present with the past condition of the Globe.

"It is suggested that the Expedition should leave this country in the latter half of the year 1872; and as its perfect organization will require much time and labour, it is desirable that suitable preparations should be commenced forthwith.

"For effectively carrying out the objects just specified, there will be required :—

"1. A Ship of sufficient size to furnish ample accommodation and storage-room for sea-voyages of considerable length and for a probable absence of four years.

"2. A Staff of Scientific Men, qualified to take charge of the several branches of investigation above enumerated.

"3. An ample supply of all that will be required for the Collection of the objects of research; for the prosecution of Physical and Chemical investigations; and for the study and preservation of the various forms of Organic Life which will be obtained.

"The Committee would propose that in making this Application to the Admiralty, the President and Council should offer their services in suggesting the Route which may appear to be most desirable for the Expedition to pursue; and also in framing Instructions for the Officers charged with the several branches of Scientific Research; with a view to facilitate the preparation by their Lordships of their general Instructions for the conduct of the Voyage to the Naval Officers commanding.

"With this object they would propose that a Committee should be appointed by the Council, which should include persons thoroughly versed in the various branches of Science to be represented in the Expedition, who should give their advice and assistance previous to and during the progress of the Expedition.

"The President and Council should also express their readiness to select and recommend to their Lordships persons qualified to be entrusted with the various branches of Scientific investigation to be represented, naming the Salaries which may appear to them commensurate with the duties to be fulfilled.

"The President and Council should also, in the opinion of this Committee, recommend that, in accordance with former precedents in regard to Expeditions of a similar character undertaken by this and other Governments, a full and complete publication of the results of the Voyage with adequate illustrations should form a part of the general plan; and that the work should be brought out as soon after the return of the Expedition as may be convenient.

"It may be well to point out to the Admiralty, that the operations of the Expedition now proposed should not dispense with such researches of a less laborious character as their Lordships might be disposed to make from time to time from either the home or the foreign stations of the British Navy."

> Resolved,—That this Report be received, and be taken into consideration at the next Meeting of Council.

December 7th, 1871.

The Report of the Committee on the subject of a Scientific Circumnavigation Voyage, received at the last Meeting, having been taken into consideration, it was

> Resolved,—That application be made to Her Majesty's Government, as recommended by the Committee, and that the following Draft of a Letter to be addressed by the Secretary to the Secretary of the Admiralty be approved:—

> > "*To the Secretary of the Admiralty.*
> > "THE ROYAL SOCIETY, BURLINGTON HOUSE,
> > "*December 8th,* 1871.

"SIR,—I am directed by the President and Council of the Royal Society to request that you will represent to the Lords Commissioners of the Admiralty that the experience of the

F

recent scientific investigations of the deep sea, carried on in European waters by the Admiralty at the instance of the Royal Society (Reports of which will be found in their 'Proceedings' herewith enclosed), has led them to the conviction that advantages of great importance to Science and to Navigation would accrue from the extension of such investigations to the great oceanic regions of the Globe. The President and Council therefore venture to submit to their Lordships' favourable consideration a proposal for fitting out an Expedition commensurate to the objects in view; which objects are briefly as follows:—

"(1) The physical conditions of the deep sea throughout all the great Ocean-basins.

"(2) The chemical constitution of the water at various depths from the surface to the bottom.

"(3) The physical and chemical characters of the deposits.

"(4) The distribution of organic life throughout the areas explored.

"For effectively carrying out these researches there would, in the opinion of the President and Council, be required—

"(1) A ship of sufficient size to afford accommodation and storage-room for sea-voyages of considerable length and for probable absence of four years.

"(2) A staff of scientific men qualified to take charge of the several branches of investigation.

"(3) A supply of everything necessary for the collection of the objects of research, for the prosecution of the physical and chemical investigations, and for the study and preservation of the specimens of organic life.

"The President and Council hope that, in the event of their recommendation being adopted, it may be possible for the Expedition to leave England some time in the year 1872; and they would suggest that as its organization will require much time and labour, no time should be lost in the commencement of preparations.

"The President and Council desire to take this opportunity of expressing their readiness to render every assistance in their power to such an undertaking; to advise upon (1) the route which might be followed by the Expedition, (2) the scientific equipment, (3) the composition of the scientific staff, (4) the instructions for that staff; as well as upon any matter connected with the Expedition upon which their Lordships might desire their opinion.

"The President and Council have abstained from any allusion to geographical discovery or hydrographical investigations, for which the proposed Expedition will doubtless afford abundant opportunity, because their Lordships will doubtless be better judges of what may be conveniently undertaken in these respects, without departing materially from the primary objects of the voyage; and they would only add their hope that, in accordance with the precedents followed by this and other countries under somewhat similar circumstances, a full account of the voyage and its scientific results may be published under the auspices of the Government as soon after its return as convenient, the necessary expense being defrayed by a grant from the Treasury.

"The President and Council desire, in conclusion, to express their willingness to assist in the preparation for such publication of the scientific results.

"I remain," &c.,

Resolved,—That the appointment of the Committee proposed in the Report be deferred until an answer has been received from the Government.

March 21st, 1872.

Read the following communication from the Admiralty :—

"ADMIRALTY, *2nd March*, 1872.

"SIR,—In reply to your Letter of the 8th of December, 1871, conveying a representation from the President and Council of the Royal Society that advantages of great importance to Science and Navigation would result from equipping an Expedition for the Examination of the Physical Conditions of the Deep Sea throughout all the Great Oceanic Basins, and for other special objects therein named,—

"2. I am commanded by my Lords Commissioners of the Admiralty to acquaint you, for the information of the President and Council, that they have had the subject under their consideration, and have decided to fit out one of Her Majesty's ships to leave England on a Voyage of Circumnavigation towards the close of the present year, in prosecution of the objects specified in your letter.

"3. I am further desired to inform you that their Lordships will be prepared to receive from the President and Council of the Royal Society any suggestions that they may desire to make on the Scientific Equipment of the Vessel, the Composition of the Civilian Scientific Staff, or any other Scientific matter connected with the Expedition upon which that body may desire to offer their opinion.

"I am, Sir,
"Your obedient Servant,
"THOS. WOLLEY."

"*The Secretary to the Royal Society.*"

> Resolved,—That the Letter from the Admiralty be referred for consideration and for report to the Council, to a Committee consisting of the President and Officers, Dr. Carpenter, Dr. Frankland, Dr. Hooker, Professor Huxley, the Hydrographer of the Admiralty, Mr. Gwyn Jeffreys, Mr. Siemens, Sir William Thomson, Dr. Wyville Thomson, Dr. Williamson, and Mr. Alfred R. Wallace, with power to add to their number.

June 20th, 1872.

In reference to the arrangements to be made for the Circumnavigatory Expedition, for which H.M.S. 'Challenger' has now been put in commission, the Committee presented the following Report to the Council, viz.:—

"The Committee suggest that the President and Council should direct a Letter to be written to the Secretary of the Admiralty to the following effect:—'That it appears desirable that the Scientific gentlemen who are to accompany the 'Challenger' Expedition should be selected at an early date and their salaries decided on, in order that they may be enabled to make the necessary arrangements for an extended absence from England.

"'The President and Council of the Royal Society therefore recommend as a fit and proper person, to superintend and be at the head of the Civilian Scientific Staff of the Expedition, Wyville Thomson, LL.D., F.R.S., &c., Regius Professor of Natural History in the University of Edinburgh; and that, as Professor Thomson will have to give up his position, with its emoluments, at Edinburgh for the time he is absent, the President and Council are of opinion that a less sum than 1,000*l.* per annum cannot properly be offered to him.

"'They propose that the other members of the Staff and their Salaries should be as follows:—

Mr. John James Wild, as Secretary to the Director and Artist . £400
Mr. John Young Buchanan, M.A., Principal Laboratory Assistant
    in the University of Edinburgh, as Chemist and Physicist 200
Mr. Henry Nottidge Moseley, B.A. (Oxon.), Radcliffe Travelling
    Fellow of Oxford University, as Naturalist 200
Dr. William Stirling, D.Sc. (Edin.), M.B., Falconer Fellow of
    the University of Edinburgh, as Naturalist 200
Mr. John Murray, as Naturalist 200

"The Committee further report that Prof. Wyville Thomson informed them that he had gone with Admiral Richards to Sheerness to examine the 'Challenger,' and that the arrangements appeared to be satisfactory in every respect."

Resolved,—That the Report of the Circumnavigation Committee be adopted, and that a communication be made to the Admiralty in terms of their recommendation.

November 14*th*, 1872.

The Council proceeded to consider the Report of the Circumnavigation Committee.

The following is the Letter from the Admiralty to which the Report refers :—

"ADMIRALTY, *August* 22*nd*, 1872.

"SIR,—With reference to my letter of the 6th instant, and to previous correspondence on the subject of the intended deep-sea exploratory Expedition, I am commanded by My Lords Commissioners of the Admiralty to acquaint you that H.M.S. 'Challenger' will probably be ready to leave this country about the end of November; and their Lordships will be glad to learn what are the precise objects of research which the President and Council of the Royal Society have in view, and in what particular portions of the Ocean such investigations may, in their opinion, be carried out with the greatest advantage to science and the best probability of success.

"2. The object of their Lordships is to frame their instructions to the Officer in command of the 'Challenger,' so far as may be possible, to meet the recommendations of the President and Council of the Royal Society.

"I am, Sir,
"Your obedient Servant,
"VERNON LUSHINGTON."

"*W. Sharpey, Esq., M.D., &c.*
"*Secretary of the Royal Society, Burlington House.*"

The Report having been considered, was adopted as follows :—

The Circumnavigation Committee have had before them the Letter from the Admiralty to the Royal Society, dated August 22, 1872, and as the Council were not in Session and the matter was pressing, they have thought it best to treat the letter as having been referred to them by the Council. They beg leave to recommend to the Council that an answer be returned to the Admiralty to the following effect :—

The principal object of the proposed Expedition is understood to be to investigate the physical and biological conditions of the great Ocean-basins; and it is recommended for that purpose to pass down the coast of Portugal and Spain, to cross the Atlantic from Madeira to the West-Indian Islands, to go to Bermuda, thence to the Azores, the Cape de Verde Islands, the Coast of South America, and across the South Atlantic to the Cape of Good Hope. Thence by the Marion Islands, the Crozets, and Kerguelen Land, to Australia and New Zealand, going southwards *en route*, opposite the centre of the Indian Ocean, as near as may be with convenience and safety to the southern Ice-barrier. From New Zealand through the Coral Sea and Torres Straits, westward between Lombok and Bali, and thence through the Celebes and Sulu Seas to Manilla, then eastward into the Pacific, visiting New Guinea, New Britain, the Solomon Islands; and afterwards to Japan, where some considerable time might be profitably spent. From Japan the course should be directed across the Pacific to Vancouver Island, then southerly through the eastern trough of the Pacific, and homewards round Cape Horn. This route will give an opportunity of examining many of the principal ocean phenomena, including the Gulf-stream and Equatorial currents; some of the biological conditions of the sea of the Antilles; the fauna of the deep water of the South Atlantic, which is as yet unknown, and the specially interesting fauna of the borders of the Antarctic Sea. Special attention shall be paid to the botany and zoology of the Marion Islands, the Crozets, Kerguelen Land, and any new groups of islands which may possibly be met with in the region to the south-east of the Cape of Good Hope. Probably investigations in these latitudes may be difficult; it must be remembered, however, that the marine fauna of these regions is nearly unknown, that it must bear a most interesting relation to the fauna of high northern latitudes, that the region is inaccessible except under such circumstances as the present, and that every addition to our knowledge of it will be of value. For the same reasons the Expedition should, if possible, touch at the Auckland, Campbell, and especially the Macquarie Islands. Particular attention should be paid to the zoology of the sea between New Zealand, Sydney, New Caledonia, and the Fiji and Friendly Islands, as it is probable that the Antarctic fauna may be found there at accessible depths. New Britain and New Ireland are almost unknown, and from their geographical position a special interest attaches to their Zoology, Botany, and Ethnology. The route through this part of the Pacific will give an opportunity of checking and repeating previous observations on the structure of coral-reefs and the growth of coral, and of collecting series of volcanic rocks. The Japan current will also be studied, and the current along the coast of California. The course from Japan to Vancouver Island and thence to Valparaiso will afford an opportunity of determining the physical geography and the distribution of life in these regions, of which at present nothing is known.

I. Physical Observations.

In crossing the great Ocean-basins, observations should be made at stations the positions of which are carefully determined, chosen so far as possible at equal distances, the length of the intervals being of course dependent on circumstances. At each station should be noted the time of the different observations, the state of the weather, the temperature of the surface of the sea, the depth, the bottom temperature determined by the mean of two Miller-Casella thermometers, the specific gravity of the surface- and bottom-water. The nature of the bottom should be determined by the use of a sounding-instrument constructed to bring up samples of the bottom, and also, if possible, by a haul of the dredge. When practicable, the amount and nature of the gases contained in the water, and the amount and nature of the salts and organic matter, should be ascertained. As frequently as possible, especially in the path of currents, serial temperature-soundings ought to be taken either with the instrument of Mr. Siemens or with the Miller-Casella thermometer, and in the latter case at intervals of 10, 50, or 100 fathoms, to determine the depth and volume of masses of moving water derived from different sources.

The simple determination of the depth of the ocean at tolerably regular distances throughout the entire voyage is an object of such primary importance that it should be carried out whenever possible, even when circumstances may not admit of dredging, or of anything beyond sounding. The investigation of various problems relating to the past history of the globe, its geography at different geological epochs, and the existing distribution of animals and plants, as well as the nature and causes of oceanic circulation, will be greatly aided by a more accurate knowledge of the contour of the sea bed.

Surface Temperature.—The surface-temperature of the sea, as also the temperature of the air as determined by the dry- and wet-bulb thermometers, should be regularly recorded every two hours during the day and night throughout the voyage.

These records should be reduced to curves, for the purpose of ready comparison: and the following points should be carefully attended to :—

1. In case of a general correspondence between the temperature of the sea and that of the air, it should be noted whether in the diurnal variation of both the sea appears to *follow* the air, or the air the sea.

2. In case of a marked discordance, the condition or conditions of that discordance should be sought in (*a*) the direction and force of the wind, (*b*) the direction and rate of movement of the ocean surface-water, (*c*) the hygrometric state of the atmosphere. When the air is very dry, there is reason to believe that the temperature of the surface of the sea is reduced by excessive evaporation, and that it may be below that of the subsurface stratum a few fathoms deep. It will be desirable therefore, that every opportunity should be taken of comparing the temperature at the surface with the temperature of the subsurface stratum,—say at every 5 fathoms down to 20 fathoms.

Temperature Soundings.—The determination of the temperature, not merely of the bottom of the ocean, over a wide geographical range, but of its various intermediate strata, is one of the most important objects of the Expedition; and should, therefore, be systematically prosecuted on a method which should secure comparable results. The following suggestions, based on the experience already obtained, in the North Atlantic, are made for the sake of indicating the manner in which time and labour may be economized in making serial soundings, in case of the employment of the Miller-Casella thermometer.

They will be specially applicable to the area in which the work of the Expedition will commence; but the thermal conditions of other areas may prove so different, that the method may need considerable modification.

The following strata appear to be definitely distinguishable in the North Atlantic:—(a) a "superficial stratum," of which the temperature varies with that of the atmosphere, and with the amount of insolation it receives. The thickness of this stratum does not seem to be generally much above 100 fathoms; and the greatest amount of heating shows itself in the uppermost 50 fathoms. (b) Beneath this is an "upper stratum," the temperature of which slowly diminishes as the depth increases down to several hundred fathoms: the temperature of this stratum, in high latitudes is considerably *above* the normal of the latitude; but in the intertropical region it seems to be considerably *below* the normal. (c) Below this is a stratum in which the rate of diminution of temperature with increasing depth is rapid, often amounting to 10° or more in 200 fathoms. (d) The whole of the deeper part of the North Atlantic, below 1,000 fathoms, is believed to be occupied by water not many degrees above 32°. With regard to this "Glacial Stratum," it is exceedingly important that its depth and temperature should be carefully determined.

It will probably be found sufficient in the first instance to take, with each deep *bottom* sounding, *serial* soundings at every 250 fathoms, down to 1,250 fathoms; and then to fill up the intervals in as much detail as may seem desirable. Thus, where the fall is very small between one 250 and the next, or between any one and the bottom, no intermediate observation will be needed: but where an abrupt difference of several degrees shows itself, it should be ascertained by intermediate observations whether this difference is sudden or gradual.

The instrument devised by Mr. Siemens for the determination of submarine temperatures is peculiarly adapted for serial measurements, as it does not require to be hauled up for each reading. It should, however, be used in conjunction with the Miller-Casella thermometer, so as to ascertain how far the two instruments are comparable: and this point having been settled, Mr. Siemens' instrument should be used in all serial soundings; and frequent readings should be taken with it, both in descending and ascending.

A question raised by the observations of the U.S. Coast Surveyors in the Florida Channel, and by those of our own surveyors in the China Sea, is the extent to which the colder and therefore heavier water may run *up hill* on the sides of declivities. The position of the Azores will probably be found very suitable for observations of this kind. Temperature-soundings should be taken at various depths, especially on their north and south slopes, and in the channels between the islands; and the temperatures at various depths should be compared with those of corresponding depths in the open ocean.

It is in the Southern Oceans that the study of ocean-temperatures at different depths is expected to afford the most important results; and it should there be systematically prosecuted. The great Ice-barrier should be approached as nearly as may be deemed suitable, in a meridian nearly corresponding to the centre of one of the three great Southern Oceans;—say to the south of Kerguelen's Land; and a line of soundings should be carried north and south as nearly as may be.

In connection with the limitation of the area and depth of the reef-building corals, it will be very important to ascertain the rate of reduction of temperature from the surface downwards in the region of their greatest activity; as it has been suggested that the limitation of living reef-builders to 20 fathoms may be a thermal one.

Wherever any anomaly of temperature presents itself, the condition of such anomaly should, if possible, be ascertained. Thus there is reason to believe that the cause of the temperature of the surface-water being below that of the subsurface stratum, in the neighbourhood of melting ice, is that the water cooled by the ice, by admixture with the water derived from its liquefaction, is also rendered less salt, and therefore floats upon the warmer and salter water beneath. Here the determination of Specific Gravities will afford the clue. In other instances a warm *current* may be found beneath a colder stratum; and the use of the "current-drag" might show its direction and rate. In other cases, again, it may happen that a warm submarine spring is discharging itself,—as is known to occur near the island of Ascension. In such a case, it would be desirable to trace it as nearly as may be to its source, and to ascertain its composition.

Movements of the Ocean.—The determination of *Surface-Currents* will, of course, be a part of the regular routine, but it is particularly desirable that accurate observations should be made along the line of sounding in the Southern Ocean, as to the existence of what has been described as a general "Southerly set" of oceanic water, the rate of which is probably very slow. It is also very important that endeavours should be made to test by the "current drag," whether any *underflow* can be shown to exist from either Polar basin towards the Equatorial region. A suitable locality for such experiments in the North Atlantic would probably be the neighbourhood of the Azores, which are in the line of the glacial flow from the North Polar Channel. The guide to the depth at which the current-drag should be suspended will be furnished by the thermometer, especially where there is any abrupt transition between one stratum and another. It would be desirable that not only the rate and direction of surface-drift, but those of the subsurface stratum at (say) 200 fathoms' depth, should be determined at the same time with those of the deep stratum.

Tidal Observations.—No opportunity of making tidal observations should be lost. Careful observations made by aid of a properly placed tide-pole in any part of the world will be valuable. Accurate measurements of the sea-level once every hour (best every *lunar* hour, *i.e.* at intervals of $1^h\ 2^m$ of solar time) for a lunar fortnight (the time of course being kept) would be very valuable information.

Beach-marks.—In reference to the interesting question of the elevation or subsidence of land, it will be very desirable, when sufficient tidal observations can be obtained to settle the mean level of the sea, that permanent beach-marks should be established, recording the date and height above such mean level. Even recording the height to which the tide rose on a certain day and time would render a comparison possible in future years.

A good determination of the mean sea-level by the simple operation of taking means may be made, in less than two days, with even a moderate number of observations *properly distributed so as to subdivide both solar and lunar days into not less than three equal parts.* Suppose, for example, we choose 8-hour intervals, both solar and lunar. Take a lunar day at $24^h\ 48^m$ solar time, which is near enough, and is convenient for division; and choosing any convenient hour for commencement, let the height of the water be observed at the following times, reckoned from the commencement:—

h.	m.	h.	m.	h.	m.
0	0	8	0	16	0
8	16	16	16	24	16
16	32	24	32	32	32

The observations may be regarded as forming three groups of three each, the members of each group being separated by 8 hours solar or lunar, while one group is separated from the next by 8 hours lunar or solar. In the mean of the nine results the lunar and solar semi-diurnal and diurnal inequalities are all four eliminated.

Nine is the smallest number of observations which can form a complete series. If the solar day be divided into m and the lunar into n equal parts, where m and n must both be greater than 2, there will be mn observations in the series; and if either m or n be a multiple of 3, or of a larger number, the whole series may be divided into two or more series having no observation in common, and each complete in itself. The accuracy of the method can thus be tested, by comparing the means obtained from the separate sub-series of which the whole is made up.

Should the ship's stay not permit of the employment of the above method a very fair determination may be made in less than a day, by taking the mean of n observations taken at intervals of the nth part of a lunar day, n being greater than 2.

Thus if $n = 3$, these observations require a total interval of time amounting to only $16^{\mathrm{h}}\ 32^{\mathrm{m}}$. The theoretical error of this method is very small, and the result thus obtained is decidedly to be preferred to the mere mean of the heights at high and low water.

The mean level thus determined is subject to meteorological influences, and it would be desirable, should there be an opportunity, to redetermine it at the same place at a different time of year. Should a regular series of observations for a fortnight be instituted, it would be superfluous to make an independent determination of the mean sea-level by either of the above methods at the same time.

Besides taking observations on the ordinary waves of the sea when at all remarkable, the Scientific Staff should carefully note the circumstances of any waves attributable to earthquakes.

Specific Gravity.—The specific gravity of the surface- and bottom-water should be carefully compared, whenever soundings are taken; and whenever serial soundings are taken, the specific gravity at intermediate depths should be ascertained. Every determination of specific gravity should be made with careful attention to temperature; and the requisite correction should be applied from the best Table for its reduction to the uniform standard of 60°. It would be well to check the most important results by the balance; samples being preserved for examination in harbour. Wherever the temperature of the surface is high,—especially, of course, in the Intertropical region,—samples should be collected at every 10 fathoms, for the purpose of ascertaining whether any effect is produced upon the specific gravity of the upper stratum by evaporation, and how far down this effect extends.

Transparency of the Water.—Observations for transparency should be taken at various depths and under different conditions by means of Mr. Siemens' photographic apparatus. As, however, the action of this depends upon the more refrangible rays, and the absorption of these and of the more luminous rays might be different, and that, in a manner varying with circumstances, such as the presence or absence of suspended matter, &c., the transparency of the sea should also be tested by lowering a white plate or large white tile to various measured depths, and noting the change of intensity and colour as it descends, and the depth at which it ceases to be visible. The state of the sky at the time should be mentioned, and the altitude of the sun, if shining, roughly measured, or if not shining, deduced from the time of day.

Relation of Barometric Pressure to Latitude.—In Poggendorff's 'Annalen,' vol. xxvi. 1832, p. 395, is a remarkable paper by Professor G. F. Schouw on the relation between the height of the barometer at the level of sea, and the latitude of the place of observation. At page 434 is a rough statement of the results of his researches, the heights being given in Paris lines.

Lat.	Barometer mercury at 0° C.
0°	337·0
10	337·5
20	338·5
30	339·0
40	338·0
50	337·0
60	335·5
65	333·0
70	334·0
75	335·5

The Expedition might contribute to the examination of this law, not only by giving especial attention to the barometer observations at about the critical latitudes 0°, 30°, 65°, 70°, but also by comparing any barometers with which long series of observations have been made at any port they may touch at, with the ship's standard barometer.

It appears probable from Schouw's paper, that certain meridians are meridians of high pressure and others of low pressure.

For comparison of barometer and measures of heights, it appears that the aneroid barometer constructed by Goldschmid of Zürich would be very useful.

It is very desirable that the state of the barometer and thermometer should be read at least every two hours.

II. CHEMICAL OBSERVATIONS.

1. Samples of sea-water should be collected for chemical analysis at the surface and at various depths, and in various conditions. Each sample should be placed in a Winchester quart glass-stoppered bottle, the stopper being tied down with tape and sealed in such a manner that the contents cannot be tampered with.

2. Portions of the same samples should be, immediately after their collection, boiled *in vacuo*, the gases collected, their volume determined as accurately as may be, and a portion, not less than one cubic inch, hermetically sealed in a glass tube, to be sent home at any time for complete analysis.

3. Frequent samples of sea-water taken at the surface, and others taken beneath as opportunity offers, should have determinations of chlorine made upon them at once, or as soon as convenient.

This operation could easily be carried on in any but very heavy weather. On the other hand, it is not thought that any trustworthy analyses of gases could be made on board ship, unless in harbour or in the calmest weather.

4. Such samples of the sea-bottom as are brought up should be carefully dried and preserved for examination and analysis.

5. The gas contained in the swimming-bladders of fishes caught near the surface and at different depths should be preserved for analysis. In each case the species, sex, and size, and especially the depth at which the fish was caught, should be stated.

III. BOTANICAL OBSERVATIONS.

The duties of a botanist in travelling are twofold, and in the case of the voyage of circumnavigation about to be undertaken by H.M.S. 'Challenger' they are of equal importance.

Of these, the one refers to forming complete collections of the plants of all interesting localities, and especially of the individual islands of oceanic groups.

The other, to making observations upon life, history, and

structure in the case of plants where special knowledge is concerned.

In the first of these the botanist must necessarily be largely helped by the assistance to be obtained on board ship from the officers and crew, working under his guidance and close supervision. When time and opportunity are wanting for making complete collections, preference should be given to the phanerogamous vegetation.

In the second he will have to depend upon his own resources, and will therefore require that the mere process of collection does not make too great demands upon his time, although in itself exceedingly important, and by no means to be neglected.

The general directions for travellers, printed in the Admiralty Manual of Scientific Inquiry, will of course be kept in view.

Especial stress must, however, be laid upon the necessity of obtaining information about the vegetation of oceanic islands. These are, in many cases, the last positions held by floras of great antiquity; and, as in the case of St. Helena, they are liable to speedily become exterminated, and therefore to pass into irremediable oblivion when the islands become occupied.

Of many that lie not far from the usual tracks of ships, absolutely nothing is known, whilst of the flora of a vast majority we possess most imperfect materials. The following are especially worth exploring; and to the list is added an indication of the least explored coast-lines of the great continents. As far as possible complete dried collections should be made, not only of each group, but of each islet of the group; for it is usually the case that the floras of contiguous oceanic islets are wonderfully different. Of those in italics the vegetation is absolutely unknown, or all but so.

1. ATLANTIC OCEAN. Cape de Verd, Tristan d'Acunha, *Fernando Noronha, Trinidad*, and *Martin Vaz* (off the Brazil coast), *Diego Ramirez*, S. Georgia. The African coast between Morocco and Senegal, the Gaboon, and Damara Land offer the most novel fields. On the American coast, Cayenne, Bahia to Cape Frio, Patagonia.

2. WEST INDIES. The Bahamas and St. Domingo and the Antilles have been very imperfectly explored, except Dominica, Trinidad, and Martinique. On the mainland, Honduras, Nicaragua, and the coast region of Mexico, the Mosquito shores and Guatemala offer rich fields for botanical research.

3. INDIAN OCEAN. The Seychelles, *Ammirantes*, Madagascar, Bourbon, *Socotra*, St. Paul's, and Amsterdam Islands, *Prince Edward's*, the *Crozets* and *Marion* groups. Of the E. African coast to the north of Natal no part is well explored, and the greater part is utterly unknown botanically.

4. PACIFIC OCEAN. 1. N. TEMPERATE. Collections are wanted from N. Japan and the Kuriles and Aleutian Islands. 2. TROPICAL. Considerable collections have been made only in the Sandwich Islands, Fiji Islands, Tahiti, and New Caledonia; from all of which more are much wanted. The Marquesas, New Hebrides, *Marshall's*, Solomon's, and *Caroline's*, together with all the smaller groups, are still less known. Of the American continent, the Californian Peninsula, Mexico, and the whole coast from Lima to Valparaiso, are but imperfectly known. Of the small islands off the coast, Juan Fernandez and the Galapagos alone have been partially botanized. 3. S. TEMPERATE. Juan Fernandez, *Masafuera*, St. Felix, and Ambroise, *Pitcairn*, *Bounty*, *Antipodes*, *Emerald*, *Macquarie* Islands.

5. INDIAN ARCHIPELAGO. Java alone is explored, and the Philippines very partially; collections are especially wanted from all the islands east of Java to the Louisiade and Solomon Archipelagos, especially Lombok and New Guinea. Siam, Cochin China, and the whole Chinese sea-board want exploration.

6. AUSTRALIA. All the tropical coasts are very partially explored.

Photographs or careful drawings of tropical vegetation often convey interesting information, and should contain some reference to a scale of dimensions.

An inquiry of much importance, for which the present Expedition affords a favourable opportunity, is that into the vitality of seeds exposed to the action of sea-water.

Observations should especially be made on the fruits and seeds of those plants which have become widely distributed throughout the tropical regions of the world, apparently without the intervention of man; but further observations on other plants of different natural orders may be of great value with reference to questions of geographical distribution.

The following Instructions have been drawn up for the botanical collectors as to objects of special attention at particular places :—

Porto Rico.—In collecting, distinguish the plants of the Savannahs from those of the mountains, which, if possible, should be ascended. The palms and tree-ferns are quite unknown; marine algæ also are wanted.

Cape de Verdes.—Make for the highest peaks, where the vegetation is peculiar and analogous to that of Madeira and the Canaries.

Fernando de Noronha.—Land if possible. Very remarkable plants are said to occur, different from those of Brazil.

Trinidad.—A complete collection is required. A tree-fern exists, but the species is unknown.

Prince Edward's Island and Crozets.—Two spots more interesting for the exploration of their vegetation do not exist upon the face of the globe. Every effort should be made to make a complete collection.

Kerguelen's Land.—A thorough exploration should be made, and the Cryptogamic plants and algæ diligently collected. The Antarctic Expedition was only there in midwinter; flowering specimens of *Pringlea* are wanted.

Auckland and Campbell Islands.—The floras should be well explored.

South Pacific and Indian Oceans.—Attend to general instructions, more especially as regards palms and large monocotyledons generally. Marine algæ are said to be scarce, and should be looked for all the more diligently. In the North Pacific, south temperate algæ are said to prevail.

Aleutian Islands.—Collections are particularly wanted.

Every effort should be made to land on islands *between*

Lat. 30° N. *and* 30° S. along the marked track (between Vancouver Island and Valparaiso), so as to connect the vegetation of the American continent with the traces of it that exist in the Sandwich Islands.

Straits of Magellan.—Cryptogams are abundant, but very partially explored.

The following additional notes have been drawn up for the more especial guidance of the botanists of the Circumnavigation:—

Phanerogams. — 1. Fleshy parasitic plants (*Balanophora, Rafflesia,* &c.) are little suitable for dissection and examination unless preserved in spirit; and the same remark applies to fleshy flowers and inflorescences generally. Dried specimens, however, are not without their value, and should always be obtained as well.

2. The stems of scandent and climbing plants are often very anomalous in their structure. Short portions of such stems should be collected when the cross-section is in any way remarkable, with the foliage, flowers, and fruit when possible. A few leaves and flowers should also be tied up between two pieces of card, and attached at once to the specimens of the stem, so as to ensure future identification.

3. Attention should be given to the esculent and medicinal substances used in various places. Specimens should be obtained, and whenever possible they should be accompanied by complete specimens of the plants from which such substances are obtained.

4. The common weeds and ruderal plants growing about ports or landing-places should not be overlooked, and, as far as practicable, trustworthy information should be recorded as to the date and circumstances of the introduction of foreign species.

5. The distribution of marine Phanerogamic plants (*Zostera, Cymodocea,* &c.) should also be noted, and specimens preserved with their latitude and longitude. Their buds and parts of fructification should be put into spirit.

6. The flowers of *Loranthaceæ* and *Santalaceæ* should be preserved in spirit, and also dried to exhibit general habit.

7. The inflorescence of Aroids should be dissected when fresh, or put into spirit. Note the placentation and position of the ovules.

8. Devote especial attention to the study of Screw-Pines and Palms when opportunity arises, even if necessary to the neglect of other things. The general habit of the plants should be sketched; the male and female inflorescence should be preserved, and also the fruit; the foliage should be dried and folded, and packed in boxes. Many fleshy vegetable objects may be "killed" by a longer or shorter immersion in spirit. They then dry up without decaying, and form useful specimens.

9. With respect to Palms, further note the height, position of the spadix, and preponderance of the sexes in both monœcious and diœcious species, also form and dimensions of leaves.

10. Surface-driftings should be examined, and any seeds or fragments of land-plants carefully noted when determinable, with direction of currents and latitude and longitude.

11. Facts are also required as to the part played by icebergs in plant-distribution. If any opportunity occurs for their examination, it would be desirable to preserve and note any vegetable material which might be found upon their surface; also to examine any rock-fragments for lichens.

12. *Ferns.*—Ferns should always, when possible, be obtained with fructification. In the case of tree-ferns, our knowledge of which, from the imperfection of material for description, is very defective, a portion of the stem sufficient to illustrate its structure should be obtained, with notes of its height; a fragment of a frond (between pieces of card) and the base of a stipes should be tied to the specimen of the stem; also a note as to whether the adventitious roots were living or dead.

The number of fronds should be counted, their dimensions taken, and the basal scales carefully preserved.

Note if tree-ferns are ever attacked by insects or fungi, and whether they form the food of any class of animals.

13. *Mosses, &c.*—Many mosses are aquatic. In the case of diœcious species of mosses, plants of both sexes should be, when possible, secured.

14. Aquatic species of *Ricciaceæ* should be looked for. Minute *Jungermanniaceæ* are found on the foliage of other plants.

15. *Podostemaceæ* are found in rocky running streams in hot countries. They have a remarkable superficial resemblance to Hepaticæ. Except at the flowering season they are altogether submerged. Specimens should be preserved in spirit as well as dried.

16. *Fungi.*—Take notes of all fleshy fungi, especially as regards colour; the spores should be allowed to fall on paper, and the colour of these noted also. The fleshy species may sometimes be advantageously immersed in spirit before preparing for the herbarium.

17. Examine the fungi which grow on ants' nests, taking care to get perfect as well as imperfect states, and to secure, if possible, specimens which have not burst their volva.

18. Look out for luminous species, and ascertain whether they are luminous in themselves, or whether the luminosity depends on decomposition.

19. Secure specimens of all esculent or medicinal fungi which are sold in bazaars, noting, if possible, the vernacular name.

20. Note any species of fleshy fungi which arise like the *Pietra Fungaja* from a mass of earth impregnated with mycelium, or from a globose resting-mass.

21. Attend especially to any fungi which attack crops, whether cereal or otherwise; and particularly gather specimens of vine-mildew and potato-mildew, should they be met with. Even common wheat-mildew, smut, &c., should be preserved.

22. In every case note date of collection, soil, and other circumstances relative to particular specimens.

23. Look after those fungi which attack the larvæ of insects.

24. In the case of the *Myxogastres*, sketches should be made on the spot of their general form, with details of microscopic appearance. It would be worth while attempting to preserve specimens for future microscopic examination by means of osmic acid.

25. *Algæ.*—Marine algæ may be found between tide-marks attached to rocks and stones, or rooting in sand, &c.; those in deeper water are got by dredging, and many are cast up after storms; small kinds grow on the larger, and some being like fleshy crusts on stones, shells, &c., must be pared off by means of a knife.

The more delicate kinds, after gentle washing, may be floated in a vessel of fresh water, upon thick and smooth writing or drawing paper; then gently lift out paper and plant together, allow some time to drip; then place on the sea-weed clean linen or cotton cloth, and on it a sheet of absorbent paper, and submit to moderate pressure—many adhere to paper but not to cloth; then change the cloth and absorbent paper till the specimens are dry. Large coarser kinds may be dried in the same way as land-plants; or are to be spread out in the *shade*, taking care to prevent contact of rain or fresh water of any kind; when sufficiently dry, tie them loosely in any kind of wrapping paper; those preserved in this rough way may be expanded and floated out in water at any time afterwards. A few specimens of each of the more delicate algæ ought to be dried on mica or glass. A note of date and locality ought to be attached to every species.

Delicate slimy algæ are best prepared by floating out on smooth-surfaced paper (known as "sketching paper"), then allowed to drip and dry by simple exposure to currents of air, without pressure.

26. Very little information exists regarding the range of depth of marine plants. It will be very desirable that observations should be made upon this subject, as opportunity from time to time presents itself.

Professor Dickie remarks, and the caution should be borne in mind:—" When the dredge ceases to scrape the bottom, it becomes in its progress to the surface much the same as a towing-net, capturing bodies which are being carried along by currents, and therefore great caution is necessary in reference to any marine plants found in it. Sea-weeds are among the most common of all bodies carried by currents near the surface or at various depths below, and from their nature are very likely to be entangled and brought up."

27. Carefully note and preserve algæ brought up in dredge in moderate depths, under 100 fathoms, or deeper. Preserve specimens *attached* to shells, corals, &c., which would indicate their being actually *in situ*, and not caught by dredge as it comes up.

28. Examine mud brought up by dredge from different depths for living Diatoms; examine also for the same purpose the stomachs of *Salpæ* and other marine animals.

29. Note algæ on ships, &c., with the submerged parts in a foul condition; also preserve scrapings of coloured crusts or slimy matter, green, brown, &c.

30. Observe algæ *floating*, collect specimens, noting latitude and longitude, currents, &c.

31. Examine loose floating objects, drift-wood, &c., for algæ. If no prominent species presents itself, preserve scrapings of any coloured crusts. Note as above.

32. It might be useful to have a few moderate-sized pieces of wood, oak, &c., quite clean at first, attached to some part of the vessel under water to be examined, say, monthly. The larger or shorter prominent algæ should be kept and noted, and crusts on such examined and preserved, with notes of the vessel's course.

33. Various instances have been mentioned by travellers of the coloration of the sea by minute algæ, as in the Straits of Malacca by Harvey; any case of this kind would be worth especial attention.

34. The calcareous algæ (*Melobesia*, &c.) are comparatively little known, and are apt to be overlooked.

35. Freshwater algæ should be collected as occasion presents. Professor Dickie states that they may be either dried like the marine kinds, or preserved in a fluid composed of 3 parts alcohol, 2 parts water, 1 part glycerine, well mixed.

36. Cases are recorded of the presence of algæ in hot springs. If such are met with, the temperature should be noted and specimens preserved.

IV. Zoological Observations.

As the Scientific Director of the Expedition is an accomplished zoologist, and has already had much experience in marine exploration, it will suffice to offer a few suggestions under this head.

The quadrant-like zone of the Pacific, which separates the northern and eastern boundaries of the Polynesian Archipelago (using "Polynesia" in its broadest sense as inclusive of "Micronesia") from the coasts of N. Asia and America, is as little explored from the point of view of the physical geographer as from that of the biologist. It would be a matter of great importance to examine the depth, and the nature of the deep-sea fauna, of this zone by taking a line of soundings and dredgings in its northern half (say between Japan and Vancouver) and in its eastern half (say between Vancouver and Valparaiso). If practicable, it would further be very desirable to explore the littoral fauna of Waihou, Easter Island, or Sala y Gomez, with the view of comparing it critically with that of the west coast of South America.

If H.M.S. 'Challenger' passes through Torres Straits, it will be very desirable to examine the littoral fauna of the Papuan shore of the straits in order to compare it with that of the Australian shore. The late Professor Jukes, in his 'Voyage of the Fly' many years ago, directed attention to this point and to its theoretical bearings.

The Hydrographic examination of "Wallace's line" in the Malay Archipelago, and of the littoral faunas on the opposite sides of that line, is of great importance, considering the significance of that line as a boundary between two Distributional provinces. And additional interest has been given to the exploration of this region by Capt. Chimmo's recently obtained sounding of 2,800 fathoms in the Celebes Sea, the mud brought up being almost devoid of calcareous organisms, but containing abundant spicula of Sponges and *Radiolaria*.

The light from any self-luminous objects met with should be examined with a prism as to its composition. The colours of animals captured should also be examined with a prism, or by aid of the microscopic spectroscope.

V. Concluding Observations.

Attention should be paid to the Geology of districts which have not hitherto been examined, and collections of minerals, rocks, and fossils should be made. Detailed suggestions as to the duties of the geologist accompanying the Expedition are unnecessary; but it seems desirable that, at all shores visited, evidence of recent elevation or subsidence of land should be sought for, and the exact nature of these evidences carefully recorded.

Every opportunity should be taken of obtaining photographs of native races to one scale; and of making such observations as are practicable with regard to their physical characteristics, language, habits, implements, and antiquities. It would be advisable that specimens of hair of unmixed races should in all cases be obtained.

Each station should have a special number associated with it in the regular journal of the day's proceedings, and that number should be noted prominently on everything connected with that station; so that in case of labels being lost or becoming indistinct, or other references failing, the conditions of the dredging or other observations may at once be forthcoming on reference to the number in the journal. All specimens procured should be carefully preserved in spirit or otherwise, and packed in cases with the contents noted; to be dealt with in the way which seems most likely to conduce to the rapid and accurate development of the scientific results of the Expedition.

A diary, noticing the general proceedings and results of each day, should be kept by the Scientific Director, with the assistance of his Secretary; and each of the members of the Scientific Staff should be provided with a notebook, in which to enter from day to day his observations and proceedings; and he should submit this diary at certain intervals to the Scientific Director, who would then abstract the results, and incorporate them, along with such additional data as may be supplied by the officers of the ship, in general scientific reports to be sent home to the Hydrographer at every available opportunity.

The Scientific Staff should be provided with an adequate set of books of reference, especially those bearing on perishable objects.

> Resolved,—That the Report of the Circumnavigation Committee, now adopted by the Council, be transmitted by the Secretary to the Secretary of the Admiralty, with the following Letter :—

In reply to your Letter of the 22nd of August, referring to the Exploratory Voyage of H.M.S. 'Challenger,' and desiring to learn, for the information of the Lords Commissioners of the Admiralty, what are the precise objects of research which the Royal Society have in view, and in what particular portions of the Ocean such investigations may, in their opinion, be carried out with the greatest advantage to science and the best probability of success, I am directed to acquaint you that the matter was carefully considered by a Committee, consisting of the President and Officers, with Dr. Allman, Dr. Carpenter, Dr. Frankland, Dr. Hooker, Prof. Huxley, the Hydrographer of the Admiralty, Mr. Gwyn Jeffreys, Mr. Siemens, Sir William Thomson, Dr. Wyville Thomson, Mr. Wallace, and Dr. Williamson. That Committee has presented a Report which has been approved by the President and Council, by whose direction I herewith transmit it to you, to be communicated to the Lords Commissioners of the Admiralty in answer to their Lordships' inquiry.

November 30th, 1872.

Read the following Letter :—

"ADMIRALTY, 27th November, 1872.

" SIR,—I am commanded by My Lords Commissioners of the Admiralty to thank you for your communication of the 22nd instant, in regard to the objects of research which the Royal Society have in view with reference to the intended voyage of

H.M.S. 'Challenger,' and to acquaint you that they are desirous of affording to the President and Council of the Royal Society, as well as the Members of the Circumnavigation Committee, an opportunity of inspecting the ship, and the arrangements made with a view to her equipment for the service she is intended to perform.

"2. My Lords therefore invite those gentlemen to proceed to Sheerness on the 6th proximo for the purpose of visiting the 'Challenger;' and a saloon carriage will be ordered to be in readiness to convey them to that port by the 10.30 A.M. train from Victoria Station.

"3. The visitors will be able to return by the 5.10 train from Sheerness, and free railway passes will be provided for them both ways. They will also be met by their Lordships' Hydrographer on the occasion.

"4. I am to request you will inform me, as soon as may be convenient, of the number of the gentlemen who will avail themselves of their Lordships' invitation, in order that the proper number of tickets may be procured.

"I am, Sir,
"Your obedient Servant,
"ROBERT HALL."

"*The Secretary of the Royal Society.*"

Before the Expedition left England, Dr. William Stirling resigned his appointment as Naturalist; and Dr. Rudolf von Willemoës-Suhm, Privat-Docent in Zoology in the University of Munich, was appointed by the Admiralty in his place, on the recommendation of the Council of the Royal Society.

APPENDIX B.

List of the Stations in the Atlantic at which Observations were taken during the Year 1873.

Station I. December 30th, 1872.—Lat. 41° 57' N., Long. 9° 42' W. Depth, 1,125 fathoms. Nature of bottom, globigerina ooze. Locality, about 40 nautical miles west of Vigo Bay.

Station II. January 13th, 1873.—Lat. 38° 10' N., Long. 9° 14' W. Depth, 470 fathoms. Grey ooze. South of Cape Espichel.

Second sounding, same day.—Lat. 38° 5' N., Long. 9° 39' W. Depth, 1,270 fathoms. Globigerina ooze.

Station III. January 15th.—Lat. 37° 2' N., Long. 9° 24' W. Depth, 1,000 fathoms. Globigerina ooze. Bottom temperature, 3°·8 C.

Second sounding, same day.—Lat. 36° 59' N., Long. 9° 14' W. Depth, 525 fathoms. Globigerina ooze. Bottom temperature, 12°·0 C.

Third sounding, same day.—Lat. 37° 2' N., Long. 9° 14' W. Depth, 900 fathoms. Globigerina ooze. Off Cape St. Vincent.

Station IV. January 16th.—Lat. 36° 25' N., Long. 8° 12' W. Depth, 600 fathoms. Globigerina ooze. About 60 miles south-east of Cape St. Vincent.

Station V. January 28th.—Lat. 35° 47' N., Long. 8° 23' W. Depth, 1,090 fathoms. Globigerina ooze. Bottom temperature, 3°·1 C. About 90 miles south-east of Cape St. Vincent.

Station VI. January 30th.—Lat. 36° 23' N., Long. 11° 18' W. Depth, 1,525 fathoms. Globigerina ooze. Bottom temperature, 1°·6 C. About 120 miles south-west of Cape St. Vincent.

Station VII. January 31st.—Lat. 35° 20' N., Long. 13° 4' W. Depth, 2,125 fathoms. Globigerina ooze. Bottom temperature,

$2°·3$ C. About 230 miles south-west of Cape St. Vincent, and 238 miles from Madeira.

Station VIII. February 12th.—At the Canaries. Depth, 620 fathoms. Dark, sandy mud and dead shells. Near the south coast of Gomera Island.

Station I. February 15th.—Lat. 27° 24′ N., Long. 16° 55′ W. Depth, 1,890 fathoms. Globigerina ooze. Bottom temperature, $2°·0$ C. About 40 miles south of Teneriffe Island.

Station II. February 17th.—Lat. 25° 52′ N., Long. 19° 14′ W. Depth, 1,945 fathoms. Globigerina ooze. Bottom temperature, $2°·0$ C. About 260 miles west of Cape Bojador.

Station III. February 18th.—Lat. 25° 45′ N., Long. 20° 12′ W. Depth, 1,525 fathoms. Rock. Bottom temperature, $2°·2$ C. About 160 miles south-west of the Island of Ferro.

Station IV. February 19th.—Lat. 25° 28′ N., Long. 20° 22′ W. Depth, 2,220 fathoms. About 20 miles south-west of Station III., and 2,428 miles from Sombrero Island, West Indies.

Station V. February 21st.—Lat. 24° 20′ N., Long. 24° 28′ W. Depth, 2,740 fathoms. Red clay. Bottom temperature, $2°·0$ C. Distance from Sombrero Island, 2,220 miles.

Station VI. February 23rd.—Lat. 23° 22′ N., Long. 27° 49′ W. Depth, 2,950 fathoms. Red clay. Bottom temperature, $2°·0$ C. Distance from Sombrero Island, 2,013 miles.

Station VII. February 24th.—Lat. 23° 15′ N., Long. 30° 56′ W. Depth, 2,750 fathoms. Red clay. Bottom temperature, $2°·0$ C. Distance from Sombrero Island, 1,841 miles.

Station VIII. February 25th.—Lat. 23° 12′ N., Long. 32° 56′ W. Depth, 2,800 fathoms. Red clay. Bottom temperature, $2°·0$ C. Distance from Sombrero Island, 1,730 miles.

Station IX. February 26th.—Lat. 23° 23′ N., Long. 35° 10′ W. Depth, 3,150 fathoms. Red clay. Bottom temperature, $1°·9$ C. Distance from Sombrero Island, 1,607 miles.

Station X. February 28th.—Lat. 23° 10′ N., Long. 38° 42′ W. Depth, 2,720 fathoms. Red clay. Bottom temperature, $1°·9$ C. Distance from Sombrero Island, 1,411 miles.

Station XI. March 1st.—Lat. 22° 45′ N., Long. 40° 37′ W.

Depth, 2,575 fathoms. Red clay. Bottom temperature, 2°·0 C. Distance from Sombrero Island, 1,307 miles.

Station XII. March 3rd.—Lat. 21° 57′ N., Long. 43° 29′ W. Depth, 2,025 fathoms. Globigerina ooze. Bottom temperature, 2°·2 C. Distance from Sombrero Island, 1,140 miles.

Station XIII. March 4th.—Lat. 21° 38′ N., Long. 44° 39′ W. Depth, 1,900 fathoms. Globigerina ooze. Bottom temperature, 1°·9 C. Distance from Sombrero Island, 1,074 miles.

Station XIV. March 5th.—Lat. 21° 1′ N., Long. 46° 29′ W. Depth, 1,950 fathoms. Globigerina ooze. Bottom temperature, 1°·8 C. Distance from Sombrero Island, 972 miles.

Station XV. March 6th.—Lat. 20° 49′ N., Long. 48° 45′ W. Depth, 2,325 fathoms. Red clay. Bottom temperature, 1°·7 C. Distance from Sombrero Island, 844 miles.

Station XVI. March 7th.—Lat 20° 39′ N., Long. 50° 33′ W. Depth, 2,435 fathoms. Red clay. Bottom temperature, 1°·7 C. Distance from Sombrero Island, 744 miles.

Station XVII. March 8th.—Lat. 20° 7′ N., Long. 52° 32′ W. Depth, 2,385 fathoms. Red clay. Bottom temperature, 1°·9 C. Distance from Sombrero Island, 642 miles.

Station XVIII. March 10th.—Lat. 19° 41′ N., Long. 55° 13′ W. Depth, 2,675 fathoms. Red clay. Bottom temperature, 1°·6 C. Distance from Sombrero Island, 472 miles.

Station XIX. March 11th.—Lat. 19° 15′ N., Long. 57° 47′ W. Depth, 3,000 fathoms. Red clay. Bottom temperature, 1°·3 C. Distance from Sombrero Island, 344 miles.

Station XX. March 12th.—Lat. 18° 56′ N., Long. 59° 35′ W. Depth, 2,975 fathoms. Red clay. Bottom temperature, 1°·6 C. Distance from Sombrero Island, 220 miles.

Station XXI. March 13th.—Lat. 18° 54′ N., Long. 61° 28′ W. Depth, 3,025 fathoms. Red clay. Bottom temperature, 1°·3 C. Distance from Sombrero Island, 115 miles.

Station XXII. March 14th.—Lat. 18° 40′ N., Long. 62° 56′ W. Depth, 1,420 fathoms. Globigerina ooze. Bottom temperature, 3°·0 C. Sombrero Island 30 miles distant.

Station XXIII. March 15th.—Close to Sombrero Island. Depth, 460 fathoms. Globigerina ooze.

Station XXIV. March 25th.—North of Culebra Island, near St. Thomas. Depth, 390 fathoms. Globigerina ooze, coral, and broken shells.

Second sounding, same day.—Depth, 625 fathoms. Globigerina ooze and coral. About 15 miles north of Culebra Island.

Station XXV. March 26th.—Lat. 19° 41′ N., Long. 65° 7′ W. Depth, 3,875 fathoms. Grey ooze. About 85 miles north of St. Thomas, and 754 miles from Bermudas.

Station XXVI. March 27th.—Lat. 21° 26′ N., Long. 65° 16′ W. Depth, 2,800 fathoms. Grey ooze. Distance from Bermudas, 654 miles.

Station XXVII. March 28th.—Lat. 22° 49′ N., Long. 65° 19′ W. Depth, 2,960 fathoms. Grey ooze. Bottom temperature, 1°·5 C. Distance from Bermudas, 566 miles.

Station XXVIII. March 29th.—Lat. 24° 39′ N., Long. 65° 25′ W. Depth, 2,850 fathoms. Red clay. Bottom temperature, 1°·7 C. Distance from Bermudas, 458 miles.

Station XXIX. March 31st.—Lat. 27° 49′ N., Long. 64° 59′ W. Depth, 2,700 fathoms. Red clay. Bottom temperature, 1°·6 C. Distance from Bermudas, 266 miles.

Station XXX. April 1st.—Lat. 29° 5′ N., Long. 65° 1′ W. Depth, 2,600 fathoms. Red clay. Bottom temperature, 1°·8 C. Distance from Bermudas, 191 miles.

Station XXXI. April 3rd.—Lat. 31° 24′ N., Long. 65° 0′ W. Depth, 2,475 fathoms. Grey ooze. Bottom temperature, 1°·7 C. Distance 53 miles from Bermudas.

Station XXXII. April 3rd.—Lat. 31° 49′ N., Long. 64° 55′ W. Depth, 2,250 fathoms. Grey ooze. Bottom temperature, 1°·8 C. Distance from Bermudas, 26 miles.

Station XXXIII. April 4th.—East of Bermudas. Depth, 435 fathoms. Coral clay.

Station XXXIV. April 21st. About 18 miles north of Bermudas. Depth, 1,375 fathoms. Grey ooze. Bottom temperature, 3°·0 C.

Station XXXV. April 22nd.—Lat. 32° 39′ N., Long. 65° 6′ W. Depth, 2,450 fathoms. Coral clay. Bottom temperature, 1°·6 C.

Second sounding, same day.—Lat. 32° 26′ N., Long. 65° 9′ W. Depth, 2,100 fathoms. Coral clay. Bottom temperature, 1°·6 C.

Third sounding, same day.—Lat. 32° 15′ N., Long. 65° 8′ W. Depth, 1,950 fathoms. Coral clay. West of Bermudas.

Station XXXVI. April 23rd.—About 13 miles south-west of Bermudas. Depth, 32 fathoms. Pebbles and stones.

Station XXXVII. April 24th. — Lat. 32° 19′ N., Long. 65° 39′ W. Depth, 2,650 fathoms. Grey ooze. Bottom temperature, 1°·6 C. Bermudas distant 43 miles east.

Station XXXVIII. April 25th. — Lat. 33° 3′ N., Long. 66° 32′ W. Depth, 2,600 fathoms. Grey ooze. Bottom temperature, 1°·8 C. Distance from Sandy Hook, near New York, 568 miles.

Station XXXIX. April 27th.—Lat. 34° 3′ N., Long. 67° 32′ W. Depth, 2,850 fathoms. Grey ooze. Bottom temperature, 1°·8 C. Distance from Sandy Hook, 490 miles.

Station XL. April 28th.—Lat. 34° 51′ N., Long. 68° 30′ W. Depth, 2,675 fathoms. Distance from Sandy Hook, 420 miles.

Station XLI. April 29th.—Lat. 36° 5′ N., Long. 69° 54′ W Distance from Sandy Hook, 324 miles.

Station XLII. April 30th.—Lat. 35° 58′ N., Long. 70° 39′ W. Depth, 2,425 fathoms. Grey ooze. Bottom temperature, 1°·8 C. Distance from Sandy Hook, 308 miles.

Station XLIII. May 1st.—Lat. 36° 23′ N., Long. 71° 51′ W. No bottom at 2,600 fathoms. Lowest temperature, 1°·8 C. Distance from Sandy Hook, 262 miles.

Station XLIV. May 2nd.—Lat. 37° 25′ N., Long. 71° 40′ W. Depth, 1,700 fathoms. Grey ooze. Bottom temperature, 1°·7 C. Distance from Sandy Hook, 209 miles.

Station XLV. May 3rd.—Lat. 38° 34′ N., Long. 72° 10′ W. Depth, 1,240 fathoms. Sand and mud. Bottom temperature, 2°·4 C. Distance from Sandy Hook, 139 miles.

Station XLVI. May 6th.—Lat. 40° 17′ N., Long. 66° 48′ W. Depth, 1,350 fathoms. Sand and mud. Bottom temperature, 2°·3 C. South of Little George Bank.

Station XLVII. May 7th.—Lat. 41° 15′ N., Long. 65° 45′ W.

H

Depth, 1,340 fathoms. Sand and mud. West of Little George Bank.

Station XLVIII. May 8th.—Lat. 43° 2' N., Long. 64° 2' W. Depth, 75 fathoms. Rock. On Le Have Bank, and about 90 miles south of Halifax.

Station XLIX. May 20th.—Lat. 43° 3' N., Long. 63° 39' W. Depth, 83 fathoms. Stones and gravel. Bottom temperature, 1°·8 C. On the eastern edge of Le Have Bank, and 651 miles north of Bermudas.

Station L. May 21st.—Lat. 42° 8' N., Long. 63° 39' W. Depth, 1,250 fathoms. Grey ooze. Bottom temperature, 2°·7 C. Distance from Bermudas, 596 miles.

Station LI. May 22nd.—Lat. 41° 19' N., Long. 63° 12' W. Depth, 2,020 fathoms. Grey ooze. Bottom temperature, 1°·5 C. Distance from Bermudas, 550 miles.

Station LII. May 23rd.—Lat. 39° 44' N., Long. 63° 22' W. Depth, 2,800 fathoms. Grey ooze. Bottom temperature, 1°·5 C. Distance from Bermudas, 456 miles.

Station LIII. May 26th.—Lat. 36° 30' N., Long. 63° 40' W. Depth, 2,650 fathoms. Grey ooze. Bottom temperature, 1°·8 C. Distance from Bermudas, 261 miles.

Station LIV. May 27th.—Lat. 34° 51' N., Long. 63° 59' W. Depth, 2,650 fathoms. Grey ooze. Distance from Bermudas, 162 miles.

Station LV. May 28th.—Lat. 33° 20' N., Long. 64° 37' W. Depth, 2,500 fathoms. Grey ooze. Position, 66 miles north of Bermudas.

Second sounding, same day.—Lat. 32° 46' N., Long. 64° 39' W. Depth, 1,775 fathoms. Rock. Bottom temperature, 1°·7 C. 32 miles north of Bermudas.

Station LVI. May 29th.—Eastward of Bermudas. Depth, 1,325 fathoms. Coral clay.

Second sounding, same day.—Depth, 1,075 fathoms. Coral clay. Bottom temperature, 3°·2 C.

Station LVII. May 30th.—Soundings round the south-west side of Bermudas Reef. First sounding, 690 fathoms. Second sounding, 1,250 fathoms. Third sounding, 1,575 fathoms. West of Bermudas.

Station LVIII. June 13th.—Lat. 32° 37′ N., Long. 64° 21′ W. Depth, 1,500 fathoms. Grey ooze. Bottom temperature, 2°·3 C. East of Bermudas.

Station LIX. June 14th.—Lat. 32° 54′ N., Long. 63° 22′ W. Depth, 2,360 fathoms. Grey ooze. Bottom temperature, 1°·7 C. Distance from Fayal, one of the Azores, 1,720 miles.

Station LX. June 16th.—Lat. 34° 28′ N., Long. 58° 56′ W. Depth, 2,575 fathoms. Grey ooze. Bottom temperature, 1°·5 C. Distance from Fayal, 1,482 miles.

Station LXI. June 17th.—Lat. 34° 54′ N., Long. 56° 38′ W. Depth, 2,850 fathoms. Grey ooze. Bottom temperature, 1°·5 C. Distance from Fayal, 1,356 miles.

Station LXII. June 18th.—Lat. 35° 7′ N., Long. 52° 32′ W. Depth, 2,875 fathoms. Grey ooze. Bottom temperature, 1°·8 C. Distance from Fayal, 1,175 miles.

Station LXIII. June 19th.—Lat. 35° 29′ N., Long. 50° 53′ W. Depth, 2,750 fathoms. Grey ooze. Distance from Fayal, 1,077 miles.

Station LXIV. June 20th.—Lat. 35° 35′ N., Long. 50° 27′ W. Grey ooze. Distance from Fayal, 1,055 miles.

Station LXV. June 21st.—Lat. 36° 33′ N., Long. 47° 58′ W. Depth, 2,700 fathoms. Grey ooze. Bottom temperature, 1°·7 C. Distance from Fayal, 960 miles.

Station LXVI. June 22nd.—Lat. 37° 24′ N., Long. 44° 14′ W. Depth, 2,750 fathoms. Grey ooze. Bottom temperature, 1°·8 C. Distance from Fayal, 780 miles.

Station LXVII. June 23rd.—Lat. 37° 54′ N., Long. 41° 44′ W. Depth, 2,700 fathoms. Grey ooze. Bottom temperature, 1°·8 C. Distance from Fayal, 606 miles.

Station LXVIII. June 24th.—Lat. 38° 3′ N., Long. 39° 19′ W. Depth, 2,175 fathoms. Grey ooze. Bottom temperature, 1°·6 C. Distance from Fayal, 496 miles.

Station LXIX. June 25th.—Lat. 38° 23′ N., Long. 37° 21′ W. Depth, 2,200 fathoms. Grey ooze. Bottom temperature, 1°·7 C. Distance from Fayal, 404 miles.

Station LXX. June 26th.—Lat. 38° 25′ N., Long. 35° 50′ W. Depth, 1,675 fathoms. Globigerina ooze. Distance from Fayal, 332 miles.

Station LXXI. June 27th.—Lat. 38° 18′ N., Long. 34° 48′ W. Depth, 1,675 fathoms. Globigerina ooze. Bottom temperature, 2°·2 C. Distance from Fayal, 284 miles.

Station LXXII. June 28th.—Lat. 38° 34′ N., Long. 32° 47′ W. Depth, 1,240 fathoms. Globigerina ooze. Bottom temperature, 2°·8 C. Distance from Fayal, 210 miles.

Station LXXIII. June 30th.—Lat. 38° 30′ N., Long. 31° 14′ W. Depth, 1,000 fathoms. Globigerina ooze. Bottom temperature, 3°·7 C. Distance from Fayal, 114 miles.

Station LXXIV. July 1st.—Lat. 38° 22′ N., Long. 29° 37′ W. Depth, 1,350 fathoms. Globigerina ooze. Distance from Fayal, 19 miles.

Station LXXV. July 2nd.—Lat. 38° 37′ N., Long. 28° 30′ W. Depth, 450 fathoms. Volcanic sand, dead shells, and corals. Between Fayal and Pico.

Station LXXVI. July 3rd.—Lat. 38° 11′ N., Long. 27° 9′ W. Depth, 900 fathoms. Globigerina ooze. Bottom temperature, 4°·2 C. South of Terceira.

Station LXXVII. July 4th.—Lat. 37° 52′ N., Long. 26° 26′ W. Depth, 750 fathoms. Rock. About 30 miles westward of San Miguel.

Station LXXVIII. July 10th.—Lat. 37° 20′ N., Long. 25° 15′ W. Depth, 1,000 fathoms. Globigerina ooze. About midway between East Point of San Miguel and the Island of Sta. Maria.

Station LXXIX. July 11th.—Lat. 36° 21′ N., Long. 23° 31′ W. Depth, 2,025 fathoms. Globigerina ooze. Bottom temperature, 1°·5 C. Distance from Madeira, 376 miles.

Station LXXX. July 12th.—Lat. 35° 3′ N., Long. 21° 25′ W. Depth, 2,660 fathoms. Globigerina ooze. Bottom temperature, 1°·8 C. Distance from Madeira, 256 miles.

Station LXXXI. July 13th.—Lat. 34° 11′ N., Long. 19° 52′ W. Depth, 2,675 fathoms. Globigerina ooze. Bottom temperature, 2°·0 C. Distance from Madeira, 171 miles.

Station LXXXII. July 14th.—Lat. 33° 46′ N., Long. 19° 17′ W. Depth, 2,400 fathoms. Globigerina ooze. Bottom temperature, 1°·8 C. Distance from Madeira, 116 miles.

Station LXXXIII. July 15th.—Lat. 33° 13′ N., Long. 18° 13′ W. Depth, 1,650 fathoms. Globigerina ooze. Bottom temperature, 2°·2 C. Distance from Madeira, 52 miles.

Station LXXXIV. July 18th.—Lat. 30° 38′ N., Long. 18° 5′ W. Depth, according to a sounding on Admiralty Chart, about 2,400 fathoms. Distance from St. Vincent, 920 miles.

Station LXXXV. July 19th.—Lat. 28° 42′ N., Long. 18° 6′ W. Depth 1,125 fathoms. Volcanic sand. Off Palma Island, Canaries.

Station LXXXVI. July 21st.—Lat. 25° 46′ N., Long. 20° 34′ W. Depth, 2,300 fathoms. Globigerina ooze. Bottom temperature, 1°·8 C. Westward of Station III., and about 600 miles from St Vincent.

Station LXXXVII. July 21st.—Lat. 25° 49′ N., Long. 20° 12′ W. Depth, 1,675 fathoms. Black coral and rock. About 4 miles north of Station III., and 600 miles from St. Vincent.

Station LXXXVIII. July 22nd.—Lat. 23° 58′ N., Long. 21° 18′ W. Depth, 2,300 fathoms. Globigerina ooze. Bottom temperature, 1°·7 C. Distance from St. Vincent, 478 miles.

Station LXXXIX. July 23rd.—Lat. 22° 18′ N., Long. 22° 21′ W. Depth, 2,400 fathoms. Globigerina ooze. Bottom temperature, 1°·8 C. Distance from St. Vincent, 362 miles.

Station XC. July 24th.—Lat. 20° 58′ N., Long. 22° 57′ W. Depth, 2,400 fathoms. Globigerina ooze. Bottom temperature, 1°·8 C. Distance from St. Vincent, 268 miles.

Station XCI. July 25th.—Lat. 19° 4′ N., Long. 24° 6′ W. Depth, 2,075 fathoms. Globigerina ooze. Bottom temperature, 1°·8 C. Distance from St. Vincent, 144 miles.

Station XCII. July 26th.—Lat. 17° 54′ N., Long. 24° 41′ W. Depth, 1,975 fathoms. Globigerina ooze. Distance from St. Vincent, 61 miles.

Station XCIII. July 27th.—Between San Antonio and St. Vincent, Cape Verde Islands. Depth, 1,070 fathoms. Volcanic sand.

Second sounding, same day.—Depth, 1,000 fathoms. Volcanic sand.

Third sounding, same day.—Depth, 465 fathoms. Volcanic sand. Bottom temperature, 6°·2 C.

Fourth sounding.—Depth, 52 fathoms. Coral.

Station XCIV. August 5th.—South-west of St. Vincent. Depth, 103 fathoms. Coral.

Second sounding, same day.—Depth, 85 fathoms. Coral and shells.

Third sounding.—Depth, 260 fathoms. Volcanic sand.

Fourth sounding.—Depth, 675 fathoms. Volcanic sand.

Fifth sounding.—Depth, 1,150 fathoms. Grey ooze.

Station XCV. August 10th.—Lat. 13° 36′ N., Long. 22° 49′ W. Depth, 2,300 fathoms. Globigerina ooze. Bottom temperature, 1°·8 C. About 90 miles south of S. Iago, Cape Verde Islands.

Station XCVI. August 11th.—Lat. 12° 15′ N., Long. 22° 28′ W. Distance from St. Paul's Rocks, 801 miles.

Station XCVII. August 13th.—Lat. 10° 25′ N., Long. 20° 30′ W. Depth, 2,575 fathoms. Globigerina ooze. Bottom temperature. 1°·8 C. West of the Bijouga Islands, West Coast of Africa.

Station XCVIII. August 14th.—Lat. 9° 21′ N., Long. 18° 28′ W. Depth, 1,750 fathoms. Globigerina ooze. Bottom temperature, 2°·0 C. West of the Isles de Los, West Coast of Africa.

Station XCIX. August 15th.—Lat. 7° 53′ N., Long. 17° 26′ W. Off Sierra Leone. Temperature sounding only taken to 500 fathoms.

Station C. August 16th.—Lat. 7° 1′ N., Long. 15° 55′ W. Depth, 2,425 fathoms. To the south-west of Sierra Leone.

Station CI. August 19th.—Lat. 5° 48′ N., Long. 14° 20′ W. Depth, 2,500 fathoms. Black mud. Bottom temperature, 1°·7 C. Off Cape Mesurado, West Coast of Africa.

Station CII. August 21st.—Lat. 3° 8′ N., Long. 14° 49′ W. Depth, 2,450 fathoms. Grey ooze. Bottom temperature, 1°·7 C. Off Cape Palmas, West Coast of Africa, and 884 miles distant from St. Paul's Rocks.

Station CIII. August 22nd.—Lat. 2° 49′ N., Long. 17° 13′ W.

Depth, 2,475 fathoms. Dark sandy mud. Bottom temperature, 1°·6 C. Distance from St. Paul's Rocks, 738 miles.

Station CIV. August 23rd.—Lat. 2° 25′ N., Long. 20° 1′ W. Depth, 2,500 fathoms. Grey ooze. Bottom temperature, 1°·7 C. Distance from St. Paul's Rocks, 568 miles.

Station CV. August 24th.—Lat. 2° 6′ N., Long. 22° 53′ W. Depth, 2,275 fathoms. Grey ooze. Bottom temperature, 1°·4 C. Distance from St. Paul's Rocks, 428 miles.

Station CVI. August 25th.—Lat. 1° 47′ N., Long. 24° 26′ W. Depth, 1,850 fathoms. Globigerina ooze. Bottom temperature, 1°·8 C. Distance from St. Paul's Rocks, 301 miles.

Station CVII. August 26th.—Lat. 1° 22′ N., Long. 26° 36′ W. Depth, 1,500 fathoms. Globigerina ooze. Bottom temperature, 2°·8 C. Distance from St. Paul's Rocks, 169 miles.

Station CVIII. August 27th.—Lat. 1° 10′ N., Long. 28° 23′ W. Depth, 1,900 fathoms. Globigerina ooze. Bottom temperature, 2°·1 C. Distance from St. Paul's Rocks, 37 miles.

Station CIX. August 29th.—Westward of St. Paul's Rocks, 26 miles. Depth, 1,425 fathoms. Globigerina ooze.

Station CX. August 30th.—Lat. 0° 9′ N., Long. 30° 18′ W. Depth, 2,275 fathoms. Globigerina ooze. Bottom temperature, 0°·9 C. Distance from Fernando Noronha, 265 miles.

Station CXI. August 31st.—Lat. 1° 45′ S., Long. 30° 58′ W. Depth, 2,475 fathoms. Globigerina ooze. Bottom temperature, 0°·2 C. Distance from Fernando Noronha, 132 miles.

Station CXII. September 1st.—Lat. 3° 33′ S., Long. 32° 16′ W. Depth, 2,200 fathoms. Globigerina ooze. Bottom temperature, 0°·5 C. About 22 miles north-east of Fernando Noronha.

Station CXIII. September 1st.—Close to Fernando Noronha. Depth, 1,010 fathoms. Rock. Bottom temperature, 2°·8 C.

Station CXIV. September 3rd.—Lat. 3° 58′ S., Long. 32° 42′ W. Depth, 820 furlongs. Rock. About 13 miles east of Fernando Noronha.

Station CXV. September 3rd.—Lat. 4° 2′ S., Long. 32° 47′ W. Depth, 2,150 fathoms. Globigerina ooze. About 19 miles south-west of Fernando Noronha.

Station CXVI. September 4th.—Lat. 5° 1′ S., Long. 33° 50′ W.

Depth, 2,275 fathoms. Globigerina ooze. Bottom temperature, 0°·7 C. About 90 miles east of Cape St. Roque.

Station CXVII. September 6th.—Lat. 5° 56′ S., Long. 34° 45′ W. Depth, 1,375 fathoms. Mud. About 130 miles north of Pernambuco.

Station CXVIII. September 8th.—Lat. 7° 28′ S., Long. 34° 2′ W. Depth, 2,050 fathoms. Mud. Bottom temperature, 1°·1 C. Off Parahyba.

Station CXIX. September 8th.—Lat. 7° 39′ S., Long. 34° 12′ W. Depth, 1,650 fathoms. Mud. Bottom temperature, 2°·3 C. Distance from Pernambuco, 47 miles.

Station CXX. September 9th.—Lat. 8° 37′ S., Long. 34° 28′ W. Depth, 675 fathoms. Mud. Distance from Cape San Antonio, Bahia, 360 miles.

Station CXXI. September 9th.—Lat. 8° 28′ S., Long. 34° 31′ W. Depth, 500 fathoms. Mud. Off Cape San Agostinho.

Station CXXII. September 10th.—Lat. 9° 5′ S., Long. 34° 50′ W. Depth, 350 fathoms. Sand.

Second sounding.—Lat. 9° 10′ S., Long. 34° 32′ W. Depth, 120 fathoms. Coral and sand.

Third sounding.—Lat. 9° 9′ S., Long. 34° 53′ W. Depth, 32 fathoms. Coral sand.

Fourth sounding.—Lat. 9° 10′ S., Long. 34° 39′ W. Depth, 400 fathoms. Sand and mud. Off Point Calvo.

Station CXXIII. September 11th.—Lat. 10° 9′ S., Long. 35° 11′ W. Depth, 1,715 fathoms. Mud. Bottom temperature, 2°·3 C. Off Maceio.

Station CXXIV. September 11th.—Lat. 10° 11′ S., Long. 35° 22′ W. Depth, 1,600 fathoms. Mud. Distance from Cape San Antonio, Bahia, 251 miles.

Station CXXV. September 12th.—Lat. 10° 46′ S., Long. 36° 2′ W. Depth, 1,200 fathoms. Mud. Off the mouth of the River San Francisco.

Station CXXVI. September 12th.—Lat. 10° 46′ S., Long. 36° 8′ W. Depth, 770 fathoms. Mud.

Second sounding.—Lat. 10° 45′ S., Long. 36° 9′ W. Depth, 700 fathoms. Mud. Off the mouth of the River San Francisco.

Station CXXVII. September 13th.—Lat. 11° 42' S., Long. 37° 3' W. Depth, 1,015 fathoms. Mud. Bottom temperature, 3°·3 C. Distance from Cape San Antonio, Bahia, about 120 miles.

Station CXXVIII. September 14th.—Lat. 13° 6' S., Long. 38° 7' W. Depth, 1,275 fathoms. Mud. Off the Cape San Antonio, Bahia.

Station CXXIX. September 30th.—Lat. 20° 13' S., Long. 35° 19' W. Depth, 2,150 fathoms. Red mud. Bottom temperature, 0°·6 C. Distance from Tristan d'Acunha, 1,572 miles.

Station CXXX. October 3rd.—Lat. 26° 15' S., Long. 32° 56' W. Depth, 2,350 fathoms. Red mud. Bottom temperature, 0°·8 C. Distance from Tristan d'Acunha, 1,235 miles.

Station CXXXI. October 6th.—Lat. 29° 35' S., Long. 28° 9' W. Depth, 2,275 fathoms. Globigerina ooze. Bottom temperature, 0°·7 C. Distance from Tristan d'Acunha, 912 miles.

Station CXXXII. October 10th.—Lat. 35° 25' S., Long. 23° 40' W. Depth, 2,050 fathoms. Globigerina ooze. Bottom temperature, 1°·1 C. Distance from Tristan d'Acunha, 561 miles.

Station CXXXIII. October 11th.—Lat. 35° 41' S., Long. 20° 55' W. Depth, 1,900 fathoms. Globigerina ooze. Bottom temperature, 1°·3 C. Distance from Inaccessible Island, Tristan d'Acunha group, 410 miles.

Station CXXXIV. October 14th.—Lat. 36° 12' S., Long. 12° 16' W. Depth, 2,025 fathoms. Globigerina ooze. Bottom temperature, 1°·6 C. About 50 miles north of Tristan d'Acunha.

Station CXXXV. October 18th.—Between Nightingale Island and Tristan d'Acunha. First sounding, depth, 1,000 fathoms. Shells and rock. Second sounding, depth, 1,100 fathoms. Third sounding, depth, 550 fathoms.

Station CXXXVI. October 20th.—Lat. 36° 43' S., Long. 7° 13' W. Depth, 2,100 fathoms. Rock. Bottom temperature, 1°·1 C. Distance from the Cape of Good Hope, 1,293 miles.

Station CXXXVII. October 23rd.—Lat. 35° 59' S., Long. 1° 34' E. Depth, 2,550 fathoms. Red mud. Bottom temperature, 0°·7 C. Distance from the Cape of Good Hope, 850 miles.

Station CXXXVIII. October 25th.—Lat. 36° 22′ S., Long. 8° 12′ E. Depth, 2,650 fathoms. Red mud. Bottom temperature, 1°·0 C. Distance from the Cape of Good Hope, 519 miles.

Station CXXXIX. October 27th. Lat. 35° 35′ S., Long. 16° 8′ E. Depth, 2,325 fathoms. Black manganese and grey mud. Bottom temperature, 0°·5 C. Distance from the Cape of Good Hope, 138 miles.

Station CXL. October 28th.—Lat. 35° 0′ S., Long. 17° 57′ E. Depth, 1,250 fathoms. Grey mud. Off the Cape of Good Hope.

CHAPTER II.

FROM PORTSMOUTH TO TENERIFFE.

Departure from England.—Rough weather in the Channel.—Lisbon.—Trawling in deep water.—Deep-water Fishes.—Surface animals.—Gibraltar.—*Cystosoma neptuni*.—Venus' Flower-basket.—*Naresia cyathus*.—The 'Clustered Sea-polype.'—Madeira.—Temperature Observations.—Meteorological Observations.—Teneriffe.

APPENDIX A.—Particulars of Depth, Temperature, and Position at the Sounding Stations between Portsmouth and Teneriffe; the Temperatures corrected for pressure.

APPENDIX B.—Comparative Table of the Indications of 'Stevenson's Mean Thermometers,' and the ordinary Maximum and Minimum Thermometers in Air, for the six months from the 1st May to the 31st October, 1873.

WE were well aware that we had many difficulties to contend with before we could get so complicated a system into full working order. There seemed, at first, to be special difficulty in dredging and taking deep-sea observations from so large a ship. The roll of the ship, her height above the water, and her want of flexibility of movement when compared with vessels which had previously been used for the purpose, raised new questions as to methods of working which it would require some time to settle, and it would likewise take some time before each of us fell into his place and laid out a line for himself, as part of a general programme. We therefore determined very early in the cruise to consider everything done during the voyage from England to the Canary Islands as tentative and introductory, and to regard the first section across the Atlantic, from Teneriffe to the West Indies, as the commencement of the true work of the expedition.

December 30*th*.—In the forenoon, the weather was much more moderate, and the first sounding was taken off the entrance of Vigo Bay at a depth of 1,125 fathoms, with a bottom of 'globigerina ooze' rich in coccoliths and coccospheres. The dredge was put over, but the sea was still too high and it came up capsized. Later in the day a second attempt was made with somewhat greater success, but the number of species procured was small; the most remarkable among them a fine specimen of *Hymenaster membranaceus* of a rich crimson colour and upwards of 70 mm. in diameter. On the 1st of January, 1873, we tried the dredge again, in water of moderate depth off Cape Mondego, but the weather was still too boisterous and the attempt was unsatisfactory; and on the 2nd, dredging at a depth of 1,975 fathoms a little to the N.W. of the Burlings, the dredge fouled something at the bottom, an unusual occurrence in such deep water, and carried away.

On the 3rd of January all our troubles were over for the time. Passing Cape Roca and the beautiful heights of Cintra, we steamed slowly up the Tagus, past the straggling suburb of Lisbon with its many-coloured villas scattered over the slopes; past the wonderful Castle of Belem, with its elegant proportions and rich ornament—a record of the skill and refined taste of the old master-masons; past the new Palace of the Ajuda, the present residence of the King—a large plain building, in a certain sense handsome, but sadly inferior in tone to the little square Keep by the side of the river.

Fig. 19.—Belem Castle, Lisbon.

About midday we were moored in the Tagus off the town. Several of us went on shore, and took up our quarters at the Hôtel Braganza, where we were very comfortable, and enjoyed greatly the splendid view over the town and river. Some went to Cintra, though we were at Lisbon at the worst season of the year for country excursions, while others spent their time in seeing what was most interesting in the city and its immediate neighbourhood, and resting after the fatigue of our earlier experiences at sea.

There is a gem of Gothic-Moresque architecture near Lisbon—the monastery and church of Santa

Fig. 20.—The Porch of Santa Maria, Belem.

Maria at Belem. One or two of us walked there one lovely Sunday afternoon. The porch of the church is Gothic, and rich beyond description. Up to the very roof of the church every pinnacle and buttress, and even the flat portions of the wall, are encrusted with ornament. The carvings are singularly easy and varied. They are executed in a pale brick-red limestone, which seems, unfortunately, to be suffering a good deal from the effects of air and rain.

The interior of the church is even more pleasing than the exterior. Here there is no excess of ornament, but simple, delicate shafts of pale grey marble support, with all the airiness of effect of Moorish architecture, a wonderfully carved and fretted Gothic roof. Service was going on when we were there. The church was cool and dim, and the clear sweet voices of the choristers rose and fell along the aisle, and seemed to linger in the roof among the sculptured palm-leaves. The high altar, with its lighted candles and vases of flowers, and the rich robes of the officiating priests, formed a warm patch of colour strongly in contrast with the cold simplicity of the grey marble. A small flock of worshippers, whose strongly-marked olive faces and picturesque attire had to us all the interest of novelty, made up for their small number by the apparent earnestness of their devotion. The whole scene struck us as being wonderfully harmonious and pleasant.

The monastery of Santa Maria of Belem, commonly called the Monastery of St. Jeronymo from its having been occupied by monks of that order, is in connection with the church, and, with it, was founded in 1499 by King Manoel the Fortunate in commemoration of the discovery of the Indies, on the spot where Vasco da Gama embarked on his first eastern voyage.

The monastery is now used as a State asylum for

upwards of a thousand poor children. Visitors are admitted to the cloisters at three o'clock, and as that hour had not arrived we were turning away from the door, when a servitor opened it and invited us to enter.

We passed through an archway into a large quadrangle with the sun shining brightly into it. Flower-

Fig. 21.—Quadrangle of the Monastery of Santa Maria, Belem.

beds occupied the centre of the square, cut out in various shapes, and separated by neat, trimly-kept gravel walks. The beds were planted with oranges and bananas, and clumps of aloes with their rich crimson spikes, alternating with the cool, white cups of the calla; while heliotrope and jasmine, and many aromatic flowering shrubs and herbs, sent up an almost oppressive fragrance into the warm still air. While we were there several beautiful hawk-moths were hovering over the flowers and dipping their long trunks into their bells.

FIG. 22.—Cloisters of the Monastery of Santa Maria, Belem.

All round the quadrangle runs a double tier of cloisters, supported by low, gracefully proportioned and richly ornamented arches—a kind of compromise between the Moorish and the Gothic.

The stone is a light pink carboniferous limestone, almost a marble, with many fossils; and in some places the elegant forms of the imbedded shells have been reproduced by the sculptor, and the nautilus and the goniatite of the elder times, and arabesques and horns of plenty, and the chubby faces of Christian cherubs, blend in the creation of the old architect like truth and fiction in the dream of a poet.

Behind the cloisters are the rooms of the seminary, and the cloisters are hung with neatly-designed programmes of the courses of study, and lists of the successful candidates for honours at various examinations. We were invited to see the school, but we declined. The pupils were at dinner, and we lingered about the silent quadrangle reluctant to leave it—it was so sweet and still. I am surprised that we do not hear more of the monastery and church of Santa Maria of Belem, for our little party, all of whom had already seen many things in all parts of the world, agreed that it was unusually pleasing.

There are many things in Lisbon to interest 'philosophers' as our naval friends call us,—not I fear from the proper feeling of respect, but rather with good-natured indulgence, because we are fond of talking vaguely about 'evolution,' and otherwise holding on to loose ropes; and because our education has been sadly neglected in the matter of cringles and toggles and grummets, and other implements by means of which England holds her place among the nations.

The buildings of a new Polytechnic School had been just completed at the time of our visit. The institution is of imposing dimensions, built in the form of a hollow square, with a quadrangular garden in the centre. It contains lecture-rooms, a consulting library, spacious and well-lighted chemical and physical laboratories, and galleries for museum purposes. The collections in mineralogy and palæontology are on the ground floor, and the zoological museum, under the able superintendence of Professor Barboza du Bocage, is lodged in four fine galleries on the upper story, one of them devoted entirely to the African fauna in which the museum is particularly rich.

Table-cases containing a fine collection of shells are placed along the centre of the rooms, and upright cases filled with stuffed mammals and birds and variously preserved reptiles and fishes are ranged along the walls. The collection of mammalia is rich in insectivora,—moles, shrews, and the like; and among the scarcer mammals are two manatees, and good examples of the Aye-Aye, and of the singular little African otter-like animal *Potamogale velox*. The collection of birds is particularly good. The specimens are well stuffed and mounted, and well arranged. This collection belongs to the present King, and was chiefly brought together by him and his brother. It contains many rarities: *Didunculus strigirostris*, a good set of birds of paradise including *Semiopteryx*, and a fine specimen of the great auk, *Alca impennis*, in excellent plumage and preservation, given to King Luiz by his brother-in-law the King of Italy. The King of Portugal is very fond of natural history, and has a good general knowledge of it. He did us the honour to visit the ship when we were at Lisbon, and expressed himself greatly interested in the expedition.

The Botanic Garden is near the Palace of the Ajuda. It seems to have been very fine at one time, for there is a good range of glass, and the handsome terraces have been laid out and decorated with statuary and fountains with some taste, and evidently at great expense; but the garden has fallen into disuse for scientific purposes, and has been allowed to get into disorder and disrepair; and only a beautiful group or two of date-palms and a splendid dragon-tree, with a head 30 feet in diameter, still maintain a trace of its former character. It is intended to lay out a piece of ground near the new Polytechnic School as a garden for teaching purposes, and I believe the old Botanic Garden will probably be merged in the grounds of the palace.

There is a very complete Meteorological and Magnetic Observatory, now under the energetic management of MM. J. C. de Brito Capello and Gama Lobo. The meteorological department is in telegraphic communication with the principal European Observatories, and the Magnetic Observatory is similar in almost all respects to that at Kew. The building is insulated for temperature, and a very complete photographic registering apparatus is in constant use. Photographs of the sun, registering the form and position of the spots, are taken daily; and an Astronomical Observatory, which is to be used for the present chiefly for observations of the sun, is in process of completion.

We enjoyed greatly our few days in Lisbon. The British Minister, the Hon. Sir Charles Murray, was most courteous in his attention. The weather was delicious, we were in the middle of the orange and lemon harvest and the air was redolent of the perfume of the golden fruit; and there was certainly little to remind us of the winter we had so lately

PLATE I. *The track of the Ship from Portsmouth to Teneriffe.*
The blue arrows indicate the direction of the currents, and the red of the winds.

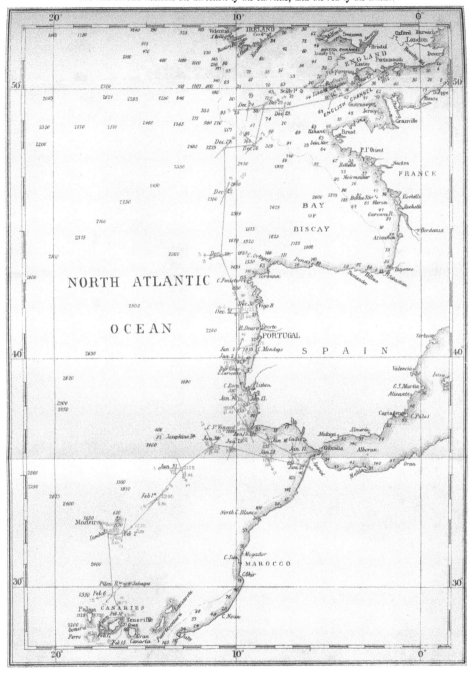

The material originally positioned here is too large for reproduction in this reissue. A PDF can be downloaded from the web address given on page iv of this book, by clicking on 'Resources Available'.

left behind us except the leafless planes with their curious pendant bullet-like seed-vessels, ranged along the Boulevards.

On the evening of the 12th of January we steamed out of the Tagus, and the next day we dredged in 470 fathoms off Setubal. The bottom was the ordinary grey ooze of the North Atlantic, and we sifted out of it many of our old acquaintances of the British area, such as *Limopsis borealis, Columbella haliæti, Dacrydium vitreum,* and many others, which confirmed us in our anticipation that we should find our deep-sea fauna very widely diffused.

We dredged again off Cape St. Vincent on the 15th in fine light weather in 525 fathoms, and brought up some of the dead coils of *Hyalonema,* each with its coating of *Palythoa,* but no perfect or living specimens; and on the following day, weary with the comparatively unproductive sifting of tons of tenacious mortar, we made our first attempt with the trawl at a depth of 600 fathoms. The experiment, as I have already said, was entirely successful. The number of individual specimens procured was now much larger, and in addition to the smaller invertebrates which were usually almost the sole produce of the dredge, several fishes were taken, and many of the larger crustaceans and echinoderms. It is of course open to question in such cases whether the fishes have come from the bottom and are to be referred to the depth indicated by the sounding, or whether they may have entered the trawl at some stage of its way to the surface. The fishes captured on the present occasion were a single specimen of *Mora mediterranea,* two of

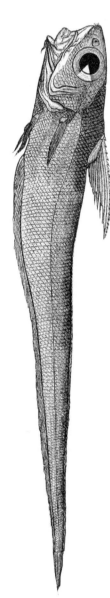

Fig. 3.—*Coryphænoides serratus*, Lowe. Half the natural size. (No. 4.)

Coryphænoides serratus (Fig. 23), and one or two small forms which were undoubtedly from the surface. The *Mora* was in a very peculiar condition; its eyes were blown nearly out of its head by the expansion of air contained probably in some spaces about the spinal cord, and its swimming-bladder was forced out at its mouth and distended almost to bursting; *Coryphænoides* had likewise the eyes forced outwards, but the distortion was not so great; all the fishes were almost denuded of scales; it is evident therefore that all must have come from a considerable depth. As *Mora* is common at moderate depths in the Mediterranean, it is more than probable that it came from some intermediate zone. *Coryphænoides* is one of a family, the MACRURIDÆ, which has yielded us by far the greater number of our deep-sea fishes; and from their peculiar appearance, from the condition in which the examples are usually found, and from the circumstance that they are not unfrequently associated with species of the genera *Melanocetus* and *Ceratias*—lophioids whose form and structure are inconsistent with a pelagic life, the balance of probability seems greatly in favour of their having been taken on the bottom.

The trawl seemed to have gone over a regular field of a delicate simple Gorgonoid, with a thin wire-like axis slightly twisted spirally, a small tuft of irregular rootlets at the base, and long exsert polyps. The stems, which were from 18 inches to 2 feet in length, were coiled in great hanks round the trawl-beam and entangled in masses in the net, and as they showed a most vivid phosphorescence of a pale lilac colour, their immense number suggested a wonderful state of things beneath—animated cornfields waving gently in the slow tidal current and glowing with a soft diffused light, scintillating and sparkling on the slightest touch, and now and again breaking into long avenues of vivid light indicating the paths of fishes or other wandering denizens of their enchanted region. The bottom in these later dredgings, anywhere in fact along the coast of Portugal at depths beyond 500 fathoms, consisted of the now well-known 'globigerina ooze,' that is to say, it was a greyish calcareous paste, soft on the surface, becoming firmer below, and made up in a great degree of the shells of foraminifera—chiefly of the genera *Globigerina* and *Orbulina*—entire or more or less broken up and disintegrated.

Along with the foraminiferous shells some other shells of much larger size enter in varying proportion into the composition of the ooze, or perhaps may be rather said to be mixed with it. These are principally shells of Pteropods, with a few of those of Heteropods, and of pelagic Gasteropods. The last of these groups, the GASTEROPODA, are well known. They include the great mass of the mollusca of the present time; for example, the whelk, the periwinkle, and the garden snail. Their shells are with few exceptions univalve and spiral, often thin and delicate, sometimes, as in the genera *Strombus*, *Fusus*, *Conus*, and many others, thick and massive, weighty accumulations of carbonate of lime secreted from the sea-water. They have a distinct head bearing organs of sense, but the character which most distinguishes them from their nearer allies is their mode of locomotion, which is by means of a long muscular plate secreting a viscid mucus running along beneath the body of the animal, and by alternate extension and contraction enabling it to creep over a solid surface. Most of these animals live on the bottom of the sea, as their organization demands. One or two only of the shell-making genera are pelagic, and the only important one of these is the genus *Janthina*, which inhabits a spiral shell, like a snail-shell, of a most lovely blue. *Janthina* floats by spreading out its 'foot' on the surface, but it is more usually found attached to the different kinds of 'Portuguese men-of-war,' *Velella*, *Physalia*, and *Porpita*, or in the mid-Atlantic, in the wandering islands of gulf-weed. At certain seasons a peculiar kind of membranous float or raft is secreted from the animal, like a crescentic piece of honeycomb with the cells filled with air. The egg-sacs, which are not unlike those of the common whelk, are attached beneath the float, and when the float is complete and the egg-sacs full the creature disengages it, and leaves the eggs to be hatched as it drifts about on the surface in the warmth and sunlight.

The shells of *Janthina* are common in the 'globigerina ooze.' They are not unfrequently cast up on the shore on the west coasts of Ireland and Scotland, and even on the Shetlands and the Færoe Islands. They are not, however, inhabitants of our Northern Seas. They are drifted along and scattered about by our beneficent ameliorator, the Gulf-stream.

The HETEROPODA are very close to the GASTEROPODA, and in most modern works on zoology they are associated with them as a sub-class. They are entirely pelagic, and as it is only under peculiar circumstances that one can stop the ship in mid-

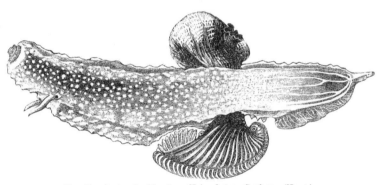

FIG. 24.- *Carinaria Atlantica*. Natural size. Surface. (No. 4.)

ocean and hunt for them, they are little known. One or two of their shells are met with in collections; one especially, *Carinaria*, a beautiful little glassy boat, which one would take at first for some form of the paper-nautilus. The shell of *Carinaria* gives no idea, however, of the form of the animal (Fig. 24), which, with one or two allied genera—such as *Pterotrachea* and *Firoloides*, which do not produce shells at all, is sometimes abundant in calm weather on the surface of the warm seas. The shell hangs below the animal, connected with it by a kind of neck, and is merely meant for the protection of some very vital organs, including the heart, the gills, and the liver. The remainder of the animal is ten times the size of the shell, and forms a large sac, usually gelatinous and very transparent, often dotted over with purple pigment spots. The front of the sac is drawn out into a long, singularly-formed snout, and near it there are bright, well-marked eyes and a pair of feelers. The posterior part of the sac is produced into a fin-like tail. Along the upper middle line of the animal in the position in which it swims in the water, the part corresponding with the 'foot' in ordinary shell-fish is raised into a high, crest-like fin. The bodies of these creatures are large, some of them not less than five or six inches in length, but like most free, floating animals, they are very soft, formed mainly of a 'connective tissue,' with little in it but sea-water. In this way their bulk is greatly increased without materially adding to their weight, and they weigh little more than an equal bulk of sea-water, and require little exertion to float or swim.

One curious result of this transparency is that we can see through the outer wall, in the most wonderful detail, all the internal arrangements—the nervous centres with the complicated organs of sense, the heart with its pulsating chambers and the blood following its course through the system and through the gills, the alimentary canal and all its accessory glands. The HETEROPODA are probably the most highly-organized group in which such transparency exists.

The shells of *Carinaria* are rare in the globigerina

ooze; but two small spiral shells belonging to animals of the same sub-class, *Atlanta peronii* and *Oxygyrus kéraudrenii*, are sometimes in such numbers as to have a sensible effect in adding to the formation. Although the Heteropod shells of the present day are insignificant in size, they played a much more important *rôle* in early times, for there seems little doubt that the great shells of the genera *Euomphalus* and *Bellerophon*, which sometimes go far to make up whole beds of limestone of the Silurian and Carboniferous periods, are to be referred to this group.

The PTEROPODA are further removed than the HETEROPODA are from the typical GASTEROPODA, and are much simpler in their structure. The head is not so markedly separated from the body, and the organs of sense are rudimentary. The body is conical and sometimes spiral, and is very usually contained in a delicate shell, sometimes spiral in form, more frequently conical or tubular; or like an ornamental flower-glass, or like a watch-pocket. The foot is modified into two wing-like appendages, one on either side of the mouth. These are frequently brightly coloured when the animal is living, and different parts of the body show iridescent blues and greens. Multitudes of these little things may now and then be seen on the surface of the water, fluttering with their wings and glittering in the sunshine; to be compared with nothing more aptly than with a congregation of the more dressy of the bombyx moths, as one sometimes comes upon them on a sunny morning just after a family of them have escaped from their chrysalises.

The Pteropods are much smaller than the larger forms among the Heteropods; the largest of the present day are not more than about an inch in length, though antediluvian species of the genus *Conularia* and its allies sometimes reached a length of nearly two feet. They make up for their small size, however, by their numbers. Everywhere in the high seas they absolutely swarm. They are not always to be taken in the towing-net, as they seem to have a habit, in the heat of the day and when there is any wind, of swimming a little way below the surface; but in a fine calm evening, no matter where, a haul of the towing-net can scarcely be made without catching many of them.

The most widely distributed species in the Atlantic seems to be *Diacria trispinosa*, with a little pocket-like shell of some weight and strength, shaded purple and white. Several species of *Cavolinia* are abundant, the largest *C. tridentata*. *Clio cuspidata*, with a fretted shell whose ornament reminds one of some of the fossil genera, is perhaps the species most frequently seen on the surface, and the one which shows the iridescent colouring with the greatest brilliancy (Fig. 25). The several species of *Styliola*, much smaller than the others, are much more numerous, and sometimes throng the towing-net with their glassy needles. *Styliola subulata, S. acicula,* and *S. virgula* are in immense abundance and very generally

Fig. 25.—*Clio pyramidata,* Browne. Slightly enlarged. Surface. (No. 4.)

distributed (Fig. 26). Some of these species sometimes reach the coast of Britain, but an indraught of northern water which includes the British Islands in a fork keeps out these oceanic things from our shores. If the British naturalist to whom these things are usually unknown in a living state will only push his towing-net work by a tug steamer, or his own or a friend's yacht, 40 or 50 miles from the west coast of Scotland or Ireland, he will get beyond the Arctic water, and will wonder, as I did only lately, at the new animal world in the shape of Pteropoda, Heteropoda, Siphonophora, and above all Polycystina and Acanthometrina in all their wonderful varieties of form and sculpture, which will suddenly burst upon him.

FIG. 26.—*Triptera columella.* Twice the natural size. (No. 4.)

The Pteropoda extend far to the northward; one, *Limacina helicina,* with a delicate but very elegant spiral shell, and another, *Clione borealis,* which belongs to the shell-less subdivision, are frequently seen by Arctic voyagers in such numbers that they actually colour the surface of the sea in patches of many square miles in extent, and they are said to form a considerable item in the food of the Greenland whale, which strains them out of the water as it passes through his mouth, with his whalebone sieve. I have dwelt on this little group because their history is not very familiar, and because I hope to show that, small as they are, they play by no means an unimportant part in some of the recent geological processes of reconstruction.

On the evening of the 17th of January we passed Cape Trafalgar of glorious memory, and sighted the light of Tarifa; and when we went on deck at sunrise the next morning we were close under the Rock of Gibraltar, the endless line of batteries and the sulky ironclads of the Channel fleet, which happened to be lying in the bay, clearing up in the increasing light; and the grand outline formed by the mountains of Seville and Granada on the one hand, and Jebel Musa and the distant range of the Atlas on the other, glowing out peak after peak in rose-colour and bronze, and then slowly subsiding into their normal shades of purple, while the blue Mediterranean stretched away without a ripple to the eastward.

The 'Challenger' remained lashed alongside the New Mole at Gibraltar for a week. The weather, although it was little past mid-winter, was warm and bright; the spring seemed already starting, and some beautiful mauve patches of almond-blossom lightened up the face of the grim old rock. The aloes were in full flower, and the 'Alaméda' and the grounds of the Governor's summer cottage were crimson with them. I do not know any plant more ornamental. The rich colour of the flower-spikes contrasts admirably with the cold grey-green of the foliage, and the rigid spear-like leaves have a thoroughly exotic look, more so than most of the plants of warmer latitudes.

January 24*th*.—A small party of us had a most pleasant excursion with Captain Phillimore, the

Captain-Superintendent of the Dockyard. We started after breakfast in the gun-boat 'Pigeon,' across the bay to Algesiras. After paying our respects to the Spanish Governor, a handsome dignified man who received us with great courtesy and returned our visit on the following day, we took a walk about the town, admired the market with its ample supply of fresh vegetables and fruit, and visited a large circus-like building, where for about a week in the year, in carnival time, bull-fights are held; and which was filled with horse-trappings, and banners and swords and small feathered spears, and all the other tawdry and horrid paraphernalia of that barbarous sport.

A splendid aqueduct, evidently built in the old times when Spain held a very different position from the one she holds now, brings abundance of water into the town from the high grounds at some distance. Near the town it crosses a valley, raised high on a long line of quaint narrow arches. We wandered up this valley for several miles to a pretty water-fall called 'La Chorrea,' where we stopped and had a delicious luncheon, the chief element in it the supply of luscious ripe oranges which we had got fresh from the trees on our way; and after sitting sketching and chatting during the heat of the day, we quietly retraced our steps to the town, where we found the 'Pigeon' awaiting our return.

The valley is well wooded, and from the peculiar character of the vegetation, in the middle of the region of evergreen shrubs and trees, there was little to remind us of winter. The greater part of our route lay through an old cork-forest, the gnarled rugged trees a considerable distance apart with greensward beneath, like the trees in an English park. The foliage of an old cork tree is dark and grey, and somewhat scanty, but along our path the rough boles and branches were everywhere beautifully relieved by great patches of the furry tawny rhizomes and light green feathery fronds of one of the hare's-foot ferns (*Davallia*), perched in the clefts of the branches or clinging in the deep grooves formed by the splitting of the outer layers of the cork.

We passed several fruit farms, most of them evidently of old standing, with orange and lemon trees 30 and 40 feet high, now bending and breaking under their load of golden fruit—eight or ten ripe oranges at the end of every branch, thrown out in splendid contrast against the dark green leaves. The spring flowers were beginning to bloom, and great spikes of *Narcissus polyanthus* sent out a delightful fragrance from the hedgerows. The whole scene was very beautiful, but the small amount of cultivation and the evident carelessness and bad management, produced the feeling of regret which seems inseparable from everything concerning the Spain of the present,—that a country naturally capable of so much should do so little.

At Gibraltar we visited the remarkable caves which penetrate the limestone rock, and on one occasion Captain Nares lit up St. Michael's cave, the largest of them, with candles and blue lights, throwing out the magnificent curtains and columns of semi-transparent stalactite in contrasts of light and shade and colouring, and producing a highly picturesque effect.

PLATE II. *Meteorological Observation*

Barometer ———— Dry Bulb Thermometer ———— Wet Bu

The arrows indicate the direction of the wind, and the num

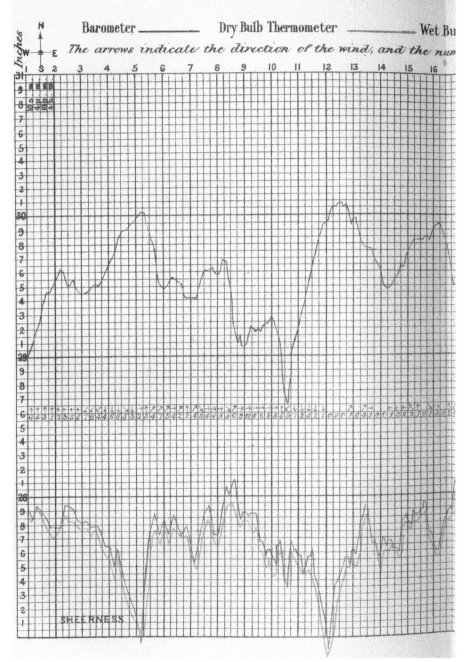

SHEERNESS

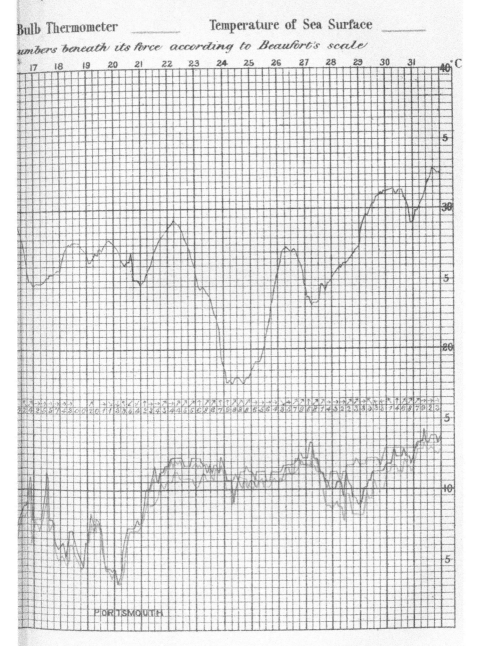
ions for the month of December, 1872.

We had not time to make any systematic exploration of the caves with a view to increasing our knowledge of the animal remains which they contain, but His Excellency, Sir Fenwick Williams of Kars, the Governor of Gibraltar, with other kind attentions, placed at our disposal a considerable collection of bones and implements which had been brought together by an officer formerly stationed at Gibraltar, and we had thus an opportunity of forming some idea of the curious succession of the inmates of the caves from pre-historic times down almost to our own.

We left Gibraltar on the 26th. We had fine moderate weather, and had a very fairly successful week, sounding, trawling, dredging, and taking temperatures, between Gibraltar and Madeira.

The trawl was again employed successfully on the 28th, at a depth of 1,090 fathoms, about 90 miles to the south-east of Cape St. Vincent. Several fishes occurred, including an example of *Macrurus atlanticus*, and a fine specimen of the rare *Halosaurus owenii*. The trawl on this occasion contained a single example of the female of a very large amphipod crustacean, briefly described under the name of *Cystosoma neptuni* by Guerin-Méneville from a single specimen obtained in the Indian Ocean. We have since taken several specimens at different stations in the Atlantic; and as a small male was in one case captured in the towing-net, there can be little doubt that, like *Phronima*, to which genus it is allied, *Cystosoma* is a pelagic animal, probably retiring during the day to a considerable depth, but occasionally coming to the very surface of the water.

K

The male example figured (Fig. 27), which is 103 mm. in length, was taken in Lat. 1° 22′ N., Long. 26° 36′ W., a little to the east of St. Paul's Rocks, where the depth was 1,500 fathoms.

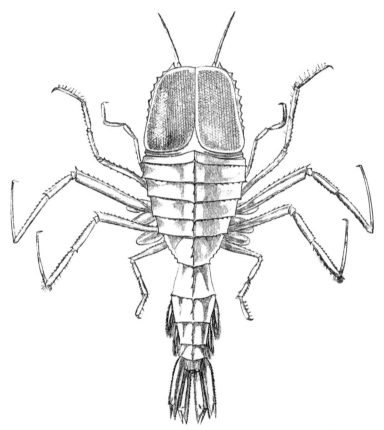

Fig. 27.—*Cystosoma neptuni*, Guerin-Meneville. Slightly reduced. (No. 107.)

The animal presents a very remarkable appearance. It is absolutely colourless and transparent, so that by transmitted light the internal organs can be perfectly seen through the test—the cephalic ganglion with the nerve-fibres running to the antennæ and the eyes; the ganglia of the double ventral cord with the filaments passing to the appendages; the heart, an elongated tube with three openings; the stomach, a large sac with a small intestine leading from its base to the excretory opening in the telson; in the female two large rose-coloured ovaries, the oviducts passing to an opening covered by two small lamellæ, at the base of the first segment of the pereion; in the male two elongated testes, their ducts opening between the appendages of the seventh segment.

The head is large and greatly inflated, and its upper surface is entirely occupied by two enormous facetted eyes reminding one of the eyes of *Æglina* among trilobites. There are two rows of spines along the lateral borders of the head, and some spines are placed round the mouth, which is in the usual position at the base of the cephalic segment on the lower surface of the body. The first pair of antennæ only are developed in either sex. The antenna consists of two joints, and is attached to the anterior margin of the head.

The parts of the mouth and the maxillipeds are very small; the two gnathopods are terminated by claws as in the Typhids, and act functionally as second and third maxillipeds.

The pereion consists of seven segments; and the pleon of five, to the two last of which the caudal appendages are attached. The five pairs of ambulatory legs are long and slender, and the three pairs of 'swimmerets' are normal. The eggs are large and few in number; some of those observed contained embryos in which nearly all the appendages were developed, showing that the young undergo no metamorphosis.

Dr. von Willemœs-Suhm, who has carefully described this singular form, has proposed to establish for the genus a family CYSTOSOMIDÆ, holding a place intermediate between the TYPHIDÆ and the PHRONOMIDÆ.

Two small specimens of *Centrostephanus longispinus* were entangled in the trawl-net, and many star-fishes, including *Archaster andromeda* and *bifrons*, *Astrogonium longimanum*, and, among other Ophiurids, some fine examples of *Ophiomusium lymani* and of an undescribed species of the same genus. The trawl seems specially suitable to the capture of Holothurids; indeed without its use we should never have imagined that animals of this group occurred so abundantly as they do, and acquired so considerable a size in deep water. Almost every haul along the coasts of Portugal and Africa yielded several species, and particularly many specimens of a remarkable form, referred to a section of the order with which we are now very familiar as inhabitants of the deeper regions of the sea. The animal is of a rich violet colour. Like *Psolus*, it has a distinct ambulatory surface with a central double row of water-feet. The body-cavity is small, but the perisom is represented by an enormously thick layer of jelly, which rises on either side of the middle-line of the back into a series of rounded lobes, each perforated for the passage of an ambulacral tube and corresponding therefore to an ambulacral foot. The upper pair of vessels of the trivium send out series of leaf-like sacs loaded with purple pigment, which fringe the ambulatory disk on either side and appear to be chiefly concerned in the function of respiration.

This haul gave us another interesting evidence of the wide geographical distribution of some of the characteristic forms of the deep-sea fauna; several examples of a species of the genus *Euplectella* were entangled in the netting of the trawl.

In the year 1841 Professor Owen gave an excellent description, in the Transactions of the Zoological Society of London, of a wonderfully beautiful sponge which had been lately received from the Philippine Islands, and which he named *Euplectella aspergillum* (Fig. 28). In the specimen described by Owen the soft gelatinous coating had been entirely removed and nothing remained except the skeleton, composed of silica and resembling an exquisitely delicate fabric woven in spun glass. The skeleton is in the form of a slightly curved tube, contracted downwards and expanding upwards to a wide circular mouth edged by an elegant frill. The mouth is closed by a wide-meshed netted lid. The walls of the tube are formed by a number of parallel longitudinal bands of glassy siliceous fibres closely united together by a cement of silica, and a series of like bands running round the tube, and thus cutting the longitudinal bands at right angles and forming a square-meshed net. The corners of the squares are then filled in with a minute irregular fret-work of siliceous tubing, and the openings in the wall of the sponge become rounded. Ornamental ridges of the same fine fret-work are arranged in irregular spirals on the outer surface, and round

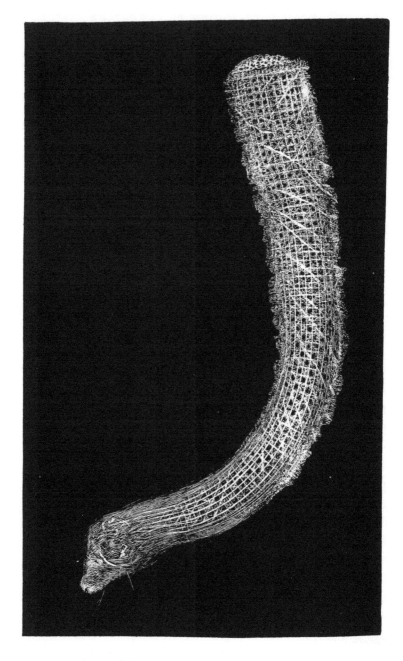

Fig. 28.—*Euplectella aspergillum*, Owen. Half the natural size.

the bottom of the tube a fringe rises of glistening threads of silica four or five inches long.

In 1859 Professor Owen described what he considered a second species of the same genus, *Euplectella cucumer*, from a beautiful specimen procured by Dr. Arthur Farre from the Seychelles. In form Dr. Farre's specimen was more inflated, more like a cucumber, as its name implies, but in minute structure the two are identical, and I believe they are merely varieties. For many years these two specimens of the two forms, one in the British Museum and the other in the possession of Dr. Farre, remained unique. A sponge existed in the gallery in the Jardin des Plantes, procured in the Moluccas during the voyage of the 'Astrolabe,' and figured but not described by Quoy and Gaimard, which had been confused with *Euplectella*, but a careful examination of the Paris specimen showed that although belonging to the same group of sponges it was generically different, and I described it under the name of *Habrodictyon*.

Some years ago some fine specimens of *Euplectella aspergillum* found their way into the London market and on account of their great beauty brought high prices. When the principal museums, and the amateurs who will have what they fancy at any price, were supplied, specimens were still to be had among the dealers, and at greatly reduced prices; and finally, when the demand for scientific purposes had nearly ceased and it was found that the number for sale was unlimited, the prices of good specimens fell from four or five pounds to ten shillings or less, and 'Venus' flower-basket' was often to be seen under a glass shade in a drawing-room, whose owner had little idea of its close relation to his familiar bath companion.

It seems that *Euplectella* is very abundant in some spots in deep water among the Philippine Islands, and particularly near the Island of Zebu. It lives partially buried in mud, which is so soft and loose as not to crush it, nor to impede in any way the assumption of its elegant form—that of a horn or a graceful bouquet-holder,—and it is supported in its position and prevented from sinking by its fringe-like root of glassy spicules. The natives get it by dragging weighted bars of wood to which fish-hooks are attached over the bottom. The sponges are pulled out of the mud by the hooks; many of them are torn and injured, but they are in sufficient numbers to give an ample supply of perfect specimens. The soft animal-matter is then removed in some way, and the skeleton is cleaned and bleached.

Until within the last few months no examples of *Euplectella* were known with the soft parts preserved, but I understand that lately spirit specimens have been received at the British Museum; and the late Professor Max Schultze of Bonn stated at a meeting of the 'Niederrheinische Gesellschaft,' on the 3rd of March last, that a fine series had been placed in his hands by Drs. Gutschow and Heuthe of the German ship-of-war 'Hertha.' As might have been anticipated in fresh specimens, the crystal framework is covered and entirely masked by a layer of grey-brown gelatinous matter, 'sarcode,' as it is technically called, which Professor Schultze describes as being very thin, and loaded with granules,

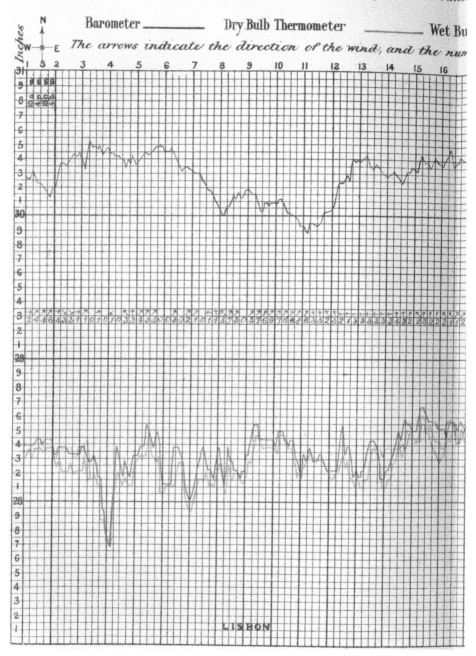

PLATE III. *Meteorological Observatio*

LISBON

tions for the month of January, 1873.

Bulb Thermometer —————— Temperature of Sea Surface ——————

umbers beneath its force according to Beaufort's scale

GIBRALTAR

pigment masses, grains of sand, and the shells of foraminifera. Even in this slimy covering, however, there is not absent the element of beauty, for a multitude of minute siliceous spicules which pervade it everywhere, and whose function seems to be to bind its particles together and to add to its consistency, present singularly elegant forms. One of the most complicated is a spicule with six very short rays, each terminated by a vase-shaped arrangement of curved and fringed siliceous plates.

All the specimens of *Euplectella aspergillum* which I have seen, with one exception—there may be others of which I am not aware—have had the spicules of the skeleton fused together into one continuous net-work, so that the fabric although fragile retains its form and will bear some handling. An examination of a portion of the skeleton under the microscope shows, however, that it was originally composed of distinct spicules. Each spicule has a very fine central tube running through its axis and all its branches have the like, and in the mature skeleton these central tubes remain, indicating the form and position of the original spicules and showing that they are only imbedded in and bound together by an external and secondary siliceous cement. The exceptional specimen is one of those in the excellent Free Museum in Liverpool. In this case the fibres and spicules have remained distinct from one another, and consequently the skeleton, although the netted lid is complete and it has all the appearance of maturity, is quite flexible and soft. Whether this be a character common to young individuals, whether it be an abnormal condition of this single specimen owing to deficiency of silica or some such cause, or whether the specimen ought to be regarded as the type of another species, I am not yet prepared to say. I have seen small specimens of the ordinary form with the spicules soldered together, apparently perfect, but short and without the lid, and I always supposed these to be the young; but from what we know of the intimate structure of the skeleton it would seem probable enough that the spicules may remain separate for a time in early life during the expansion of the creature and the development of its form.

The fine species (Fig. 29) for which I propose the name *Euplectella suberea*, of which three specimens, all unfortunately more or less injured, were taken in the trawl, forms a hollow cylinder about 25 centimetres in length by 5 centimetres in diameter. The walls are composed as in *E. aspergillum* of a fundamental square-meshed siliceous net-work, bands of spicules running longitudinally from end to end of the sponge, and transverse bands intersecting these at right angles. The spicules are in some cases straight and smooth, frequently four projecting knobs ranged round the centre of the shaft of the spicule show that in essential form the spicule is six-rayed, and often one of the side rays is strongly developed and projects to a distance of half an inch or more from the surface of the sponge. The spicules are all free from one another, and those composing the bands can easily be teased asunder with a pair of needles. In this species, as in *E. aspergillum*, the corners of the square meshes are filled up,—a pale-brown corky-looking substance reducing them

CHAP. II.] *FROM PORTSMOUTH TO TENERIFFE.* 139

to round tube-like holes and rising into spirally

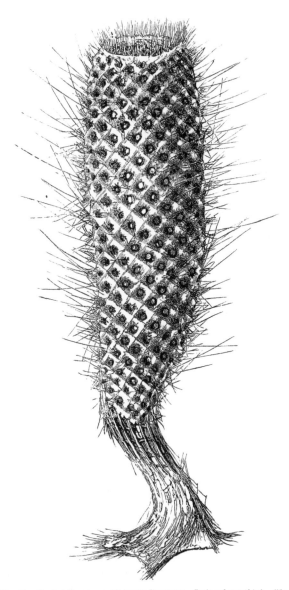

Fig. 29.—*Euplectella suberea*, Wyville Thomson. Reduced one-third. (No. 5.

arranged ridges between them; but the ridges, instead of having a continuous glassy skeleton, have their soft substance supported by a multitude of delicate six-rayed separate spicules interspersed with the usual minute siliceous stars and rosettes. The sponge is hirsute with sheaves of feathered spicules which project from the crests of the spiral ridges, and a series of like sheaves of great length replace round the mouth the fretted frill of the Philippine Islands form. The mouth is closed by a very delicate net-work of a gelatinous substance supported by sheaves of fine needles. The correspondence in form between its ultimate spicules and those of *E. aspergillum* appeared to be so close that when I first saw this sponge I suspected that it might turn out to be the same thing under different conditions. I am now however convinced that the two species are entirely distinct. *Euplectella suberea* is not the only species of the genus living in the Atlantic. Fragments of at least two others have occurred to us, but they are too imperfect for description.

The tube of the Philippine examples of *Euplectella aspergillum* is very frequently inhabited by one, sometimes by a pair of a 'commensal' decapod crustacean; indeed, the association is so constant that only a few years ago a paper was written to show that the sponge was a wonderful siliceous habitation constructed by the crab! It is singular that while *Palythoa fatua* is as constantly associated with examples of *Hyalonema* from the coast of Portugal as with those from Japan, no commensal crustacean has been found in any of the Atlantic specimens of *Euplectella*.

We were greatly interested in adding one of these singular sponge-forms to the fauna of Europe. *Euplectella* belongs to a very special group of sponges which have been called the HEXACTINELLIDÆ, from the circumstance that the siliceous spicules throughout the whole family appear to be six-rayed. This fundamental form is often curiously masked—one, two, three, or four of the rays being frequently suppressed; but when this is the case some branching or splitting of the central canal, or some symmetrical arrangement of projections in the ornament of the spicule, is sure not only to refer it to its ground-form, but to give some clue to the particular kind of suppression or modification which has taken place.

The HEXACTINELLIDÆ are a beautiful family. Besides *Euplectella*, which is perhaps the flower, it contains *Hyalonema*, the glass-rope sponge of the Atlantic and the North Pacific, *Aphrocallistes*, another beautiful lacey fabric of flint, *Holtenia, Rossella, Poliopogon*, and many other wonderful genera. The group belongs specially to the fauna of the deep sea, and they seem to thrive best among the elements of nascent limestones. They are an old family, abounding in many graceful shapes in the beds of chalk and greensand of the south of England, but until lately the fossil 'ventriculites' were supposed to be extinct, and the discovery of their descendants living in the modern chalk-beds of the Atlantic was one of the most interesting of the many corroborative evidences in favour of the view of the 'continuity of the chalk.' We had another successful haul of the trawl on the 30th in 1,525 fathoms, a little farther to the south-west of Cape St. Vincent. The chief prize on this occasion was a Bryozoon of singular beauty, and differing very widely in structure and habit from all previously discovered members of the class (Fig. 30). A straight transparent stem, like the stem of a claret-glass, 60 mm. in length and 5 mm. wide at the base, contracting to half that width at the top, rises from a tuft of fibrous roots and supports on its summit a very graceful cup formed of branches, which, in their general character, somewhat resemble those of the cœnœcium of the common *Bugula neritina*. The backs of the zoœcia are turned outwards and the openings towards the interior of the cup, and a large avicularium is attached to the wall of each cell. The bases of the branches are connected together to a height of about 10 mm. by a transparent membrane. The cell-bearing portion of the cœnœcium ends beneath in a curiously abrupt way at the side of the top of the stem, with which it does not seem to be in any way continuous. The stem passes into the membrane supporting the base of the cup, and the stem and membrane, and the cell-bearing branches, are so different from one another in appearance, that one was inclined to make sure in the first place that he was dealing with a single organism, and not with some singular case of 'commensal' association. We got, however, many specimens of the same species in all stages of growth in the deep water of the mid-Atlantic, and also one or two allied forms which appeared to lead up to it. The transparent stem may probably homologate with the stem of *Loxosoma*, but the branches of the

Fig. 30.—*Naresia cyathus*, Wyville Thomson. Slightly enlarged. (No 6.)

zoœcia and the polypides are certainly those of a normal cheilostomatous Bryozoon not far from *Bugula*. This remarkable form was dedicated to Captain Nares under the name of *Naresia cyathus*, as an early recognition of the confidence and esteem with which he had already inspired the civilian scientific staff. *Naresia* in the form and structure of its cup certainly presents a curious resemblance to the Cambrian *Dictyonema*. This resemblance must, however, be only superficial, for *Dictyonema* passes with many points of accordance equally remarkable into the graptolites, which we can scarcely doubt must be referred to a very different group.

There were six specimens of a beautiful little sea-urchin, with a small purple body and long white serrated spines, somewhat like those of the 'piper' of the Shetland fishermen (*Cidaris hystrix*). There is however an anatomical character in this little urchin which removes it very widely from *Cidaris*, and gives it to some of us a tremendous fancy value. The character is a very small one. Instead of having at the top of the shell a rosette of ten plates, five of them perforated to lodge the eyes, and five for the passage of the tubes of the ovaries, this little urchin has eleven plates in the rosette—an additional one, large, crescent-shaped, and without a perforation (Fig. 31). This is entirely contrary to the usage of all the 'regular' urchins of modern times; but when we go back to the time of the chalk, we find a very compact and characteristic little family, the SALENIADÆ, with the additional plate in the same position, and I agree with Professor A. Agassiz, who has referred a specimen of a species either the same as the one we dredged off the coast of Spain, or closely allied to it, dredged by

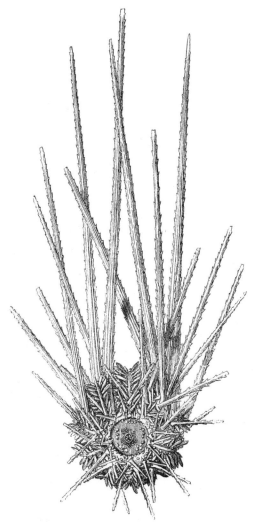

Fig. 31.—*Salenia varispina*, A. Ag. Four times the natural size. (No. 6.)

Count Pourtales in the Strait of Florida, to the chalk genus *Salenia*, under the name of *Salenia varispina*.

The same haul gave a large, handsome urchin, radiant with mauve and white bands springing from the centre of the disk. This was a fine new species of the genus *Phormosoma* which I have described

Fig. 32.—*Salenia varispina*, A. Ag.   Showing the structure of the apical disk.

elsewhere, and shown to represent the genus *Echinothuria* of the chalk, and to belong to a family which were supposed to have become extinct with the close of Mezozoic times.

The present form, which I will call *Phormosoma uranus* (Fig 33), corresponds with *Phormosoma placenta*, a species taken in the 'Porcupine' in 450

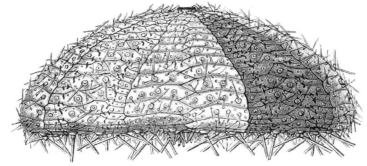

Fig. 33.—*Phormosoma uranus*, Wyville Thomson.   Natural size. (No. 6.)

fathoms off the Butt of the Lews, in having the calcareous pavement of the perisom continuous, no membranous spaces being left between the plates, which are imbricated like the slates on the roof of a house, the shell being perfectly flexible. One marked character, however, of the northern species was that the lower (oral) surface of the test was quite

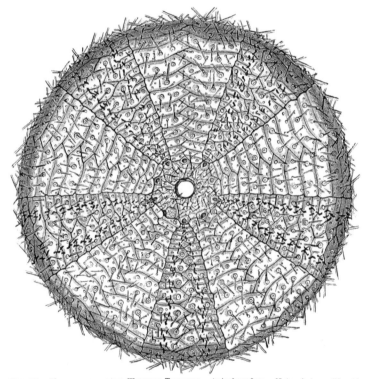

Fig. 34.—*Phormosoma uranus*, Wyville Thomson. Apical surface. Natural size. (No. 6.)

different in character from the apical surface, the pore areæ and the water-feet on the lower surface being reduced to insignificance, and the test uniformly studded with tubercles for primary spines surrounded by enormous areolæ giving attachment to masses of muscle quite out of proportion to the size of the spines. In the only specimen which we procured, the spines on the oral surface were broken and incomplete, but I remarked on this curious disproportion when briefly noticing the species.[1] After going over the siftings from the 'Porcupine' dredgings, Dr. Carpenter sent me, with a number of other things *incertæ sedis*, one or two peculiar conical calcareous bodies jointed on short stalks. These I recognised as a part of some unknown echinoderm, what part I was at a loss to conceive. Curiously enough, a couple of days ago (June 13th, 1874) we

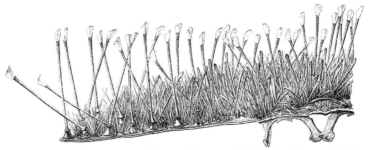

Fig. 35.—*Phormosoma hoplacantha*, Wyville Thomson. Southern Sea between Australia and New Zealand. Portion of the ventral surface of the test. Reduced one-third.

dredged in 400 fathoms 100 miles to the eastward of Sydney, a splendid specimen a foot in diameter of another species of *Phormosoma* (Fig. 35), all the primary spines of the oral surface tipped with calcareous cones, identical in structure with those found on the Scottish coast, but considerably larger. All the bases of the cones had been bevelled off towards their outer aspect, and they had evidently been used for vigorous locomotion over firm ground. The large areolæ and spine-muscles were again present, and

[1] The Depths of the Sea. London: Macmillan and Co., 1873. P. 161.

the object of their excessive development was now obvious.

In *Phormosoma uranus* there is but little difference in character between the upper and the lower surfaces of the test, and the species thus holds a place intermediate between the genera *Phormosoma* and *Calveria*.

January 31*st*.—The trawl was put over in 2,125 fathoms about midway between Cape St. Vincent and Madeira. Since our first essay with the trawl we had been employing it at gradually increasing depths, and our experience was that on perfectly smooth ground without rocks it worked with even greater certainty than the dredge, always falling in the right position and scarcely ever fouling in any way. The operation took a little longer than dredging, for the trawl being of lighter material and exposing a much larger surface to the water takes somewhat longer to go down, and offers greater resistance in coming up. The strain, measured by the accumulator, is however probably not greater, for as the trawl is not constructed to bring up the material of the bottom in any quantity, the weight of its contents is usually much less.

The number of living animals certainly diminishes after we reach a certain point, with increasing depth; and even the wide-mouthed trawl rarely brings up a heavy freight from depths beyond 2,000 fathoms. Such captures, however, frequently countervail their scantiness by their interest and value. This haul, for example, gave us with a few star-fishes and holothurids a fine specimen of the 'clustered sea-polype,' *Umbellularia grœnlandica* (Fig. 36), an animal of great interest, historical as well as scientific. Twelve gigantic alcyonarian polyps, each with eight long fringed arms, terminate in a close cluster a calcareous stem 90 centimetres high.

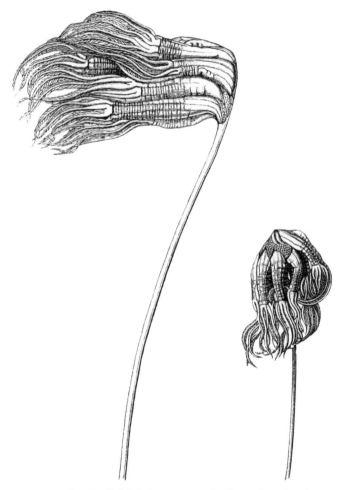

Fig. 36.—*Umbellularia grœnlandica*, L. Natural size. (No. 7.)

Two specimens of this fine species were brought from the coast of Greenland early in last century by Captain Adriaanz, who found them adhering to the sounding line brought up from about 300 fathoms water. They were described by M. Christlob Mylius in 1754, and one of them was afterwards described in the Philosophical Transactions by Ellis, who formed the erroneous notion, much more intelligible when one has seen a specimen, that he had discovered in *Umbellularia* the living representative of the 'lily encrinite.' The two specimens described by Mylius and Ellis were lost, and for a century the animal was never seen. A year ago two specimens were taken in deep water during the expedition of the Swedish frigates 'Ingegerd' and 'Gladan' to the Northern Ocean, and will shortly be described by M. Lindahl, the naturalist who accompanied the expedition.*

When taken from the trawl the polyps and the membrane covering the hard axis of the stem were so brightly phosphorescent, that Captain Maclear found it easy to determine the character of the light by the spectroscope. It gave a very restricted spectrum sharply included between the lines b and D.

On Monday, the 3rd of February, we dredged off the Desertas, and in the afternoon put down the trawl. The bottom, however, proved to be too rocky and the trawl fouled and was lost,—to our great regret, for it was well made and well balanced, and had been successful in every cast, even to the greatest depths.

Next morning we anchored in the roadstead of Funchal.

We spent two days only at Madeira, and we can of course add nothing to the many excellent descriptions

* M. Lindahl's memoir 'Om Pennatulid-slägtet Umbellula' was read before the Royal Swedish Academy in February, 1874, and is now published (February, 1875).

which have been given of the Island. We, apparently in common with all others, were struck with its exceeding beauty, and particularly with the wonderful profusion of flowers. Patches of glorious colour were visible from the ship even at this early season, and these when looked at more closely resolved themselves into astonishing festoons and garlands of flowering creepers, hanging across the streets and clothing and mantling over every wall and trellis.

There was a fine *Bougainvillea* in Captain Phillimore's garden at Gibraltar which greatly excited our admiration; but round Funchal it was everywhere, in every shade of colour from a brick-red through rich crimsons and violets to a pale delicate mauve. What a pity it is that this singularly ornamental plant will not as yet stand the climate of England! From its brilliancy not-depending upon the flower, but upon a bunch of *bracts* or flower-leaves, it stands a long time, scarcely varying in effect from early in December to the middle or end of August. Second only to the *Bougainvillea* in decorating the verandas and trellises in Madeira are several species of *Bignonia*, particularly *B. venusta*, which runs out into long wreaths, clustering round every available projection and glorifying it with its trusses of golden bells. Many of our party went off at once on horseback to the hills, whilst others found enough to enjoy during the few hours of our stay, in the gardens and walks in the neighbourhood of the town.

Madeira is very rich in land-shells, which are particularly interesting owing to the singular position in which these Atlantic islands stand as to the source and extension of their land faunæ. Some of the naturalists of our party took the opportunity of going over a very instructive collection of the landshells which has been made by the Rev. R. Boog Watson, during his residence.

On the morning of the 5th of February we left Madeira and stood for the Canaries. We had a capital breeze all the way, force = 5—6, from the north-east. We only stopped to sound once in 1,970 fathoms, about half-way; and we made on an average eight knots an hour, so that we found ourselves off Teneriffe early on the morning of the 7th.

We anchored in the bay of Santa Cruz, and remained there a couple of days, and sent off a tenting party, consisting of several of the civilian scientific staff and one or two of the naval officers, to the Canâdas, the mountain plateau at an elevation of about 7,000 feet, 5,000 feet below the summit of the Peak. With the friendly assistance of the British Consul their arrangements were all satisfactorily made. They went across the island to Orotava, whence they commenced the ascent of the mountain. They spent four nights camped on the high grounds. It was too early in the season to attempt the ascent of the Peak, and rather too early for natural-history work; the chief object was to get into the *modus operandi* of such expeditions in preparation for the future. Still collections were made, both in zoology and in geology, and the party were greatly interested in the wonderful atmospheric effects from their camping-ground above the lower stratum of clouds. The ship weighed anchor on the morning of Monday the 10th, and took a sounding and dredging cruise among the islands. We dredged between Teneriffe and Palma and passed close under Palma, near enough to see the outline and bounding wall of its wonderful crater; past Gomera, a fine rugged island, the coast showing the usual alternation,—deep ravines and lava spurs covered with vines and maize. Past Hierro, or Ferro, another fine bold island, interesting as having been long regarded as the western point of the old world, and on that account chosen as the zero in reckoning longitude. The zero line passing through Hierro is still used in some countries, though the meridian of Greenwich is now almost universally employed for practical purposes, such as the construction of charts. The position of Hierro appears to have determined the line of division of the world on maps into an eastern and a western hemisphere.

We had splendid moonlight during our cruise, and although the Peak remained almost constantly shrouded in mist during the day, it shone out at night as a dazzling white cone through the rifts in the clouds.

We returned and anchored in Santa Cruz Bay on the morning of Thursday the 13th.

Owing to the uncertain and boisterous weather the temperature observations taken during this early part of the cruise were not so numerous or so complete as we could have wished. The bottom temperatures are given in Appendix A. of this chapter. Those off the coast of Portugal corresponded very closely with the observations made in the 'Porcupine' in 1870 and in the 'Shearwater' in 1871.

A serial sounding was taken on the 15th January off Cape St. Vincent, which showed a great uniformity in the temperature at that season for the first 200 fathoms:—

Surface	16°·1 C.
25	15·5
50	15·5
75	15·3
100	14·2
125	13·6
150	12·8
175	12·8
200	12·5

and another near the Island of Gomera on the 12th of February, at intervals of 10 and 20 fathoms, gave the same result:—

Surface	18°·3 C.	120		16°·3 C.
10	18·0	140		15·7
20	17·9	160		15·0
30	17·9	180		14·3
40	17·8	200		14·2
50	17·8	220		13·7
60	17·7	240		13·3
70	17·7	260		13·1
80	17·6	280		12·7
90	17·5	300		12·2
100	16·6			

A sounding taken at the same station at intervals of 100 fathoms down to 1,000, showed an entire modification in the distribution of the band of abnormally warm water between 300 and 800 fathoms, which presents so marked a feature in the temperature sections off the west coast of Europe:—

Surface	18°·9 C.	600	7°·2 C.
100	15·0	700	6·4
200	13·2	800	5·5
300	10·8	900	4·7
400	9·0	1,000	3·4
500	7·6	1,620—bottom	2·3

Observations made at subsequent periods in the same district gave a much better insight into the nature of this change.

Throughout the year meteorological observations were taken with great care and regularity. The barometer, the wet- and dry-bulb thermometer in air, the temperature of the sea-surface, the direction and force of the wind, and the proportion of cloud and the general state of the weather, were registered not less than once in two hours by the officers of the watch. Whenever there was any marked peculiarity in atmospheric condition or in the temperature of the sea, or any suspicion that there might be cause for such, observations were made hourly and often half-hourly.

It must be remembered, however, that owing to the ship constantly changing her position and passing into different latitudes and being subjected to different local conditions, these observations have not the kind of value of a series of observations taken at a single spot, but must be regarded as isolated observations, each good only for the position and date at which it was taken. As such, however, their importance can scarcely be overrated in their bearings upon the discussion of the questions regarding the movement of air and water, and the conditions of climate affecting the maintenance and distribution of living beings, which it was our mission to attempt to solve. This of course does not apply to series of observations extending over a considerable time at one place, such as those taken in harbour at Bermudas, Halifax, &c.

The number of separate observations was so great,

amounting during the twelve months from the 1st December 1872 to the corresponding date in 1873 to upwards of 50,000, and would have occupied so much space in a tabular form, that it has been thought preferable in this preliminary sketch to reduce some of the general results to diagrams. In a series of twelve plates, accordingly, the results are given for the several months. The darker vertical lines on the plates indicate the hour of 4 A.M. for each day, and the three intermediate lighter lines the hours of 10 A.M., 4 P.M., and 10 P.M. respectively. The spaces between the transverse lines have the value of $\frac{1}{20}$th of an inch on the barometer scale referred to a column on the left side of the diagram; and to $\frac{1}{20}$th of a degree centigrade referred to a corresponding scale on the right. The black curved line indicates the variations in atmospheric pressure reduced from bi-hourly observations taken with a mercurial barometer. Later in the cruise we had out from Messrs. Elliott of London a self-registering aneroid barometer, a beautiful instrument, but somewhat too delicate in its construction for use on board a ship. Owing probably to a trace of friction, the aneroid was usually very slightly behind the mercurial barometer in its indications, but otherwise the close correspondence between the two curves was most satisfactory as an evidence of the care and accuracy with which the barometrical observations were taken.

The red line gives the variations of the temperature of the air projected from observations of a standard mercurial thermometer housed in the usual way. An instrument devised by Mr. Thomas Stevenson of Edinburgh with a view of eliminating any transient and accidental changes of temperature, and thus arriving at a more natural mean for the day, was observed twice daily, at 6 A.M. and at 6 P.M., by Mr. Tizard. A maximum and a minimum registering thermometer have their bulbs immersed in a small flask filled with brine and suspended on gimbals. The whole of the contents of the flask must thus be affected by a change of temperature before that change can be registered by the thermometers. The size of the flask is so adjusted that there is ample time for this to occur with the gradual and normal cycle of daily temperature, while spasmodic oscillations are neutralized before they are recorded.

The general result of this arrangement is that the maximum stands somewhat higher in Stevenson's than in the ordinary maximum thermometer, while in the minimum there is very little difference. A comparative table of the indications of the common registering thermometers in air and of 'Stevenson's mean thermometers,' for six months, is given in Appendix B. to this chapter.

The yellow line indicates the rise and fall of the wet-bulb thermometer.

The blue line, perhaps the most interesting to us of the whole, gives the temperature of the surface of the sea,—ascertained by bringing up some water at bi-hourly intervals in a draw-bucket, and testing it with a standard thermometer. The course of this blue line on the diagrams gives an example of the advantage of some such plan of graphic representation, where it can be employed, in giving a rapid and vivid impression to the eye of phenomena which it would be somewhat difficult to realize from the data presented in a tabular form.

When there is no interfering cause, and where the sea surface is free to assume the local temperature conditions, the normal position of the blue line may perhaps be stated as markedly below the mean of the red line, but a little above its greatest depressions. The wide elevations and depressions of the red line correspond with changes of wind and alternations of bright and cloudy weather, and these principal deviations of the red line from a straight path the blue line follows slowly but upon the whole closely. The daily oscillations which are the most prominent on most of the diagrams indicate of course the difference between the warmest hour of the day and the coolest hour of the night. The sea-surface absorbs radiant heat during the day which it does not entirely dissipate during the night, so that while its maximum never rises nearly so high as that of the air at midday, its minimum for the coldest period of the night does not fall so low, and the daily mean which it tends to take is just a little below that of the air.

It is rarely, however, that the water of the sea-surface is at liberty to assume the local conditions of temperature. Owing to the universal system of oceanic circulation, the water at any one point forms necessarily part of an indraught or current moving from a warmer or a colder source, and unless this movement be inappreciably slow, the temperature observed is not simply that of the station of observation, but is greatly affected by that of the region from which the water is moving. When the current is warm—and this is usually the case in dealing with surface movements, since the cold return indraughts move, unless in exceptional cases, on the bottom—the blue line rises high above the mean of the red one, and where the phenomenon is extreme, as in the case of the Gulf-stream and the Guinea current, this divergence is most striking.

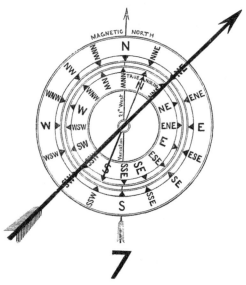

Fig. 37.—Diagram showing the direction and force of the wind (direction south-west by Compass; force = 7 by Beaufort's scale) at midnight, January 1st, 1873.

The direction of the wind is given by a line of arrows flying with the wind. In the Atlantic where the deviation of the compass is comparatively small, the direction is referred to the magnetic north.

The force of the wind is indicated by numbers according to Admiral Beaufort's scale. This scale ranges from 1 to 12. I add a Table, giving what must be only regarded as rough approximations to the somewhat indefinite values of the numbers—in ordinary language, in velocity, and in pressure.

No. on Beaufort's Scale.	Equivalent Expression.	Velocity per Hour in Miles.	Pressure on the Square Foot in Pounds.
1	Light Air	1·5	$\frac{1}{100}$
2	Light Breeze	2	$\frac{2}{100}$
3	Gentle Breeze	4·5	$\frac{1}{10}$
4	Moderate Breeze	8	$\frac{3}{10}$
5	Fresh Breeze	12·5	$\frac{7}{10}$
6	Strong Breeze	18	1·5
7	Moderate Gale	24·5	2·8
8	Fresh Gale	32	4·8
9	Strong Gale	40·5	7·4
10	Whole Gale	50	11·5
11	Storm	61	17
12	Hurricane	72	24

GIBRALTAR, FROM THE SEA.

APPENDIX A.

Particulars of Depth, Temperature, and Position, at the Sounding Stations between Portsmouth and Teneriffe; the Temperatures corrected for Pressure.

Date.	Depth in Fathoms.	Bottom Temperature.	Surface Temperature.	Position.
1872. Dec. 30	1,125	...	12°·8 C.	Lat. 41° 58′ N. Long. 9° 42′ W.
1873. Jan. 1	325	...	13·9	Lat. 40° 25′ N. Long. 9° 38′ 30″ W.
,,	730	...	13·9	Lat. 40° 24′ N. Long. 9° 45′ W.
2	1,975	...	13·6	Lat. 39° 55′ N. Long. 10° 5′ W.
13	470	...	11·7	Lat. 38° 10′ N. Long. 9° 14′ W.
,,	1,270	..	13·9	Lat. 38° 5′ N. Long. 9° 39′ W.
14	84	...	13·9	Lat. 38° 31′ N. Long. 9° 31′ W.
,,	280	...	13·9	Lat. 38° 28′ N. Long. 9° 35′ W.
,,	580	10°·9 C	13·9	Lat. 38° 26′ N. Long. 9° 38′ W.
,,	1,290	...	13·9	Lat. 38° 22′ 30″ N. Long. 9° 44′ W.
,,	1,475	2·6	14·2	Lat. 38° 14′ 25″ N. Long. 9° 49′ 42″ W.
,,	1,380	2·9	14·2	Lat. 38° 9′ 43″ N. Long. 9° 48′ W.
,,	1,800	2·2	13·9	Lat. 37° 56′ N. Long. 10° 8′ W.
15	1,000	3·9	15·3	Lat. 37° 1′ 45″ N. Long. 9° 23′ 45″ W.
,,	525	12·0	15·6	Lat. 36° 58′ 50″ N. Long. 9° 14′ 20″ W.
,,	900	...	15·0	Lat. 37° 2′ N. Long. 9° 14′ W.
16	600	...	15·6	Lat. 36° 25′ N. Long. 8° 12′ W.
28	1,090	3·1	15·6	Lat. 35° 47′ N. Long. 8° 23′ W.
29	2,500	...	15·0	Lat. 36° 13′ N. Long. 10° 7′ W.
30	1,525	1·6	14·4	Lat. 36° 23′ N. Long. 11° 18′ W.
31	2,125	2·0	15·6	Lat. 35° 20′ N. Long. 13° 4′ W.
Feb. 1	2,250	2·0	15·6	Lat. 34° 4′ N. Long. 14° 18′ W.
2	2,225	1·8	17·2	Lat. 32° 43′ N. Long. 15° 52′ W.
,,	670	8·0	17·2	{ R. T. Bugio 8° 30′, R. T. Deserta. { L. T. Bugio 18° 30′, R. T. Bugio.
,,	1,150	3·4	17·2	{ R. T. Bugio N. 7′ W. (true). { L. T. Burio 9° 35′, R. T. Bugio.
,,	930	6·0	17·2	{ L. T. Bugio 33° 0′, R. T. Bugio. { L. T. Chao 21° 10′, L. T. Bugio

FROM PORTSMOUTH TO TENERIFFE.

Date.	Depth in Fathoms.	Bottom Temperature.	Surface Temperature.	Position.
1873. Feb. 2	1,500	2°·4 C.	17°·2 C	Lat. 32° 27′ N. Long. 16° 40′ 30″ w.
3	1,150	3·4	17·2	Sail Rock 37° 28′, R. T. Bugio. Fora Lt. 46° 50′, Sail Rock.
,,	790	6·9	17·2	Same $\begin{Bmatrix} 29° \ 40' \\ 40° \ 30' \end{Bmatrix}$ objects.
,,	490	...	17·2	L. T. Madeira φ. R. T. Madeira 59° 30′, R. T. Bugio. φ 133° 0′, R. T. Madeira.
6	1,975	1·6	17·0	Lat. 29° 19′ N. Long. 16° 38′ w.
10	975	4·6	17·8	St. Andre Tr. 52° 33′, R. T. Teneriffe. Pico di Teyde 27° 50′, St. Andre Tr.
,,	278	...	17·8	Mole Lt. 137° 45′, Antequerra Pt. Aboona Pt. 61° 10′, Mole Lt. Guadamete 40° 0′, ditto.
,,	630	7·0	17·8	Mole Lt. 119° 0′, Antequerra Pt. Pico di Teyde 26° 55′ Mole Lt.
,,	560	7·2	17·8	St. Andre Tr. 60° 5′, Lt. Ho. Pico di Teyde 13° 0′, St. Andre Tr.
,,	78	...	17·8	Antequerra Pt. φ Roquette Pt. φ 77° 40′, Anaga Rk.
,,	179	...	...	Lt. Ho. 34° 0′, R. T. Teneriffe. Antequerra Pt. 27° 40′, Lt. Ho.
,,	640	7·4	17·8	Anaga Rk. 34° 0′, R. T. Teneriffe. L. T. Teneriffe 28° 50′ Anaga Rk.
,,	1,390	3·0	17·8	Anaga Rk. 41° 10′, Madeira Pt. L. T. Teneriffe 13° 15′, Anaga Rk.
11	1,750	2·3	17·2	Gomera Pk. 84° 50′, Palma Pt. Pico di Teyde 54° 40′ Gomera Pt.
,,	1,340	3·0	18·3	L. T. Gomera 28° 55′, R. T. Gomera. High part of R. T. Palma 133° 50′, L. T. Gomera.
,,	1,620	2·3	18·3	L. T. Ferro 23° 37′, R. T. Ferro φ Salme Rk. R. T. Gomera, 137° 30′, L. T. Ferro.
12	620	...	18·3	L. T. Gomera 71° 0′, R. T. Gomera. R. T. Palma 54° 20′, L. T. Gomera.

APPENDIX B.

Comparative Table of the Indications of Stevinson's Mean Thermometers, and the ordinary Maximum and Minimum Thermometers in Air, for the Six Months from the 1st May to the 31st October, 1873.

Date. 1873.	Latitude.	Longitude.	6 A.M.				6. P.M.			
			In Air.		In Water.		In Air.		In Water.	
			Max.	Min.	Max.	Min.	Max.	Min.	Max.	Min.
	° ′	° ′	Deg. Cent.	Deg. Cent.	Deg. Cent.	Deg. Cent.	Deg. Cent.	Deg. Cent.	Deg. Cent.	Deg. Cent.
May 1	36 23 N.	71 51 W.	17·1	16·2	16·5	16·0	20·0	16·5	18·7	16 0
2	37 25 N.	71 40 W.	18·1	14·4	18·0	14·5	16·3	14·5	15·3	14·5
3	38 34 N.	72 10 W.	15·3	12·3	15·0	12·2	13·2	6·5	11·3	8·0
4	39 13 N.	71 20 W.	7·4	4·7	6·5	5·0	8·1	5·3	7·8	4·0
5	39 50 N.	69 14 W.	10·4	7·0	9·7	7·0	12·3	9·7	11·5	9·0
6	40 17 N.	66 48 W.	10·9	5·7	9·0	6·0	12·2	5·7	10·0	6·0
7	41 15 N.	65 45 W.	9·0	4·2	9·7	4.5	9·1	5·0	7·5	4·7
8	43 2 N.	64 2 W.	5·7	2·3	6·0	3·0	9·2	2·8	7·5	2·5
9	At Halifax.	Nova Scotia	6·2	2·9	6·7	3·5	Not registered.			
10	Ditto.	Ditto.	17·8	3·6	15·0	1·0	12·6	5·9	10·0	5·8
11	Ditto.	Ditto.	7·1	4·8	7·0	5·5	8·2	5·7	7·2	5·2
12	Ditto.	Ditto.	7·9	6·6	7·5	5 5	11·6	6·8	10·5	7·5
13	Ditto.	Ditto.	9·0	6·2	9·0	6·5	Not registered.			
14	Ditto.	Ditto.	18·4	7·3	16·0	7·0	13·7	8·7	12·5	8·0
15	Ditto.	Ditto.	11·6	4·4	12·0	5·5	10·0	4·7	9·0	5·2
16	Ditto.	Ditto.	9·5	4·0	9·0	4·5	16·2	7·3	14·0	5·0
17	Ditto.	Ditto.	12·3	4·6	12·7	5·5	9·5	4·2	8·0	5·0
18	Ditto.	Ditto.	5·7	3·6	5·0	4·0	13·4	4·4	12·0	4·2
19	Ditto.	Ditto.	11·8	6·8	12·0	7 0	11·2	4·2	9·5	6·5
20	43 3 N.	63 39 W.	6·2	3·2	7·2	3·2	10·3	4·1	8·0	3·2
21	42 10 N.	63 39 W.	8·3	4·7	8·0	5·0	11·7	5·4	10·0	5·0
22	41 19 N.	63 11 W.	11·4	8·3	10·2	8·0	16·1	11·1	15·0	9·0
23	39 44 N.	63 22 W.	19·5	14·7	18 5	13·0	22·0	18·7	20·7	17·2
24	38 32 N.	63 36 W.	21·7	20·1	21·0	20·0	22.4	20 8	22·0	19·5
25	37 9 N.	62 30 W.	22·3	19·2	22·0	19.2	23.2	19·6	22·5	18·0

FROM PORTSMOUTH TO TENERIFFE.

Date. 1873.	Latitude.	Longitude.	6 A.M.				6 P.M.			
			In Air.		In Water.		In Air.		In Water.	
			Max.	Min.	Max.	Min.	Max.	Min.	Max.	Min.
	° '	° '	Deg. Cent.	Deg. Cent.	Deg. Cent.	Deg. Cent.	Deg. Cent.	Deg. Cent.	Deg. Cent.	Deg. Cent.
May 26	36 30 N.	63 40 W.	22·8	20·8	22·5	21·0	24·4	20·7	23·2	21·5
27	34 51 N.	63 59 W.	23·1	20·6	23·0	21·0	22·6	20·6	22·1	21·5
28	33 20 N.	64 37 W.	22·6	20·6	22·1	21·0	24·4	20·9	23·0	21·0
29	Off Bermudas.		23·9	22 0	23·2	22·0	Not registered.			
30	Ditto.		25·7	22·3	24·7	22·7	26·6	22·8	25·0	23·0
31	Ditto.		25·8	22·6	25·0	22·5	25·5	24·4	25·0	24·0
June 1	At Bermudas.		26·0	22·3	25·0	22·3	25·8	22·0	24·0	22·0
2	Ditto.		22·8	20·3	23·1	20·1	23·2	20·3	22·1	20·1
3	Ditto.		23·3	19·6	22·0	19·5	23·4	20·1	22·8	20·0
4	Ditto.		23·1	18·9	22·0	19·0	24·4	19·5	23·7	19·0
5	Ditto.		23·4	20·5	23·5	20·5	26·4	20·9	25·5	20·5
6	Ditto.		25·0	22 8	25·5	23·0	26·7	23·1	26·0	23·0
7	Ditto.		25·0	22·9	25·2	23·2	25·5	22·1	25·0	22·5
8	Ditto.		24·5	22·8	24·7	23·0	25·0	21·6	23·7	22·0
9	Ditto.		22·3	18·9	22·0	19·5	23·2	18·2	22·1	19·0
10	Ditto.		22·1	19·7	22·0	20·0	24·2	20·3	23·0	20·0
11	Ditto.		22·6	20·4	23·0	20·5	23·9	20·6	22·8	20·6
12	Ditto.		23·9	20·3	22·8	20 5	25·0	21·4	24·0	20·7
13	Ditto.		25·0	22·0	26·5	21·7	25·1	22·9	26·5	20·0
14	32 52 N.	63 34 W.	24·1	23·0	24·1	23·0	25·3	23·1	24·8	22 8
15	33 41 N.	61 28 W.	25·0	22·9	24·8	22·8	24·7	22·9	24·0	22·5
16	34 27 N.	58 56 W.	23·4	21·7	23·5	21·6	25·0	21·7	23·8	22 0
17	34 54 N.	56 38 W.	23·4	21·8	23·5	21·8	24·7	21·8	24·0	22·0
18	35 6 N.	52 50 W.	23·2	21·7	23·0	21·8	24·5	21·1	23·0	21 8
19	35 29 N.	50 53 W.	24·1	19 3	23·0	19·5	23·7	19·6	22·5	19·5
20	35 35 N.	50 27 W.	22·4	19·7	22·2	20·0	24·5	20·0	23·5	20·0
21	36 22 N.	48 37 W.	23·7	22·0	23·5	22·0	23·9	21·4	23·0	21·8
22	37 19 N.	45 6 W.	23·7	21·2	23·0	21·2	23·9	21·4	22·8	21·2
23	37 54 N.	41 38 W.	22·9	20·7	22·8	20·5	24·4	20·6	23·0	20·5
24	38 3 N.	39 19 W.	22·0	20·6	22·0	20·5	23·3	20 6	22·5	20·5
25	38 23 N.	37 21 W.	22·3	20·7	22·0	21·0	23·3	20·6	22·8	20·5
26	38 25 N.	35 50 W.	22·3	20·6	22·5	20·5	23·3	20·9	22·5	20·6
27	38 18 N.	34 48 W.	22·7	20 7	22·5	20·8	24·7	20·8	23·0	21·0
28	38 32 N.	33 17 W.	23·9	19·7	23·0	20·2	23·4	20·3	22·5	20·2
29	37 47 N.	31 2 W.	22·5	19 2	22·0	19·5	22·3	19·5	21·5	20·0
30	38 30 N.	31 14 W.	20·3	19·0	20·2	19·0	22·3	19·1	21·0	19·0
July 1	Off Fayal.		21·7	19·3	21·2	19·1	22·3	19·6	21·8	19·5
2	At Fayal.		21·7	18·1	22·0	19·0	22·9	19·2	21·8	18·6
3	38 11 N.	27 9 W.	21·4	19·5	21·5	19·5	23·0	20·0	22·0	19·8
4	37 47 N.	26 9 W.	22·3	19·1	22·0	20·0	22·9	20·0	21·8	20·0
5	At San Miguel Azores.		22 6	19 4	22·0	20·0	24·5	19·7	23·0	19·0
6	Ditto.		22·9	19·7	23·0	20·0	24·0	20·0	23·0	20·7
7	Ditto.		22·6	20·4	23·0	20·5	25·3	21·4	23·5	20·5
8	Ditto.		22·3	20 6	23·5	21·0	...	...	...	..
9	Ditto.		24·0	19·5	23·0	19·8	23·4	19·5	22·5	20·0
10	Off San Miguel Island.		21·2	20·0	21·8	20·1	25·3	20·2	23·6	20·2
11	36 21 N.	23 31 W.	22·9	20·7	23·2	20·8	25·1	20·8	23·9	20·8

166   THE ATLANTIC.   [CHAP. II.

Date. 1873.	Latitude.	Longitude.	6 A.M.				6 P.M.			
			In Air.		In Water.		In Air.		In Water.	
			Max.	Min.	Max.	Min.	Max.	Min.	Max.	Min.
	° ′	° ′	Deg. Cent.	Deg. Cent.	Deg. Cent.	Deg. Cent.	Deg. Cent.	Deg. Cent.	Deg. Cent.	Deg. Cent.
July 12	35 8 N.	21 33 W.	22·4	19·8	23·0	20·0	24·0	20·1	23·0	20·2
13	34 17 N.	20 10 W.	23·4	19·7	23·0	20·0	25·5	20·0	25·0	20·2
14	33 46 N.	19 17 W.	23·2	20·0	24·1	20·1	25·8	20·2	24·2	20·5
15	33 11 N.	18 13 W.	23·4	20·2	24·0	20·5	23·4	20·2	23·0	20·5
16	At Madeira.		22·3	19·7	23·0	19·5	25·5	19·7	23·2	19·5
17	Ditto.		25·5	19·7	23·5	19·2	...	...	...	...
18	30 49 N.	17 59 W.	24·0	19·2	23·0	19·0	24·0	20·2	23·0	20·5
19	28 42 N.	18 7 W.	22·5	20·0	23·0	20·5	23·1	20·1	24·0	20·5
20	27 0 N.	19 38 W.	22·3	20·5	23·0	20·5	23·4	20·5	22·8	20·5
21	25 51 N.	20 15 W.	22·3	20 8	22·0	20·7	25·0	21·4	23·5	21·2
22	24 7 N.	21 18 W.	22·6	21·4	23·0	21·7	25·5	22·9	24·2	22·0
23	22 18 N.	22 2 W.	24·0	22·0	24·0	22·0	25·0	22·3	24·3	22·0
24	20 58 N.	22 57 W.	24·0	22·3	24·0	22·0	25·0	22·5	24·1	22·0
25	19 11 N.	24 7 W.	23·7	22·3	23·7	22·0	25·8	22·6	25·0	22·5
26	17 54 N.	24 41 W.	24·2	22·0	24·8	22·0	25·4	22·3	24·2	22·2
27	Off St. Vincent Island.		24·2	23·0	24·0	23·0	26·1	22·9	25·5	23·0
28	At St. Vincent.		25·0	22·9	25·5	23·0	...	...	...	...
29	Ditto.		27·2	23·0	26·0	23·0	26·8	23·4	26·0	23·0
30	Ditto.		25·0	22·9	25·5	23·0	27·2	22·9	26·0	23 0
31	Ditto.		24·4	22·9	25·0	23·0	25·5	22·9	24·8	22·8
Aug. 1	At Porto Grande.		24·4	23·0	24·2	22·8	27·2	23·2	26·0	23·0
2	St. Vincent.		25·0	23·0	25·8	23·0	26·9	23·3	26·0	23·0
3	Ditto.		25·4	23·6	26·0	23·5	27·8	23·6	26·6	23·6
4	Ditto.		25·0	23·4	25·8	23·0	30·0	23·4	28·0	23·0
5	Ditto.		25·6	23·6	27·0	23·8	28·3	24·3	26·7	24·7
6	15 43 N.	24 15 W.	26·7	24·2	26·5	24·0	26·9	24·4	26·7	24·2
7	At Porto Praya.		25·6	23·8	26·2	24·1	27·8	23·9	26·5	22·0
8	Ditto.		26·1	24·2	26·2	24·0	27·2	24·4	26·6	24·0
9	Ditto.		25·8	23·9	26·0	24·0	28·6	23·9	27·2	23·5
10	13 58 N.	23 5 W.	26·4	24·2	27·0	24·0	26·4	24·4	26·0	24·2
11	12 30 N.	22 38 W.	25·6	23·4	25·7	23·3	26·7	24·0	25·7	24·2
12	11 59 N.	21 12 W.	25·6	23·9	25·5	24·8	28·1	24·9	26·7	24·8
13	10 25 N.	20 30 W.	26·7	24·6	26·8	24·5	27·9	24·2	26·7	24·5
14	9 21 N.	18 28 W.	26·1	24·7	26·2	24·7	26·9	24·4	26·1	25·0
15	8 25 N.	18 2 W.	25·6	22·5	26·0	23·2	25·4	21·7	25·0	22·3
16	7 3 N.	16 3 W.	25·6	24·4	25·0	23·8	27·2	24·4	26·3	24·2
17	6 44 N.	16 42 W.	25·8	24·2	26·0	24·0	26·2	24·0	26·0	24·1
18	6 11 N.	15 57 W.	25·6	23·3	25·5	24·2	25·8	24·4	25·5	24·5
19	5 48 N.	14 20 W.	25·6	23·6	25·0	24·0	27·5	24·7	27·0	24·5
20	4 29 N.	13 52 W.	26·1	23·0	26·5	24·0	28·3	24·2	27·5	24·8
21	3 8 N.	14 49 W.	26·1	23·3	27·0	23·5	26·2	23·3	25·5	23·8
22	2 49 N.	17 13 W.	24·4	23·0	25·0	23·0	26·0	23·2	25·2	23·2
23	2 25 N.	20 1 W.	24·7	23·3	25·0	21·0	26·7	23·3	26·0	23·8
24	2 13 N.	22 21 W.	25·0	23·0	25·0	23·0	25·6	23·3	25·2	23·2
25	1 47 N.	24 26 W.	25·0	23·4	25·0	23·7	26·7	23·3	26·0	23·8
26	1 22 N.	26 36 W.	25·6	23·6	25·5	23·8	28·0	23·9	26·1	24·2
27	1 7 N.	28 48 W.	26·1	23·9	26·2	23·8	27·2	23·9	26·2	24·2

CHAP. II.] *FROM PORTSMOUTH TO TENERIFFE.* 167

Date. 1873.	Latitude.	Longitude.	6 A.M. In Air. Max. Deg. Cent.	6 A.M. In Air. Min. Deg. Cent.	6 A.M. In Water. Max Deg. Cent.	6 A.M. In Water. Min Deg Cent.	6 P.M. In Air. Max. Deg. Cent.	6 P.M. In Air. Min. Deg. Cent.	6 P.M. In Water. Max. Deg. Cent.	6 P.M. In Water. Min. Deg. Cent.
Aug. 28	At St. Paul's Rocks.		25·6	24·0	25·8	24·0	28·4	23·9	27·8	24·0
29	Ditto.		26·1	24·4	27·0	24·5	28·0	24·0	26·8	24·5
30	0 4 N.	30 20 w.	26·1	23·9	26·5	23·9	27·2	23·6	26·0	23·9
31	2 6 s.	31 4 w.	25·3	22·9	25·5	23·9	27·5	23·9	26·7	24·0
Sept. 1	At Fernando Noronha.		26·1	24·4	26·8	24·5	28·0	24·6	27·0	24·8
2	Ditto.		26·4	24·2	27·2	24·2	28·0	23·3	27·5	24·0
3	Ditto.		23·9	21·9	26·0	22·0	26·7	22·9	25·2	22·5
4	5 1 s.	33 50 w.	25·6	23·6	25·5	24·0	25·8	23·0	25·0	23·5
5	4 45 s.	33 7 w.	25·4	23·3	25·0	23·0	26·1	22·5	25·5	23·5
6	5 54 s.	34 39 w.	25·6	24·2	25·2	24·0	26·7	23·8	25·8	23·8
7	6 38 s.	34 33 w.	25·1	23·8	25·0	24·0	26·1	23·9	25·6	23·5
8	7 39 s.	34 12 w.	25·8	23·4	25·2	23·5	26·9	23·6	26·0	24·5
9	8 33 s.	34 30 w.	25·4	23·6	25·2	23·5	27·2	23·3	26·2	23·8
10	9 10 s.	34 49 w.	25·3	22·2	25·5	22·5	26·7	21·9	25·5	22·0
11	10 11 s.	35 22 w.	25·3	23·6	25·5	23·6	27·8	24·2	26·2	23·8
12	10 46 s.	36 8 w.	26·1	23·3	26·0	23·5	26·4	23·4	25·8	23·5
13	11 52 s.	37 10 w.	25·3	22·2	25·0	22·2	25·0	22·2	24·8	22·5
14	Off Bahia.		24·4	22·8	24·5	22·7	28·3	22·8	25·7	22·5
15	At Bahia.		24·4	22·8	25·0	23·0	26·1	23·0	25·5	21·5
16	Ditto.		25·3	22·8	25·2	23·0	28·6	22·8	27·0	23·0
17	Ditto.		25·7	23·9	26·5	24·0	24·7	21·4	25·0	21·5
18	Ditto.		24·2	20·8	24·0	20·5	23·6	21·0	23·0	20·5
19	Ditto.		23·9	21·9	23·2	22·0	24·6	21·7	23·5	21·7
20	Ditto.		23·9	21·8	23·2	22·2	26·1	24·4	25·0	21·7
21	Ditto.		24·7	22·2	24·3	22·1	26·7	22·3	25·2	22·0
22	Ditto.		25·0	22·8	25·0	22·5	26·4	22·6	25·5	21·5
23	Ditto.		24·9	22·5	25·2	22·5	27·5	21·9	25·0	22·5
24	Ditto.		25·1	22·7	25·0	22·5	27·3	22·7	25·5	23·0
25	Ditto.		25·8	22·9	25·7	23·0	26·9	23·0	25·7	23·0
26	13 45 s.	37 59 w.	25·3	23·6	25·5	23·5	27·8	23·6	26·0	23·5
27	14 51 s.	37 1 w.	25·8	23·6	25·0	23·0	26·9	23·5	26·0	23·2
28	17 7 s.	36 50 w.	25·0	23·0	25·0	22·8	27·8	23·0	26·2	23·0
29	19 6 s.	35 40 w.	25·4	22·5	25·0	22·5	26·2	22·5	25·5	22·5
30	20 13 s.	35 19 w.	24·4	21·7	25·0	21·5	26·1	21·7	24·5	21·5
Oct. 1	22 15 s.	35 37 w.	23·9	21·1	24·0	21·0	24·4	21·1	24·2	21·0
2	24 43 s.	34 17 w.	22·2	20·3	23·0	20·3	23·9	20·1	23·8	20·5
3	26 15 s.	32 56 w.	22·2	19·6	23·0	19·2	23·9	19·4	23·0	18·0
4	27 43 s.	31 3 w.	22·2	18·9	22·5	19·0	23·3	18·9	23·0	19·0
5	29 1 s.	28 59 w.	21·1	17·8	22·5	18·0	23·9	17·8	22·8	18·2
6	29 35 s.	28 9 w.	21·7	17·8	22·5	17·8	22·5	17·8	20·5	17·8
7	29 11 s.	26 25 w.	20·0	16·1	20·0	15·8	18·3	15·9	17·8	16·0
8	31 22 s.	26 54 w.	16·9	13·9	17·0	14·0	16·9	14·2	16·0	14·0
9	33 57 s.	24 33 w.	15·8	13·0	16·0	12·7	13·9	12·2	13·8	12·0
10	35 25 s.	23 40 w.	14·7	12·5	14·8	12·0	16·1	12·5	15·2	13·0
11	35 41 s.	20 55 w.	13·3	11·4	13·6	11·2	15·6	11·9	14·0	11·5
12	36 10 s.	17 52 w.	14·4	11·1	14·0	11·2	11·9	8·9	12·0	9·2
13	36 7 s.	14 27 w.	9·7	6·6	10·0	7·0	9·7	7·2	9·0	7·5

Date. 1873.	Latitude.	Longitude.	6 A.M.				6 P.M.			
			In Air.		In Water.		In Air.		In Water.	
			Max.	Min.	Max.	Min.	Max.	Min.	Max.	Min.
			Deg. Cent.	Deg. Cent.	Deg. Cent.	Deg. Cent.	Deg. Cent.	Deg. Cent.	Deg. Cent.	Deg. Cent.
	° ′	° ′								
Oct. 14	36 12 s.	12 18 w.	9·4	5·6	9·0	5·5	11·5	6·8	10·0	6·8
15	At Tristan d'Acunha.		9·2	7·8	9·8	7·8	12·8	8·3	12·0	8·0
16	At Inaccessible.		11·1	9·2	11·5	9·0	12·5	8·9	11·5	8·5
17	At Nightingale.		10·6	9·4	11·0	8·5	13·3	9·4	11·5	9·0
18	Between Nightingale and Tristan.		11·7	9·4	11·8	9·5	12·8	9·6	11·7	9·2
19	37 5 s.	9 40 w.	11·7	9·4	...	10·0	13·8	9·4	13·0	9·8
20	36 43 s.	7 13 w.	11·9	10·6	11·5	10·0	13·8	11·9	12·8	10·0
21	36 47 s.	4 14 w.	12·8	11·9	12·5	10·8	12·8	9·4	11·0	8·2
22	35 57 s.	0 15 w.	10·0	8·6	...	8·0	10·8	9·4	...	8·0
23	35 59 s.	1 34 e.	10·8	10·6	...	9·5	13·0	11·2	...	9·5
24	36 2 s.	5 27 e.	12·8	12·5	...	11·0	13·9	12·8	...	12·0
25	36 22 s.	8 12 e.	12·8	11·1	...	10·2	12·8	11·4	...	10·5
26	35 59 s.	11 43 e.	13·3	12·0	...	11·5	13·3	11·0	...	10·5
27	35 35 s.	16 8 e.	11·7	9 9	...	9·8	14·6	12·2	...	11·0
28	Off Cape of Good Hope.		13·3	12·3	.	11·0	17·2	15·0	...	12·2
29	In Simon's Bay.		15·3	13·8	...	12·8	21·7	19·7	...	13·0
30	Ditto.		20·6	13·3	...	...	...	...	...	...
31	Ditto.		21·9	16·4	...	14·0	20·0	15·6	...	16·0

CHAPTER III.

TENERIFFE TO SOMBRERO.

The First Deep-sea Section.—*Leiosoma limicolum.*—A Grove of Deep-sea Coral.—*Poliopogon amadou.*—Red Clay.—Phosphorescence.—Surface Fauna.—Blind Crustaceans.—Fishes' Nests.—The Paucity of the Higher Forms of Life.—Deep-sea Annelids.—The Structure and Mode of Formation of Globigerina Ooze.—The Habits of the Living *Globigerina.*—*Orbulina universa.*—*Pulvinulina.*—'Coccoliths' and 'Rhabdoliths.'—The Origin and Extension of the 'Red Clay.'—Radiolarian Ooze.—The Use of the Tow-net. — The Vertical Distribution of Temperature throughout the Section.—Specific Gravities.

APPENDIX A.—Table of Temperatures observed between Teneriffe and Sombrero Island.

APPENDIX B.—Table of Specific Gravities observed between Teneriffe and Sombrero Island.

ALTHOUGH important observations had been taken and many interesting additions had been made to our knowledge of the fauna of the deep sea during the earlier part of the voyage, the regular work of the Expedition can only be said to have commenced with the section across the Atlantic from Teneriffe to Sombrero. It had taken all our time, up to our departure from the Canary Islands, to put the machinery into working order, to settle the direction and scope of the parts to be assigned to the various members of the Staff, and to devise among us a satisfactory routine of work. At Santa Cruz the old journals were closed, and the numbering of the stations and the other entries were commenced afresh with some alterations, the result of additional experience. As this first ocean section may be taken as a fair sample of the occupation of our working days for now upwards of two years, I will give a detailed account of our proceedings, even at the risk of making this chapter somewhat technical.

We left Santa Cruz on the evening of Friday the 14th of February. The weather was bright and pleasant, with a light breeze—force equal to about 5—from the north-east. Our course during the night lay nearly westward, and on the morning of the 15th we sounded about 75 miles from Teneriffe and 2,620 miles from Sombrero Island the nearest point in the Virgin Group in 1,891 fathoms, with a bottom of grey globigerina-ooze mixed with a little volcanic detritus. The average of two Miller-Casella thermometers gave a bottom temperature of 2° C.

The dredge was put over at 9 A.M., but came up empty some hours later, the rope having apparently fouled when paying out.

The ship was steamed round to obtain the deviation of the compass using the true bearing of the Peak of Teneriffe; and this very important manœuvre may possibly have had something to do on this occasion with the miscarriage of the dredging, always a delicate operation at such depths when there is any drift. During the day Lieutenant Bethell took a series of temperatures, at intervals of 100 fathoms down to 1,000, with Mr. Siemens' resistance deep-sea temperature apparatus. A

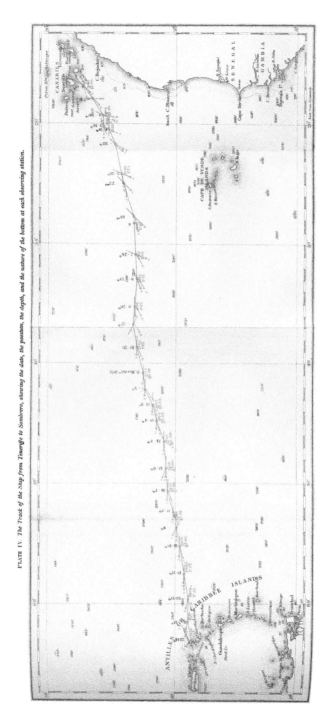

PLATE IV. The Track of the Ship from Teneriffe to Sombrero, showing the date, the position, the depth, and the nature of the bottom at each observing station.

The material originally positioned here is too large for reproduction in this reissue. A PDF can be downloaded from the web address given on page iv of this book, by clicking on 'Resources Available'.

Miller-Casella thermometer was attached to the cable at every 100 fathoms, so that the one method of determination might check the other. The sounding gave the following result:—

Depth.	Temperature by Siemens' apparatus.	Temperature by Miller-Casella thermometer.
Fathoms.		
100	15°·3 C.	16°·5 C.
200	13·6	13·2
500	8·0	8·0
700	6·6	—
800	5·5	5·1
1000	5·4	4·0

This result appeared to be on the whole satisfactory for a first trial, and it was Mr. Bethell's impression that with a little more practice in the use of Sir William Thomson's marine galvanometer, the instrument employed in observing the indications, it might be possible to arrive at considerable accuracy.

The slip water-bottle which was used by Dr. Meyer and Dr. Jacobsen in the German North Sea Expedition of the summer of 1872 was sent down to the bottom, and Mr. Buchanan determined the specific gravity of the bottom water to be 1·02584 at a temperature of 17°·9 C., the specific gravity of surface water being 1·02648 at a temperature of 18°·5 C.

All Sunday, the 16th, we spent sailing with a light air from the northward, and by Monday morning we had made about 130 miles from our previous sounding. The dredge was put over at 5.15 A.M.

with 2,700 fathoms rope, and a weight of 2 cwt. 300 fathoms before the dredge. A sounding was taken at 7 A.M. with the 'Hydra' machine and 2 cwt.; a slip water-bottle and two thermometers, Nos. 49 and 40, being sent down along with it. The sounding instrument gave a depth of 1,945 fathoms, with a bottom of grey globigerina-ooze containing many large foraminifera. The mean of the two thermometers was 2° C., and the specific gravity of the bottom water 1·02527 at 18°·3 C., that of the surface water being 1·02629 at 19°·6 C.

After steaming up to the dredge once or twice, hauling-in was commenced at 1.30 P.M., and the dredge came up at 3.30 half full of compact yellowish ooze. The ooze was carefully sifted, but nothing was found in it with the exception of foraminifera, the otolites of fishes, the dead shells of pteropods, and one mutilated specimen of what appears to be a new Gephyrean. This animal was examined by Dr. v. Willemœs-Suhm, who found that it shows a combination of the characters of the Sipunculacea and the Priapulacea. As in the former group, the excretory orifice is near the mouth, in the anterior part of the body; while, as in the latter, there is no proboscis and there are no tentacles. The pharynx is very short, and is attached to the walls of the body by four retractor muscles. The pharynx shows six to seven folds ending in a chitinous border. The mouth is a round aperture, beset with small cuticular papillæ. The perisom is divided into four muscular bands, the surface layer showing a tissue of square meshes in each of which there are four to five sense-bodies. For the reception of this singular species Dr. v. Willemœs-Suhm proposed to establish the genus *Leioderma*, which will represent a family intermediate between the sipunculids and the priapulids :—Pharynx attached to the body-wall by four retractors; no tentacles, no proboscis, and no pharyngeal teeth; excretory opening in the anterior part of the body near the mouth.

On the 18th we sounded at 9 A.M. lat. 25° 45′ N , long. 20° 12′ W , 160 miles S.W. of the Island of Ferro, and 50 miles to the west of the station of the day before, in 1,525 fathoms. The 'Hydra' tube brought up no bottom, and we sounded again with a depth of 1,520 fathoms, and again no bottom. It thus seemed that we had got upon hard ground, and as the sounding of the following day gave 2,220 at a distance of only 19 miles, we had evidently struck the top of a steep rise. The dredge was lowered at 10 A.M. with 2,220 fathoms of line and 2 cwt. leads 300 fathoms before the dredge. At 5.30 P.M. the dredge was hauled up, and contained a few small pieces of stone resembling the volcanic rocks of the Canary Islands, and some large bases of attachment and branches of the calcareous axis of an alcyonarian polyp allied to *Isis*. Some of the larger stumps were nearly an inch in diameter; the central portion very compact, and of a pure white colour; the surface longitudinally grooved, and of a glossy black. The pieces of the base of the coral which had been torn off by the dredge were in one or two cases several inches across and upwards of an inch thick, forming a thick crust from which the branches of the coral sprang. The crust was black on the surface, showing a fine regular granulation, and a fracture through the crust was of a uniform dark brown colour and semi-crystalline.

The singular observation was afterwards made by Mr. Buchanan that this black or brown substance which encrusted the coral and appeared to pass into and to form its bases of attachment consisted of almost pure black oxide of manganese. The whole of the coral was dead, and appeared to have been so for a long time. It was so fresh in its texture that it was scarcely possible to suppose that it was sub-fossil, although from the comparatively great depth at which it was found, and the many evidences of volcanic action over the whole of this region, one could scarcely avoid speculating whether it might not have lived at a higher level and been carried into its present position by a subsidence at the sea-bottom.

Attached to the branches of the coral there were several specimens of a magnificent sponge belonging to the HEXACTINELLIDÆ. One specimen, consisting of two individuals united together by their bases, is 60 centimetres across, and has very much the appearance of the large example of the tinder-fungus attached to the trunk of a tree (Fig. 38). Both surfaces of the sponge are covered with a delicate network of square meshes closely resembling that of *Hyalonema*, and formed by spicules of almost the same patterns. The sponge is bordered by a fringe of fine spicules, and from the base a large brush of strong, glassy, anchoring spicules project, fixing it to its place of attachment. The form of the barbed end of the anchoring spicules is as yet unique among sponges. Two wide, compressed flukes form an anchor very much like that of one of the skin-spicules of *Synapta*. The sponge when brought up was of a

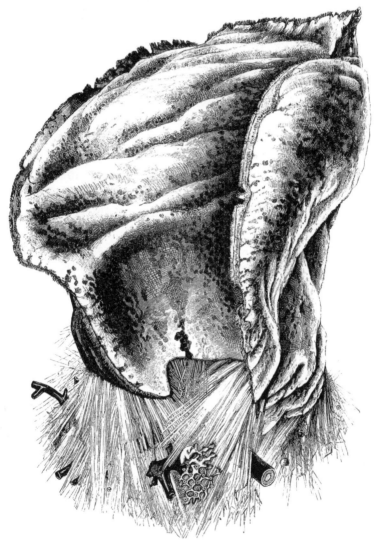

Fig. 38.—*Poliopogon amadou*, Wyville Thomson. One-third the natural size. (No. 3.)

delicate cream colour. It was necessary to steep it

in fresh water to free it from salt, and the colour changed to a leaden grey. A number of small examples of the sponge, some of them not much beyond the condition of gemmules, were found attached to the larger specimens and to branches of the coral, so that we have an opportunity of studying the earlier stages of its development.

For this sponge, which forms the type of a new genus, I propose the name *Poliopogon*[1] *amadou*. It comes nearest to *Hyalonema*, from which it differs chiefly in external form and habit, in the smaller amount of development and concentration of the large anchoring root-spicules, and in the form of these spicules and of the spicules of the outer fringe. In other particulars, as in design and structure of the outer network, the form of the spicules which make up the structure of the substance of the sponge, and the form of the double grapnel of the sarcode, the two genera approach one another very closely.

Among the branches of the coral there were many specimens of an ophiurid apparently of an undescribed species, and there was found one disk and a portion of a mutilated arm of a small *Brisinga* whose characters do not entirely correspond with those of either of the species hitherto defined.

Attached to the sponge were two examples of a fine Annelid which Dr. v. Willemœs-Suhm referred to the family AMPHINOMIDÆ, sub-family Euphrosyninæ, with many of the characters of the genus *Euphrosyne*. The body is 12 mm. long and 5 mm. broad, and consists of fifteen segments. The surface of the head is covered with a caruncle extending over

[1] Πολιὸς, white, and πώγων, a beard.

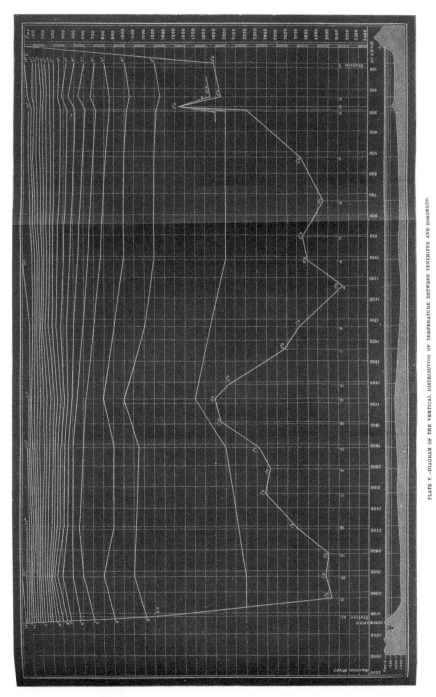

PLATE V.—DIAGRAM OF THE VERTICAL DISTRIBUTION OF TEMPERATURE BETWEEN TENERIFFE AND SOMBRERO.

The material originally positioned here is too large for reproduction in this reissue. A PDF can be downloaded from the web address given on page iv of this book, by clicking on 'Resources Available'.

the anterior segments, and the whole surface of the body is clothed with milk-white two-branched setæ, which radiate over each segment like a fan. It cannot be ascertained without careful dissection—which has not as yet been possible—whether this is the type of a new genus or a *Euphrosyne* with very small branchiæ.

Besides the species mentioned this rich haul yielded several bryozoa, one or two corals, and one or two small sponges.

On the following day we sounded in 2,220 fathoms, on the opposite side of the ridge. This was one of the rare cases in which the 'Hydra' sounding machine did not disengage its weights, and we consequently lost the instruments sent down with it—a water-bottle, two thermometers, and a pressure gauge designed by Mr. Buchanan and on trial as to its efficiency. A series of temperatures were taken from the surface to 1,500 fathoms at intervals of 100 fathoms:—

Surface	19°·5 C.	800 fathoms	5°·6 C.
100 fathoms	17·2	900 ,,	4·7
200 ,,	13·7	1000 ,,	4·6
300 ,,	11·0	1100 ,,	3·8
400 ,,	9·5	1200 ,,	3·5
500 ,,	7·6	1300 ,,	3·1
600 ,,	6·5	1400 ,,	2·8
700 ,,	6·2	1500 ,,	2·6

The dredge was not used, but, as is our custom whenever the rate of the ship is such as to make it practicable, a large tow-net was put out astern. We procured a number of specimens of the remarkable surface annelid *Alciope*, with large, deeply-pigmented eyes; and a number of the delicate, transparent crustaceans of the so-called genera *Ericthus* and *Alima,* which have been latterly shown by Claus to be larval stages of various Squillidæ.

Towards mid-day in hot, calm weather the tow-net when used on the surface is usually unsuccessful. It seems that the greater number of pelagic forms retire during the heat of the day to the depth of a few fathoms, and come up in the cool of the evening and in the morning, and in some cases in the night. The larger phosphorescent animals are frequently abundant during the night round the ship and in its wake, while none are taken in the net during the day. Mr. Moseley has been specially engaged in working up the developmental stages of *Pyrosoma,* and the intimate structure of the tissues and organs of some of the surface groups, whose extreme transparency renders them particularly suitable for such researches.

Feb. 21.—Up to 2.15 P.M. going on under all plain sail at the rate of six knots an hour before the N.E. trades, force 3 to 4. Sounded in the afternoon in 2,740 fathoms, the 'Hydra' machine bringing up a small quantity of yellowish globigerina-ooze. The thermometers registered a bottom temperature of 2° C., and the bottom water taken with the slip water-bottle gave a specific gravity of 1·02623 at 20° C., the surface water having a specific gravity of 1·02619 at 20°·5 C. The dredge was put over at 5 P.M. with 3,400 fathoms of line, and was kept down till 1 o'clock A.M. on the following morning, the ship drifting slowly. Our position at noon on the 21st was about 500 miles S.W. of Teneriffe, lat. 24° 22' N., long. 24° 11' W., Sombrero Island S. 58 W. 2,220 miles. Work began early on the 22nd, and the dredge, which had begun its ascent at 1.15 A.M., came up at 5.45 half full of a yellowish ooze, which was not so tenacious as usual, and on the whole singularly poor in higher living things. A careful and laborious sifting of the whole mass gave us three small living molluscs, referred to the genera *Arca* (Fig. 39),

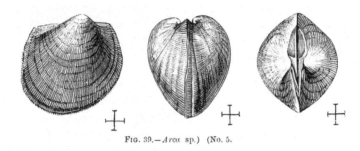

Fig. 39.—*Arca* sp.) (No. 5.

Limopsis (Fig. 40), and *Leda* (Fig. 41); and two bryozoa apparently undescribed. Foraminifera were abundant, many examples of miliolines being of unusually large size. Some beautiful radiolarians were

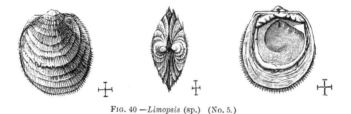

Fig. 40 —*Limopsis* (sp.) (No. 5.)

sifted out of the mud. These may have been taken into the dredge on its way up, or more probably they may have lived on the surface or in intermediate water and have sunk to the bottom after death, since they consist of continuous fenestrated shells of silica.

A series of temperature soundings were taken after the dredge came up, with the following result :—

Surface	20°·0 C.	900 fathoms	4°·7 C.
100 fathoms	17·0	1000 ,,	4·0
200 ,,	13·9	1100 ,,	3·5
300 ,,	11·2	1200 ,,	3·0
400 ,,	9·4	1300 ,,	2·5
500 ,,	7·7	1400 ,,	2·7
600 ,,	6·8	1500 ,,	2·0
700 ,,	6·0	2740 ,,	2·0
800 ,,	5·2		

On Sunday the 23rd we continued our course, going before the N.E. trades at an average rate of

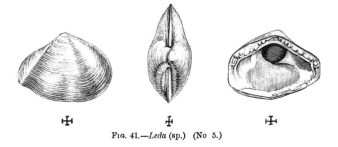

Fig. 41.—*Leda* (sp.) (No 5.)

seven knots an hour; the sky and the sea gloriously bright and blue as ever. Several flying-fish were observed, and many large examples of *Physalia*, although there was a good deal of surface motion. At 4 P.M. we shortened sail and sounded in 2,950 fathoms with the 'Hydra' machine and 3 cwt. The chamber of the 'Hydra' contained a small quantity of reddish ooze with a few foraminifera. The corrected temperature at 2,950 fathoms was 2° C.

The following day we sounded in 2,750 fathoms, and found the bottom still more unpromising. It

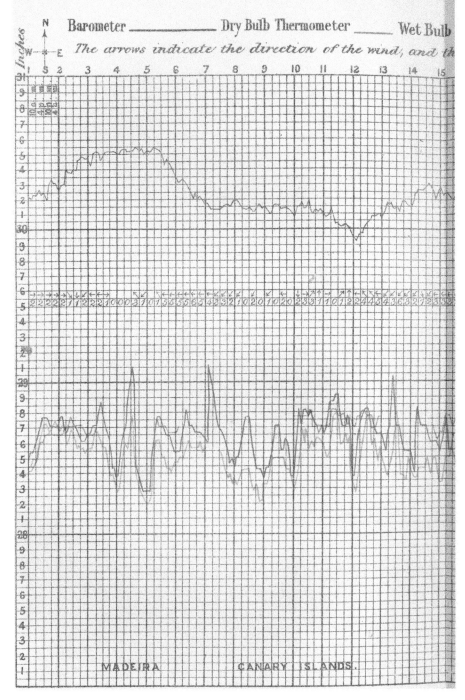
PLATE VI. *Meteorological Observation*

ions for the month of February, 1873.

lb Thermometer ———— Temperature of Sea Surface ————

the numbers beneath its force according to Beaufort's scale

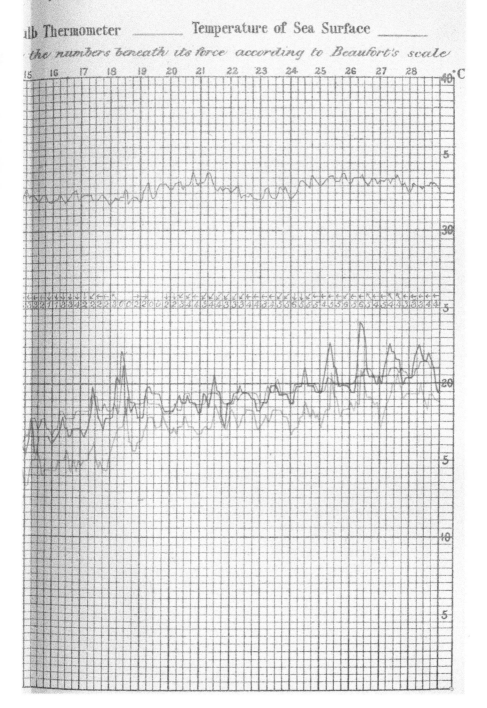

was of a decided brown colour, and contained very few definite organisms, although it gave to a rough analysis a considerable quantity of carbonate of lime. The bottom temperature was 2° C., the temperature of the surface being 20° C.

On Tuesday the 25th a small dredge was lowered at 6.30 A.M. with 3,500 fathoms of line (2,500 fathoms of 2½-inch rope and 1,000 of 2-inch), and 2 cwt. leads attached 300 fathoms in advance. At 7.30 we sounded in 2,800 fathoms, with a bottom of the same reddish ooze, and a temperature of 2° C. A series of temperatures were taken at intervals of 100 fathoms down to 1,000, the results agreeing closely with those of the previous series:—

Surface		19°·5 C.	600 fathoms		6°·3 C.
100 fathoms		19 · 0	700 ,,		5 · 5
200 ,,		14 · 9	800 ,,		5 · 2
300 ,,		11 · 1	900 ,,		4 · 3
400 ,,		9 · 3	1000 ,,		4 · 0
500 ,,		6 · 7	2800 ,,	bottom	2 · 0

At 5.15 P.M. the dredge came up clean and empty. It had either never reached the bottom owing to some local current or the drift of the ship, or else everything had been completely washed out of it on its way to the surface. The bottom water gave a specific gravity of 1·02504 at 19°·6 C., that of the surface being 1·02617 at 21°·3 C. While sounding the current-drag was tried, and indicated a slight north-westerly current.

As the attempt to drag on the previous day had been unsuccessful, it was determined to repeat the operation with every possible precaution on the 26th. The morning was bright and clear, and the swell, which had been rather heavy the day before, had gone down considerably. A sounding was taken about 10 o'clock A.M. with the 'Hydra' machine and 4 cwt. The sounding was thoroughly satisfactory, a sudden change of rate in the running out of the line indicating in the most marked way when the weight had reached the bottom. During the sounding a current-drag was put down to the depth of 200 fathoms, and it was then ascertained that, by means of management and by meeting the current by an occasional turn of the screw, the ship scarcely moved from her position during the whole time the lead was running out. The depth was 3,150 fathoms; the bottom a perfectly smooth red clay, containing scarcely a trace of organic matter—merely one or two minute granular masses. The thermometers indicated a bottom temperature of 1°·9 C.

The small dredge was sent down at 2.15 P.M. with two hempen tangles; and, in order to ensure its reaching the bottom, we attached to the iron bar, below the dredge, which is used for suspending the tangles, a 'Hydra' instrument with detaching weight of 3 cwt. Two additional weights of 1 cwt. each were fixed to the rope 200 fathoms before the dredge. 3,600 fathoms of rope were payed out—1,000 fathoms 2 inches in circumference, and the remainder (2,600 fathoms) $2\frac{1}{2}$ inches. The dredge came up at 10.15 P.M. with about 1 cwt. of red clay.

This haul interested us greatly. It was the deepest by several hundred fathoms which had yet been taken, and, at all events coincidently with this great increase in depth, the material of the bottom was totally different from what we had been in the habit of meeting with in the depths of the Atlantic. For a few soundings past the ooze had been assuming a darker tint, and showed on analysis a continually lessening amount of calcareous matter, and, under the microscope, a smaller number of foraminifera. Now, calcareous shells of foraminifera were entirely wanting, and the only organisms which could be detected after washing over and sifting the whole of the mud with the greatest care, were three or four tests of foraminifera of the cristellarian series, made up apparently of particles of the same red mud. The shells and spines of surface animals were almost entirely wanting; and this is the more remarkable as the clay-mud was excessively fine, remaining for days suspended in the water, looking in colour and consistence exactly like chocolate, indicating therefore an almost total absence of movement in the water of the sea where it is being deposited. When at length it settles, it forms a perfectly smooth red-brown paste, without the least feeling of grittiness between the fingers, as if it had been levigated with extreme care for a process in some refined art. On analysis it is almost pure clay, a silicate of alumina and the sesquioxide of iron, with a small quantity of manganese.

The weight beneath the dredge had certainly the desired effect in this case of bringing the dredge rapidly to the bottom. The stem of the 'Hydra' machine had gone deep into the mud, and was bent, apparently by the weight falling upon it. The dredge had taken a deep scoop of mud, and the line had entangled itself in a coil of 20 or 30 fathoms over the second weight, showing that it had not been exposed to any current.

On the 27th we proceeded steadily on our course, and on the 28th we sounded in the morning in 2,720 fathoms with No. 1 line, to which were attached a 'Hydra' machine with 4 cwt. disengaging weights, a slip water-bottle, and two thermometers. At a distance of 500 fathoms from the bottom a stopcock water-bottle was bent on to the line. The tube of the 'Hydra' was filled with a red mud containing a considerably larger proportion of carbonate of lime than in the previous sounding, and a few shells of calcareous foraminifera. As the bottom presented so little difference from the former, which we had found so nearly destitute of animal life, it was thought unnecessary to spend the time and great labour necessarily involved in dredging, and we proceeded under sail after taking a series of temperature observations:—

Surface	22°·2 C.	400 fathoms	8°·7 C.
10 fathoms	22·1	500 "	6·5
20 "	21·9	600 "	5·2
30 "	22·1	700 "	4·8
40 "	21·7	800 "	4·6
50 "	21·8	900 "	4·0
60 "	21·8	1000 "	3·5
70 "	21·4	1100 "	3·3
80 "	19·5	1200 "	3·1
90 "	19·7	1300 "	2·6
100 "	19·5	1400 "	2·7
200 "	14·3	1500 "	2·6
300 "	11·5	2720 " bottom	1·9

On Saturday, March 1, we sounded in the morning in 2,575 fathoms with the 'Hydra' and 3 cwt., the slip water-bottle, and one thermometer—the stopcock water-bottle being again attached at 500 fathoms from the 'Hydra.' - The bottom was ooze still containing a large proportion of silicate of alumina, but with much more calcareous matter and many more minute shells than in the previous sounding; showing that we were passing from the region of dead red clay which occupies the extreme deep water in this place. The small dredge was sent down at 9.55 A.M. with 3,000 fathoms of rope, 2,000 fathoms of which were 2-inch rope and new; 1 cwt. was attached 300 fathoms before the dredge, and later in the day an additional cwt. was slipped down the line. At 5.20 P.M. the dredge came up bottom upwards and quite empty. A little mud on the netting showed that it had been at the bottom, and as the double chain to which the dredge is immediately attached was twisted up into a close spiral, we judged that the *bouleversement* of the dredge had plainly been caused by the twist in the new line. A series of temperature observations were taken at intervals of 20 fathoms from the surface to a depth of 200 fathoms, in order to determine the depth to which the temperature of the water is affected by direct radiation. The following is the result :—

Surface . . . . 22°· 2 C.	120 fathoms . .	18°· 0 C.
20 fathoms . . . 22 · 2	140 ,, . . .	—
40 ,, . . . 22 · 2	160 ,, . . .	16 · 3
60 ,, . . . 22 · 2	180 ,, . . .	15 · 15
80 ,, . . . 21 · 8	200 ,, . . .	14 · 7
100 ,, . . . 19 · 9		

The specific gravity of the bottom water (2,575 fathoms) was 1·02459 at 21°·6 C., that of the surface water being 1·02581 at 22°·4 C.

On Sunday, March 2, we saw the first patches of

gulf-weed drifting past the ship, and flying-fish were abundant. Our position at noon was lat. 22° 30′ N., long. 42° 6′ W., Sombrero Island distant 1,224 miles. At night the phosphorescence of the sea was particularly brilliant, the surface scintillating with bright flashes from the small crustaceans, while large cylinders and globes of lambent light, proceeding probably from *Pyrosoma* and some of the Medusæ, glowed out and slowly disappeared in the wake of the vessel at a depth of a few feet.

The next morning we sounded at 7 A.M. in 2,025 fathoms with No. 1 line, the 'Hydra' machine and 3 cwt., a slip water-bottle, and one thermometer; a stopcock water-bottle was bent on at 925 fathoms from the bottom. The corrected bottom temperature was 1°·9 C., the temperature of the surface being 22°· 8 C. During the morning the naturalists were out in a boat with the tow-net, and they brought back a number of fine examples of *Porpita*, several of *Glaucus atlanticus*, some shells of *Spirula* bearing groups of a small stalked cirriped, and many large radiolarians. One of the *Spirula* shells was covered with a beautiful stalked infusorian.

The dredge was lowered at 9.30 A.M., and hauled in at 4.30 P.M., unfortunately again entangled in a coil of the rope, and empty. A small quantity of mud attached to the netting of the dredge was found to contain a large proportion of the shells of foraminifera, especially *Globigerina* and *Orbulina*. The mud was again of a pale grey colour, and consisted chiefly of calcic carbonate.

We proceeded in the evening under all plain sail. The soundings on the chart in advance of us seemed to indicate an extensive rise, with a depth of water averaging not much more than 1,700 fathoms, and it was determined to dredge again on the following day.

On the morning of the 4th of March we sounded in lat. 21° 38′ N., long. 44° 39′ W., in 1,900 fathoms, with No. 1 line, the 'Hydra' and 3 cwt., the slip water-bottle, and a thermometer. The bottom was grey ooze, as on the day before, and the bottom temperature 1°·9 C. The dredge was put over at 8 A.M. It was intended to attach a 'Hydra' tube with disengaging weight a little below the bottom of the dredge; the weight slipped, however, close to the surface, and the dredge was lowered in the ordinary way with 1 cwt. 500 fathoms in advance. The dredge came up about 4 o'clock with a small quantity of ooze containing some red clay, a large proportion of calcareous débris, and many foraminifera, chiefly *Orbulina* and *Pulvinulina*.

Warped in the hempen tangles there was a fine specimen of a handsome decapod crustacean, having all the principal characters of the family Astacidæ, but different from the typical decapods in the total absence of eye-stalks and eyes. Dr. v. Willemœs-Suhm gave this interesting deep-sea form such a preliminary examination as was possible in the absence of books of reference. I abstract from his notes. *Willemœsia*[1] *leptodactyla*, n.g. and sp. The single

[1] In my first notice of its capture, in a letter published in NATURE in May 1873, this crustacean was described by Dr. v. Willemœs-Suhm under the name of *Deidamia leptodactyla*. In NATURE of the 19th of October, 1873, Mr. Grote, one of the curators of the museum in Buffalo, U.S., pointed out that *Deidamia* was already occupied by a genus of Sphingidæ, and proposed for the crustacean the generic name *Willemœsia*, which is accordingly adopted.

specimen procured (Fig. 42), which is a male, is 120 mm. in total length and 33 mm. in width across the base of the cephalo-thorax, which is 60 mm. in length. Three rows of spines, one in the middle line and one on each side, run along the cephalo-thorax, which is divided by a transverse sulcus into an anterior and a posterior part, the former occupied by a central gastric and lateral hepatic regions, and the latter by a central cardiac and lateral branchial regions. The abdomen, which consists as usual of seven segments, has the central series of spines of the cephalo-thorax continued along the middle line. The sixth segment bears the caudal appendages, and in the seventh, the telson, we find the excretory opening. The lateral borders of the body, and all the appendages with the exception of the first pair of ambulatory legs, are edged with a close and very beautiful fringe of hair of a pale yellow colour.

There are two pairs, the normal number, of antennæ, one pair of mandibles, two pairs of maxillæ, three pairs of maxillipeds, five pairs of ambulatory legs, and five pairs of swimmerets. As most of the appendages differ from those usually met with in the Astacidæ only in detail, it is only necessary to mention that the interior antennæ have two flagella, one of which is very long, longer than the external flagellum of the external pair.

The form of the first pair of ambulatory legs is singularly elegant. They are 155 mm. in length—considerably longer than the body; they are very slender, and end in a pair of very slender denticulated chelæ, with a close, velvet-like line of hairs along their inner edges. The rest of the ambulatory legs

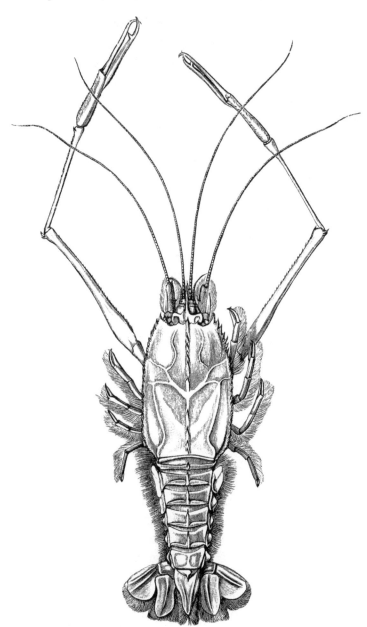

Fig. 42.—*Willemœsia leptodactyla*, v. Willemœs-Suhm. Natural size. (No. 13.)

are much shorter, and all bear chelæ. The specimen captured being a male the first pair of swimmerets are somewhat modified. The four other pairs of swimmerets, which are 33 mm. in length, bear each two narrow swimming processes richly fringed with hair, and a short flagellum.

The absence of eyes in many deep-sea animals and their full development in others is very remarkable. I have mentioned ('The Depths of the Sea.' p. 176), the case of one of the stalk-eyed crustaceans, *Ethusa granulata*, in which well-developed eyes are present in examples from shallow water. In deeper water, from 110 to 370 fathoms, eye-stalks are present, but the animal is apparently blind, the eyes being replaced by rounded calcareous terminations to the stalks; in examples from 500 to 700 fathoms in another locality, the eye-stalks have lost their special character, have become fixed, and their terminations combine into a strong pointed rostrum. In this case we have a gradual modification, depending apparently upon a gradual diminution and final disappearance of solar light. On the other hand, *Munida*, from equal depths, has its eyes unusually developed and apparently of great delicacy. Is it possible that in certain cases, as the sun's light diminishes, the power of vision becomes more acute, while at length the eye becomes susceptible of the stimulus of the fainter light of phosphorescence? The absence of eyes is not unknown among the Astacidæ. *Astacus pellucidus* from the Mammoth Cave is blind and from the same cause—the absence of light; but morphologically the eyes are not entirely wanting, for two small abortive eye-stalks still remain in the position in which eyes are developed in all normal decapods. In *Willemœsia* no trace whatever remains either of the organs of sight or of their pedicels.

Two specimens of a species of the genus *Leda* were sifted out of the mud, and two minute gasteropods, one a prettily ornamented species, allied apparently to *Solarium* (Fig. 43.) These and the other deep-sea

FIG. 43.—*Solarium* (sp.). Greatly enlarged. (No 13.)

mollusca will shortly be placed in the hands of Mr. Gwyn Jeffreys for determination. Several specimens were found on this occasion of two bryozoa of the *Farciminaria* group, which have turned up more than once from great depths, frequently in considerable quantity, entangled in the swabs or on the outside of the dredge net. They are frequent where the bottom is of smooth ooze, and from the appearance and structure of the bunches of horny tubes which form their roots it would seem that they are in a certain sense free, merely anchored in the surface of the soft calcareous mud.

Among the bryozoa, at first sight closely resembling them in form and habit, were one or two specimens of an extremely delicate and beautiful siliceous sponge; one, apparently, of the aberrant stalked Esperiadæ. Several species of this curious little group occurred in deep water in the Atlantic; they will be described on a future occasion. A serial temperature sounding was taken, the intervals being 100 fathoms from the surface to a depth of 1,500 fathoms:—

Surface	22°·2 C.	900 fathoms	3°·6 C.
100 fathoms	18·9	1000 ,,	3·4
200 ,,	15·5	1100 ,,	3·0
300 ,,	12·4	1200 ,,	2·5
400 ,,	9·6	1300 ,,	2·5
500 ,,	6·7	1400 ,,	2·3
600 ,,	5·2	1500 ,,	2·2
700 ,,	4·7	1900 ,, bottom	1·9
800 ,,	4·1		

The specific gravity of the bottom water was 1·02517 at 22°·1 C., that of the surface water being 1·02579 at 22°·8. C.

During the following night we made about 100 miles, and on the morning of the 5th we were a little to the south-west of a sounding of 1,875 fathoms (Lieut. Berryman) on the chart, so that we were still on a comparatively elevated plateau. A trawl with a 22 feet beam was sent down at 9 A.M., and we sounded at 10 A.M. in 1,950 fathoms. A slip water-bottle accompanied the 'Hydra' sounding machine to the bottom, and a stopcock water-bottle was bent on to the line at 1,000 fathoms from the bottom. On hauling up, the strong brass cylinder of the stopcock water-bottle was found collapsed and crumpled like a piece of paper. This was not a matter of surprise, for it was already suspected that the valves had closed before the bottle reached the surface from having been scarcely sufficiently tightened up. The slip water-bottle had also miscarried, some mud having got into the valve and prevented its closing fully.

The sample in the tube of the 'Hydra' was ooze containing many foraminifera; the thermometer registered a temperature of 1°·9 C.

The trawl was hauled in at 5 P.M. The beam was broken through the middle and otherwise strangely torn and crushed by the combined action of the pressure to which it had been subjected and the strain of pulling it up rapidly through three miles of water. The wood was driven in and compressed so as to reduce the diameter of the beam by half an inch, and the knots projected a quarter of an inch on all sides. In the bread-bag chamber at the end of the trawl there was a little mud, full of large foraminifera and the otolites of fishes, the finer débris having been washed through the canvas; and sticking to the net were several examples of the two bryozoa previously mentioned, and a very perfect young specimen of the remarkable form which we dredged off the coast of Portugal, *Naresia cyathus*. Our position at noon was lat. 21° 0′ N., long. 46° 30′ W., Sombrero Island distant 972 miles.

On Thursday the 6th we sounded in 2,325 fathoms, sending down a thermometer and the slip water-bottle. The bottom-temperature registered was 1°·7 C., and the specific gravity of the sample of water was 1·02470 at 21° C., that of the surface water being 1·02556, at 23°·3 C.

The bottom was a yellowish ooze with a very large proportion of red clay, and a corresponding decrease in the amount of carbonate of lime and in the number of shells of foraminifera; a considerable approach to the character of the mud from Station 7 on the 24th of February, when we were approaching the bed of fine red clay. Serial soundings were taken at the usual intervals down to 1,500 fathoms :—

Surface	22°·5 C.	900 fathoms	4°·0 C.
100 fathoms	19·4	1000 ,,	3·3
200 ,,	14·9	1100 ,,	3·2
300 ,,	11·9	1200 ,,	2·9
400 ,,	8·5	1300 ,,	2·7
500 ,,	5·9	1400 ,,	2·5
600 ,,	5·0	1500 ,,	2·6
700 ,,	4·6	2325 ,, bottom	1·7
800 ,,	4·3		

A good deal of gulf-weed drifted past during the day, and we were struck by the circumstance that a large proportion of it, instead of assuming its usual loose graceful arrangement in the water, was in the form of round compact balls a little larger than cricket-balls. A boat was sent off to collect some, and about half a dozen closely-twined bundles were brought on board. On pulling them to pieces and examining them we found that the bundles were bound together by very strong transparent strings of the viscid secretion of the singular pelagic fish *Antennarius marmoratus,* and that the spaces among the fronds were filled with its eggs. Several young examples of this grotesque little animal (Fig. 44) were found among the gulf-weed; with many crustaceans, several of the nudibranchiate mollusca characteristic of the gulf-weed fauna, such as *Scillæa pelagica* and *Glaucus atlanticus,* and many planarians.

On the following morning the dredge was put over at 8 A.M., and line veered to 3,000 fathoms; and at 10 o'clock we sounded in 2,435 fathoms, sending down the slip water-bottle and a thermometer. The thermometer registered 1°·7 C., and the sample of the bottom in the 'Hydra' tube was still redder and more unpromising than in the sounding of the day before.

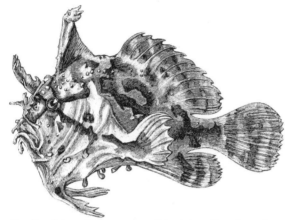

FIG. 44.—*Antennarius marmoratus.* Natural size. From the surface.

The dredge came up at 4.15 P.M. with a small quantity of red mud, in which we detected only one single but perfectly fresh valve of a small lamellibranchiate mollusc (Fig. 45). In the mud there were

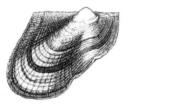 

FIG. 45.—*Avicula* (sp.). Greatly enlarged. (No. 16.)

also some sharks' teeth of at least two genera, and a number of very peculiar black oval bodies about an inch long, with the surface irregularly reticulated, and within the reticulations closely and symmetrically granulated; the whole appearance singularly like that of the phosphatic concretions which are so abundant in the green-sand and trias. My first impression was that both the teeth and the concretions were drifted fossils, but on handing over a portion of one of the latter to Mr. Buchanan for examination, he found that it consisted of almost pure peroxide of manganese.

The character both of the exterior and interior of the nodule strongly recalled the black base of the coral which we dredged in 1,530 fathoms on the 18th of February; and on going into the matter, Mr. Buchanan found not only that the base of the coral retaining its external organic form had the composition of a lump of pyrolusite, but that the glossy black film covering the stem and branches of the coral gave also the reaction of manganese. There seemed to be little doubt that it was a case of slow substitution, for the mass of peroxide of manganese forming the root showed on fracture in some places the concentric layers and intimate structure of the original coral. The coral, where it was unaltered, had the ordinary composition, consisting chiefly of calcic carbonate. Water was obtained by the slip water-bottle from 300 fathoms with a specific gravity of 1·02510 at 21°·1 C.; from 400 fathoms of 1·02475 at 20°·9 C.; from 500 fathoms of 1·02619 at 20°·5 C.; and at 200 fathoms of 1·2515 at 21°·6 C. The water from the bottom (2,435 fathoms) had a specific gravity of 1·02576 at 22° C., and the surface water of 1·02526 at 24°·4 C.

On Saturday the 8th a sounding was taken in 2,385 fathoms, and the tube brought up a sample of mud of a bright, light chocolate colour, with a mere trace of calcic carbonate—nearly a pure red clay.

We were laying our course so as to include a sounding of Lieut. Lee in lat. 19° 2′ 36″ N., long. 59° 33′ 20″ W., of 3,300 fathoms, and it seemed that we were gradually passing off the plateau—which Captain Nares has called, in recognition of the vessel from which its position was first determined, the 'Dolphin Rise,'—into the depression indicated by the deep sounding, and that again we had a change in the nature of the bottom coincident with increase in depth. A series of temperatures was taken, with the results tabulated:—

Surface	23°·3 C.	900 fathoms		4°·2 C.
100 fathoms	21·0	1000 „		3·6
200 „	16·0	1100 „		3·4
300 „	12·5	1200 „		3·2
400 „	8·1	1300 „		—
500 „	7·4	1400 „		2·5
600 „	5·6	1500 „		2·5
700 „	5·0	2385 „	bottom	1·9
800 „	4·7			

Some of our party, using the towing-net and collecting gulf-weed on the surface from a boat, brought in a number of things beautiful in their form and brilliancy of colouring, and many of them strangely interesting for the way in which their glassy transparency exposed the working of the most subtle parts of their internal machinery; and these gave employment to the microscopists in the dearth of returns from the dredge. Our position was now lat. 19° 57′ N., long. 53° 26′ W.; Sombrero distant 558 miles.

Sunday was a lovely day. The breeze had fallen off somewhat, and the force was now only from 2 to 3. The sky and sea were gloriously blue, with here and there a soft grey tress on the sky, and a gleaming white curl on the sea. A pretty little Spanish brigantine, bright with green paint and white sails, and the merry, dusky faces of three or four Spanish girls, came in the morning within speaking distance and got her longitude. She had been passing and repassing us for a couple of days, wondering doubtless at the irrelevancy of our movements, shortening sail, and stopping every now and then in mid-ocean with a fine breeze in our favour. On Monday morning we parted from our gay little companion. We stopped again to dredge and she got far before us, and we saw with some regret first her green hull and then her white sails pass down over the edge of the world.

The sounding on Monday the 10th gave 2,675 fathoms, with a bottom of the same red clay with very little calcareous matter. The bottom temperature was 1°·6 C., that of the surface being 23°·3 C. The smaller dredge was sent over at 7.15 A.M. with 3,000 fathoms of line, four hempen tangles, and a leaden weight of 28 lbs. about three fathoms below the dredge. The dredge was hauled up at 4.50 P.M. with only a very small quantity of red mud sticking about the chain and the mouth of the dredge. There could be no doubt from the appearance of the dredge-bag that it had contained a quantity of the perfectly smooth and uniform clay-mud such as had been brought up in the sounding tube, and that the greater part had been washed out in hauling up. A small fish, as yet undetermined, but with the peculiarity of having eyes so small as to be nearly microscopic, was found in one of the corners of the dredge-bag. It is very possible, however, that it was taken into the dredge on the way up. We had been struck for some time past with the singular absence of the higher forms of life. Not a bird was to be seen from morning till night. A few kittiwakes (*Larus tridactylus*) followed the ship for the first few days after we left Teneriffe, but even these had disappeared. A single petrel (*Thalassidroma pelagica*) was seen one day from one of the boats on a towing-net excursion, but we had not yet met with one of the southern sea-birds. For the last day or two some of the large sea-mammals and fishes had been visible. A large grampus (*Orca gladiator*) had been moving round the ship and apparently keeping up with it. Some sharks hung about seeking what they might devour, but we had not succeeded in catching any of them. Lovely dolphins (*Coryphœna hippurus*) passed in their varying iridescent colouring from the shadow of the ship into the sunshine, and glided about like living patches of rainbow. Flying-fish (*Exocetus evolans*) became more abundant, evidently falling a prey to the dolphins, which are readily deceived by a rude imitation of one of them, a white spinning bait, when the ship is going rapidly through the water.

The following is the result of a temperature sounding at intervals of 50 fathoms from the surface to 800 :—

Surface	23°·3 C.	
50 fathoms	23·0	
100 ,,	21·1	
150 ,,	17·8	
200 ,,	15·4	
250 ,,	13·6	
300 ,,	12·3	
350 ,,	10·3	
400 ,,	8·6	
450 fathoms		7°·1 C.
500 ,,		5·8
550 ,,		5·2
600 ,,		5·0
650 ,,		4·6
700 ,,		4·6
800 ,,		4·2
2675 ,, bottom		1·6

On Tuesday the 11th we pursued our course during the forenoon at the rate of from six to seven knots, with a light breeze, force 3 to 4. At 2 p.m. shortened and furled sails, and got up steam to sound, and at 3 o'clock sounded in 3,000 fathoms. The bottom was a pale, chocolate-coloured mud, with some gritty particles, but very little carbonate of lime, and no foraminifera. The bottom temperature, registered from the mean of two thermometers, was $1°·3$ C. On the following morning, March 12, the small dredge, with the bag lined throughout with bread-bag, was lowered early in the morning, with a slip water-bottle and two thermometers. This sounding was almost exactly on the spot of Lieut. Lee's sounding of 3,300 fathoms mentioned above, and gave a somewhat less depth, the 'Hydra' tube returning filled with red clay-mud from a depth of 2,975 fathoms. The bottom had very much the character of that of the very deep depression which we encountered earlier in the cruise, only the mud was not quite so homogeneous, containing a number of gritty particles. It was, however, on the whole very smooth, and contained scarcely a trace of carbonate of lime. The thermometer registered a bottom temperature of $1°·6$ C., and the sample of bottom water had a specific gravity of 1·02416 at $22°·6$ C., that of the surface having a specific gravity of 1·02470 at $24°·9$ C. A series of temperature observations was taken at intervals of 100 fathoms from the surface to 1,500 fathoms :—

Surface	$24°·0$ C.	300 fathoms . . . $9°·7$ C.
100 fathoms . . .	20 ·0	400 „ . . . 7 ·0
200 „ . . .	15 ·6	500 „ . . . 5 ·3

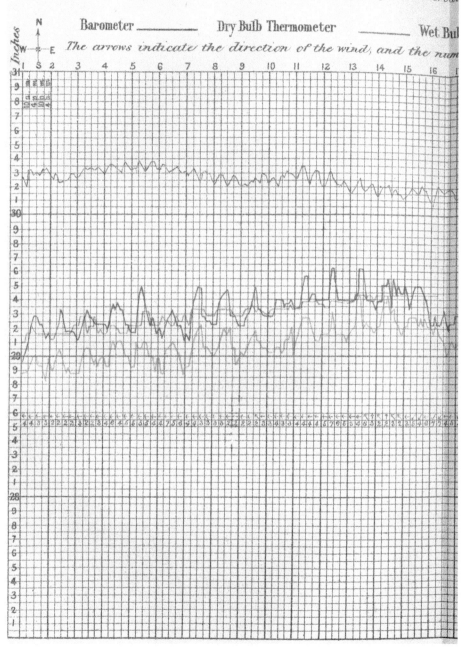

PLATE VII. *Meteorological Observat*

Barometer ———— Dry Bulb Thermometer ———— Wet Bulb

The arrows indicate the direction of the wind, and the num

rvations for the month March, 1873.

Bulb Thermometer　　　　　Temperature of Sea Surface ─────

numbers beneath its force according to Beaufort's scale

ST THOMAS

600 fathoms	5°·0 C.	1200 fathoms			2°·7 C.
700 ,,	4·6	1300 ,,			2·8
800 ,,	4·0	1400 ,,			2·7
900 ,,	3·5	1500 ,,			2·5
1000 ,,	3·3	2975 ,,	bottom		1·6
1100 ,,	2·9				

The dredge-line was veered to over 4,000 fathoms, nearly 5 statute miles. The dredge came up at about half-past 5 o'clock, full of red mud of the same character as that brought up by the sounding machine. Entangled about the mouth of the dredge and embedded in the mud were many long cases of a tube-building annelid, evidently formed out of the gritty matter which occurs, though sparingly, in the clay. The tubes with their contents were handed over to Dr. v. Willemœs-Suhm, who found the worms to belong to the family Ammocharidæ (Claparède and Malmgren), closely allied to the *Maldania* or *Clymenidæ*, all of which build tubes of sand or mud. The largest specimens dredged are 120 mm. in length by 2 mm. in width. The head is rounded, with a lateral mouth. There is no trace of cephalic branchiæ. The worm consists of only from 17 to 20 segments; the first few of these are very long—about 17 mm.; while those of the posterior portion of the body are only 5 mm. in length. The segments are not divided from one another; but the *tori uncinigeri*, which are occupied by the hair-like setæ, and the elevations bearing small *uncini*, indicate the beginning of a new segment. The number of small hooks on the *tori uncinigeri* is very large.

Claparède has calculated that *Owenia filiformis*, to which this species is nearly allied, is provided with

150,000 hooks wherewith to attach itself in its tube.

There is a pair of glands in each of the segments, from the second to the seventh. The position and structure of these has been described by Claparède in the genus *Owenia*, in which, however, there are only four pairs. Most of the specimens examined are females, and contain many eggs.

There is no doubt that this annelid is closely allied to the genus *Owenia*, but it differs from it in the absence of cephalic branchiæ. Malmgren has, however, already proposed the name of *Myriochele* for a form in which this absence of branchiæ occurs. The description of the northern form on which Malmgren's genus is founded is not at hand, so that it is impossible in the meantime to determine whether the two forms are identical or specifically distinct.

As bearing upon some of the most important of the broad questions which it is our great object to solve, I do not see that any capture which we could have made could have been more important and more conclusive than that of this annelid. The depth was 2,975, practically 3,000 fathoms—a depth at which all the conditions which might be expected to militate against the existence of animal life must have attained their full force. The nature of the bottom, from the absence of a due proportion of carbonate of lime, was very unfavourable to higher animal life, and yet this creature, which is closely related to the *Clymenidæ*, a well-known shallow-water group of high organization, is abundant and fully developed. It is fortunate in possessing such attributes as to make it impossible even to suppose that it may have been taken during the passage of the dredge to the surface, or have entered the dredge-bag in any other illegitimate way; and its physiognomy and habits are the same as those of allied forms from moderate depths. It appears, in fact, conclusive proof that the conditions of the bottom of the sea to all depths are not only such as to admit of the existence of animal life, but are such as to allow of the unlimited extension of the distribution of animals high in the zoological series, and closely in relation with the characteristic faunæ of shallower zones.

On Thursday the 13th our position at noon was lat. 18° 54′ N., long. 61° 28′ W. A sounding was taken in 3,025 fathoms, and observations were taken with the current-drag. The 'Hydra' tube brought up a nearly uniform red clay with a small trace of carbonate of lime. One of the two thermometers was broken, and in the second, which registered 1°·3 C., the mercury had parted. It seems probable, however, that this occurred after the index had registered the true minimum, as the temperature was precisely what might have been anticipated. The water in the slip water-bottle was muddy. There was a fresh breeze in the evening, which, however, fell off during the night, and we did not make so much way as we expected.

On the forenoon of the 14th we were still 35 miles from land, and we sounded in 1,420 fathoms. The bottom had altered greatly in character; it now consisted chiefly of calcareous foraminifera of many species, mixed with a considerable proportion of the broken spicules of siliceous sponges. The bottom temperature registered was 3° C. The water-bottle was accidentally broken in taking in, so that that observation was lost. A series of temperature soundings was taken with the usual intervals :—

Surface	24°·5 C.	800 fathoms		——
100 fathoms	20·6	900 ,,		4°·0 C.
200 ,,	16·2	1000 ,,		3·5
300 ,,	——	1100 ,,		3·3
400 ,,	8·7	1200 ,,		3·1
500 ,,	6·6	1300 ,,		3·0
600 ,,	5·8	1420 ,,	bottom	3·0
700 ,,	4·8			

As we were now within sight of land, and all our results were evidently modified by its immediate proximity, we regarded our first deep-sea section as completed.

It will be seen from the foregoing account, with the accompanying diagrams, that we ran a continuous section across nearly the widest part of the North Atlantic; and that along this line we established 22 observing stations, at distances varying a little according to circumstances, but averaging 120 miles apart. At each of these stations one sounding at least was taken, and, except in one single instance, where the weights of the 'Hydra' sounding instrument failed to disengage, an ample specimen of the bottom was brought up. The samples of the bottom were carefully labelled and preserved. In some cases one and in many cases two Miller-Casella thermometers registered the bottom temperature. A specimen of bottom water, amounting to about two litres, was brought up on each occasion, except on the one already mentioned where the instrument was lost, and on two others where the valves did not completely close: its specific gravity was determined, and the water was either subjected to further analysis or retained for future investigation.

In thirteen cases a dredge measuring 54 inches in length of opening and 15 in width, and weighing 137 lbs., or in very deep water one somewhat smaller, was lowered; and in nine instances, notwithstanding the great depths and the extreme difficulty of the operation, brought up a sample of the bottom usually weighing 1 cwt. or more; and what we could scarcely doubt was a fair representation of the fauna of the ground which it had gone over at the bottom.

At almost every station a serial temperature sounding was taken, the temperature being ascertained at certain stated intervals, usually at intervals of 100 fathoms from the surface to 1,500. In many cases samples of water were brought up from intermediate depths for examination, and in every case the surface temperature of the sea was taken, the temperature of the air with dry- and wet-bulb thermometers, and the amount of atmospheric pressure.

Every single operation, whether of sounding or dredging, was conducted from beginning to end by Captain Nares, and in every case the conditions required were determined with an amount of care which left no reasonable doubt of their accuracy within very narrow limits of error. I should therefore say, with reference to this first section, that the results were thoroughly satisfactory.

In the length of the section at the foot of Plate V. one centimetre division represents 100 nautical miles, so that 1 mm. corresponds with 10 miles. In order to make the differences in depth perceptible, and at the same time to avoid too great an amount of exaggeration, this proportion has been multiplied in a vertical direction twenty-five times; so that while 1 mm. represents 10 miles in distance, 2·5 mm. represent one mile or 1,000 fathoms in depth or height.

A reference to this section shows that the bottom of the Atlantic, along a line which corresponds roughly with the Tropic of Cancer, presents very much the same character which it does further north—that of a plateau showing comparatively gentle undulations on a large scale. The section does not differ very materially from the general outline given in some of the latest atlases of physical geography,—for example, in Plate XLI$^{c.}$ of Stieler's Hand Atlas; and it confirms upon the whole, to a remarkable degree, the soundings of Lieut.-Commanding Lee and Lieut.-Commanding Berryman of the U.S. Navy in the surveying ship 'Dolphin,' which have furnished nearly all the data for this particular region.

After passing over about 80 miles of volcanic mud and sand, products of the disintegration of the volcanic rocks of the Islands of the Canary group, the first four soundings, to a distance of 300 miles from Santa Cruz at depths varying from 1,525 to 2,220 fathoms, yielded 'globigerina-ooze' of the usual character. This 'modern chalk' consists first of all of a creamy surface-layer made up of little else than the shells, most of them almost entire, of *Globigerina*, *Pulvinulina*, and *Orbulina* with a relatively small proportion of finely-divided matter consisting chiefly of coccoliths and rhabdoliths, and a still smaller proportion of the spines and tests of radiolarians, and fragments of the spicules of sponges. Mixed with these there are usually a considerable number of the dead shells of pteropods of the genera *Cleodora, Diacria, Cavolinia, Triptera,* and *Styliola* in a more or less mutilated and disintegrated condition; and living among the ooze, at all events at moderate depths, there are scattered examples of many foraminifera of the crystellarian and milioline groups, and the sponges, corals, star-fishes and higher invertebrates, which, with a few fishes belonging to certain well-defined families, complete the fauna of the region. Next we have a layer an inch or two in thickness, somewhat more firm in consistence, in which most of the shells of all kinds are more or less broken up and their fragments cemented together by a calcareous paste the result of the complete disentegration of many of them; and beneath this a nearly uniform calcareous paste, coloured grey by decomposed organic matter and containing whole and fragmentary shells only sparsely scattered through it. Excellent samples showing the gradual passage from one condition into the other are often brought up in the tube of the sounding-machine.

Since the time of our departure, Mr. Murray has been paying the closest attention to the question of the origin of this calcareous formation, which is of so great interest and importance on account of its anomalous character and its enormous extension. Very early in the voyage, he formed the opinion that all the organisms entering into its composition at the bottom are dead, and that all of them live abundantly at the surface and at intermediate depths, over the globigerina-ooze area, the ooze being formed by the subsiding of these shells to the bottom after death.

This is by no means a new view. It was advocated by the late Professor Bailey, of West Point, shortly after the discovery, by means of Lieut. Brooke's ingenious sounding-instrument, that such a formation had a wide extension in the Atlantic. Johannes Müller, Count Pourtales, Krohn, and Max Schultze, observed *Globigerina* and *Orbulina* living on the surface; and Ernst Haeckel, in his important work upon the Radiolaria, remarks "that we often find upon, and carried along by, the floating pieces of sea-weed which are so frequently met with in all seas, foraminifera as well as other animal forms which habitually live at the bottom. However, setting aside these accidental instances, certain foraminifera, particularly in their younger stages, occur in some localities so constantly, and in such numbers, floating on the surface of the sea, that the suspicion seems justifiable that they possess, at all events at a certain period of their existence, a pelagic mode of life, differing in this respect from most of the remainder of their class. Thus Müller often found in the contents of the surface-net off the coast of France, the young of *Rotalia*, but more particularly *Globigerinæ* and *Orbulinæ*, the two latter frequently covered with fine calcareous tubes, prolongations of the borders of the fine pores through which the pseudopodia protrude through the shell. I took similar *Globigerinæ* and *Orbulinæ* almost daily in a fine net at Messina, often in great numbers, particularly in February. Often the shell was covered with a whole forest of extremely long and delicate calcareous tubes projecting from all sides, and probably contributing essentially to enable these little animals to float below the surface of the water by greatly increasing their surface, and consequently their friction against the water, and rendering it more difficult for them to sink."[1] In 1865 and 1866 two papers were read by Major Owen, F.L.S., before the Linnean Society, "On the Surface-fauna of Mid-Ocean." In these communications the author stated that he had taken foraminifera of the genera *Globigerina* and *Pulvinulina* living, in the tow-net on the surface, at many stations in the Indian and Atlantic Oceans. He described the special forms of these genera which were most common, and gave an interesting account of their habits; proposing for a family which should include *Globigerina* with *Orbulina* as a sub-genus, and *Pulvinulina*, the name Colymbitæ, from the circumstance that, like the radiolaria, these foraminifera are found on the surface after sunset, 'diving' to some depth beneath it during the heat of the day. Our colleague Mr. Gwyn Jeffreys, chiefly on the strength of Major Owen's papers, maintained that certain foraminifera were surface-animals, in opposition to Dr. Carpenter and myself.[2] I had formed and expressed a very strong opinion on the matter. It seemed to me that the evidence was conclusive that the foraminifera

---

[1] *Die Radiolarien.* Eine Mongraphie von Dr. Ernst Haeckel. Berlin, 1862. Pages 166, 167.

[2] "Mr. Jeffreys desires to record his dissent from this conclusion, since (from his own observations, as well as those of Major Owen and Lieutenant Palmer) he believes *Globigerina* to be exclusively an *Oceanic* Foraminifer inhabiting only the superficial stratum of the sea."—*Preliminary Report of the Scientific Exploration of the Deep Sea*, Proceedings of the Royal Society, No. 121, page 443.

which formed the globigerina-ooze lived on the bottom, and that the occurrence of individuals on the surface was accidental and exceptional; but after going into the thing carefully, and considering the mass of evidence which has been accumulated by Mr. Murray, I now admit that I was in error; and I agree with him that it may be taken as proved, that all the materials of such deposits (with the exception of course of the remains of animals which we now know to live at the bottom at all depths, and which occur in the deposit as foreign bodies) are derived from the surface.

Mr. Murray has combined with a careful examination of the soundings, a constant use of the tow-net, usually at the surface, but also at depths from ten to a thousand fathoms; and he finds the closest relation to exist between the surface-fauna of any particular locality and the deposit which is taking place at the bottom. In all seas, from the equator to the polar ice, the tow-net contains *Globigerinæ*. They are more abundant, and of a larger size, in warmer seas; several varieties attaining a large size, and presenting marked varietal characters, are found in the intertropical area of the Atlantic. In the latitude of Kerguelen they are less numerous and smaller, while further south they are still more dwarfed, and only one variety, the typical *Globigerina bulloides*, is represented. The living *Globigerinæ* from the tow-net are singularly different in appearance from the dead shells we find at the bottom (Fig. 46). The shell is clear and transparent, and each of the pores which penetrate it is surrounded by a raised crest, the crest round adjacent pores coalescing into a roughly

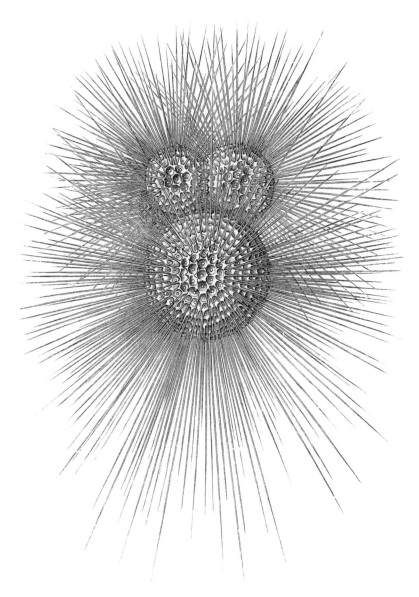

Fig. 46.—*Globigerina bulloides*, from the surface.

hexagonal network, so that the pore appears to lie at the bottom of a hexagonal pit. At each angle of this hexagon the crest gives off a delicate flexible calcareous spine, which is sometimes four or five times the diameter of the shell in length. The spines radiate symmetrically from the direction of the centre of each chamber of the shell, and the sheaves of long transparent needles, crossing one another in different directions, have a very beautiful effect. The smaller inner chambers of the shell are entirely filled with an orange-yellow granular sarcode; and the large terminal chamber usually contains only a small irregular mass, or two or three small masses run together, of the same yellow sarcode stuck against one side, the remainder of the chamber being empty. No definite arrangement, and no approach to structure, was observed in the sarcode; and no differentiation, with the exception of bright yellow oil-globules, very much like those found in some of the Radiolarians, which are scattered apparently irregularly in the sarcode; and usually one very definite patch of a clearer appearance than the general mass, coloured vividly with a carmine solution; and the presence of scattered particles of bioplasm was indicated by minute spots here and there throughout the whole substance, which received the dye.

When the living *Globigerina* is examined under very favourable circumstances, that is to say, when it can be at once placed under a tolerably high power of the microscope in fresh still sea-water, the sarcodic contents of the chambers may be seen to exude gradually through the pores of the shell, and spread out until it forms a kind of flocculent fringe round the shell, filling up the spaces among the roots of the spines and rising up a little way along their length. This external coating of sarcode is rendered very visible by the oil-globules, which are oval and filled with intensely-coloured secondary globules, and are drawn along by the sarcode, and may be seen with a little care following its spreading or contracting movements. At the same time an infinitely delicate sheath of sarcode containing minute transparent granules but no oil-globules, rises on each of the spines to its extremity, and may be seen creeping up one side and down the other of the spine with the peculiar *flowing* movement with which we are so familiar in the pseudopodia of *Gromia* and of the Radiolarians. If the cell in which the *Globigerina* is floating receive a sudden shock, or if a drop of some irritating fluid be added to the water, the whole mass of sarcode retreats into the shell with great rapidity, drawing the oil-globules along with it, and the outline of the surface of the shell and of the hair-like spines is left as sharp as before the exodus of the sarcode.

Major Owen (*op. cit.*) has referred the *Globigerina* with spines to a distinct species, under the name of *G. hirsuta*. I am inclined rather to believe that all *Globigerinæ* are, to a greater or less degree, spiny when the shell has attained its full development. In specimens taken with the tow-net the spines are very usually absent; but that is probably on account of their extreme tenuity; they are broken off by the slightest touch. In fresh examples from the surface the dots indicating the origin of the lost spines may almost always be made out with a high power. There never are spines on the *Globigerinæ* from the

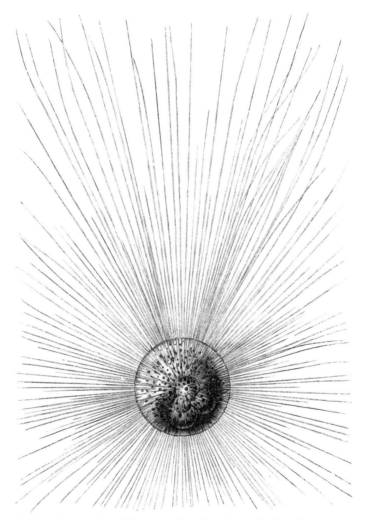

Fig. 47.—*Orbulina universa*, D'Orbigny. From the surface. Fifty times the natural size.

bottom, even in the shallowest water. Two or three very marked varieties of *Globigerina* occur; but I certainly do not think that the characters of any of them can be regarded as of specific value.

There is still a good deal of obscurity about the nature of *Orbulina universa,* an organism which occurs in some places in large proportion in the globigerina-ooze. The shell of *Orbulina* (Fig. 47) is spherical, usually about ·5 mm. in diameter, but it is found of all smaller sizes. The texture of the mature shell resembles closely that of *Globigerina,* but it differs in some important particulars. The pores are markedly of two different sizes, the larger about four times the area of the smaller. The larger pores are the less numerous; they are scattered over the surface of the shell without any appearance of regularity; the smaller pores occupy the spaces between the larger. The crests between the pores are much less regular in *Orbulina* than they are in *Globigerina;* and the spines, which are of great length and extreme tenuity, seem rather to arise abruptly from the top of scattered papillæ than to mark the intersections of the crests. This origin of the spines from the papillæ can be well seen with a moderate power on the periphery of the sphere. The spines are hollow and flexible; they naturally radiate regularly from the direction of the centre of the sphere; but even in specimens which have been placed under the microscope with the greatest care, they are usually entangled together in twisted bundles. They are so fragile that the weight of the shell itself, rolling about with the motion of the ship, is usually sufficient to break off the whole of the spines and leave only the papillæ projecting from its surface, in the course of a few minutes. In some examples, either those in process of development, or a series showing a varietal divergence from the ordinary type, the shell is very thin and almost perfectly smooth, with neither papillæ nor spines, nor any visible structure except the two classes of pores, which are constant.

The chamber of *Orbulina* is often almost empty; even in the case of examples from the surface, which appear from the freshness and transparency of the shell, to be living, it is never full of sarcode; but it frequently contains a small quantity of yellow sarcode stuck against one side, as in the last chamber of *Globigerina*. Sometimes, but by no means constantly, within the chamber of *Orbulina* there is a little chain of three or four small chambers singularly resembling in form, in proportion, and in sculpture, a small *Globigerina;* and sometimes, but again by no means constantly, spines are developed on the surface of the calcareous walls of these inner chambers, like those on the test of *Globigerina*. The spines radiate from the position of the centre of the chambers and abut against the inside of the wall of the *Orbulina*. In a few cases, the inner chambers have been observed apparently arising within or amidst the sarcode adhering to the wall of the *Orbulina*.

Major Owen regards *Orbulina* as a distinct organism, nearly allied to *Globigerina*, but differing so far from it as to justify its separation into a special sub-genus. He considers the small inner chamber of *Orbulina* to represent the smaller chamber of *Globigerina*, and the outer wall as the equivalent of the large outer chamber of *Globigerina*, developed in this form as an investing chamber. Count Pourtales, Max Schultze, and Krohn, on the other hand, believe, on account of the close resemblance in structure between the two shells, their constant association, and the undoubted fact that an object closely resembling a young *Globigerina* is often found within *Orbulina*, that the latter is simply a special reproductive chamber budded from the former, and capable of existing independently. I am rather inclined to the latter view, although I think much careful observation is still required to substantiate it; and some even of our own observations would seem to tell somewhat in the opposite direction. Although *Orbulina* and *Globigerina* are very usually associated, they are so in different proportions in different localities; and in the icy sea to the south of Kerguelen, although *Globigerina* was constantly taken in the surface-net, not a single *Orbulina* was detected. Like *Globigerina*, *Orbulina* is most fully developed and most abundant in the warmer seas.

Associated with these forms, and, like them, living on the surface—and dead, with their shells in various stages of decay, at the bottom, there are two very marked species or varieties of *Pulvinulina*, *P. menardii* and *P. micheliniana*. The general structure of *Pulvinulina* resembles that of *Globigerina*. The shell consists of a congeries of from five to eight chambers arranged in an irregular spiral. As in *Globigerina*, the last chamber is the largest; the inner smaller chambers are usually filled with yellow sarcode; and, as in *Globigerina*, the last chamber is frequently nearly empty, a small irregular mass of sarcode only occupying a part of the cavity. The walls of the chambers are closely and minutely perforated. The external surface of the wall is nearly smooth, and in particularly well-preserved tow-net specimens spines may be detected closely resembling those of *Globigerina* and *Orbulina*, but more thinly scattered and apparently somewhat more delicate. *Pulvinulina menardii*, an example of which is here figured in the condition in which it is usually met with in the ooze (Fig. 48), has a large discoidal depressed shell, consisting of a series of flat chambers overlapping one another, like a number of coins laid down somewhat irregularly, but generally in a spiral; each chamber

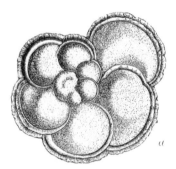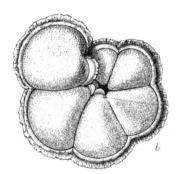

Fig. 48.—*Pulvinulina menardii*, D'Orbigny, *a*, the upper, *b*, the under surface. Thirty times the natural size. Dead shells from the bottom, at a depth of 1,900 fathoms.

is bordered by a distinct somewhat thickened solid rim of definite width. On the lower surface of the shell the intervals between the chambers are indicated by deep grooves. The large irregular opening of the final chamber is protected by a crescentic lip, which in some specimens bears a fringe of spine-like papillæ. This form is almost confined to the warmer seas. It is very abundant on the surface, and still more so during the day at a depth of ten to twenty fathoms in the Mid-Atlantic; and it enters into the composition of the very characteristic 'globigerina-ooze' of the 'Dolphin Rise' in almost as large proportion as *Globigerina*. *Pulvinulina micheliniana* is a smaller variety; the upper surface of the shell is flattened as in *P. menardii*, but the chambers are conical and prolonged downwards, so that the shell is deeper and somewhat turbinate. The two species usually occur together, but *P. micheliniana* has apparently a much wider distribution than *P. menardii;* for while the latter was limited to the region of the trade-winds and the equatorial drift current, and was found rarely if at all to the south of the Agulhas current, the former accompanied us southward as far as Kerguelen Land. Both forms of *Pulvinulina*, however, are more restricted than *Globigerina;* for even *P. micheliniana* became scarce after leaving the Cape, and the wonderfully pure calcareous formation in the neighbourhood of Prince Edward Island and the Crozets consists almost solely of *Globigerina bulloides*, and neither species of *Pulvinulina* occurred to the south of Kerguelen Land.

Over a very large part of the 'globigerina-ooze' area, and especially in those intertropical regions in which the formation is most characteristically developed, although the great bulk of the ooze is made up of entire shells and fragments of shells of the above-described foraminifera, there is frequently a considerable proportion (amounting in some cases to about twenty per cent.) of fine granular matter, which fills the shells and the interstices between them, and forms a kind of matrix or cement. This granular substance is, like the shells, calcareous, disappearing in weak acid to a small insoluble residue; with a low microscopic power it appears amorphous, and it is likely to be regarded, at first sight, as a paste made up of the ultimate calcareous particles of the disintegrated shells; but under a higher power it is found to consist almost entirely of 'coccoliths' and 'rhabdoliths.' I need scarcely enter here into a detailed description of these singular bodies, which have already been carefully studied by Huxley, Sorby, Gümbel, Haeckel, Carter, Oscar Schmidt, Wallich, and others. I need only state that I believe our observations have placed it beyond a doubt that the 'coccoliths' are the separated elements of a peculiar calcareous armature which covers certain spherical bodies (the 'coccospheres' of Dr. Wallich). The rhabdoliths are the like elements of the armature of extremely beautiful little bodies, of which two forms are represented in Figs. 49 & 50, which have been first observed by Mr. Murray and naturally called by him 'rhabdospheres.' Coccospheres and rhabdospheres live abundantly on the surface, especially in warmer seas. If a bucket of water be allowed to stand over night with a few pieces of thread in it, on examining the threads carefully many examples may usually be found attached to them; but Mr. Murray has found an unfailing supply of all forms in the stomachs of *Salpæ*.

What these coccospheres and rhabdospheres are, we are not yet in a position to say with certainty; but our strong impression is that they are either Algæ of a peculiar form, or the reproductive gemmules or the sporangia of some minute organism, probably an Alga; in which latter case the coccoliths and rhabdoliths might be regarded as representing in position and function the 'amphidisci' on the surface of the gemmules of *Spongilla,* or the spiny facets on the zygospores of many of the Desmideæ. There are many forms of coccoliths and rhabdoliths, and many of these are so distinct that they evidently indicate different species. Mr. Murray believes, however, that only one form is met with on one sphere ; and that,

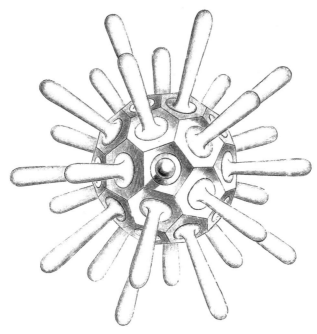

Fig. 49.—A ' Rhabdosphere.' From the surface. Five hundred times the natural size.

in order to produce the numerous forms figured by Haeckel and Oscar Schmidt, all of which, and many additional varieties, he has observed, the spheres must vary in age and development, or in kind. Their constant presence in the surface-net, in surface-water drawn in a bucket, and in the stomachs of surface-animals, sufficiently proves that, like the ooze-forming foraminifera, the coccoliths and rhabdoliths, which enter so largely into the composition of the recent deep-sea calcareous formations, live on the surface and at intermediate depths, and sink to the bottom after death. Coccospheres and rhabdospheres have a very wide, but not an unlimited, distribution. From the Cape of Good Hope they rapidly decreased in number on the surface and

Fig. 50.—A 'Rhabdosphere.' From the surface. Two thousand times the natural size.

at the bottom as we progressed southwards. The proportion of their remains in the globigerina-ooze near the Crozets and Prince Edward Island was comparatively small; and to this circumstance the extreme cleanness and the unusual appearance of being composed of *Globigerinæ* alone was probably mainly due. We found the same kind of ooze, nearly free from coccoliths and rhabdoliths, in what may be considered about a corresponding latitude in the north, to the west of Faröe.

The next seven soundings, extending along the section to a distance of about 1500 miles from Teneriffe, and at depths varying from 3150 to 2575 fathoms, are marked on the chart 'red clay.' According to our present experience, the deposit of globigerina-ooze is limited in the open oceans, such as the Atlantic, the Southern Sea, and the Pacific, to water of a certain depth, the extreme limit of the pure characteristic formation being placed at a depth of somewhere about 2250 fathoms.

Crossing from these shallower regions occupied by the ooze into deeper soundings, we find universally that the calcareous formation gradually passes into, and is finally replaced by, an extremely fine pure clay, which occupies, speaking generally, all depths below 2500 fathoms, and consists, almost entirely, of a silicate of the red oxide of iron and alumina. The clay is often mixed with other inorganic matter, particularly with particles, graduating up to the size of large nodules, of peroxide of manganese; and in volcanic regions, or in their neighbourhood, with fragments of pumice. The transition is very slow, and extends over several hundred fathoms of increasing depth; the shells gradually lose their sharpness of outline and assume a kind of 'rotten' look and a brownish colour, and become more and more mixed with a fine amorphous red-brown powder, which increases steadily in proportion until the lime has almost entirely disappeared. This brown matter is in the finest possible state of subdivision, so fine that when, after sifting it to separate any organisms it might contain, we put it into jars to settle, it remained for days in suspension.

In indicating the nature of the bottom on the charts, we came from experience, and without any theoretical consideration, to use three terms for soundings in deep water. Two of these, 'gl. oz.' and 'r. cl.', were very definite, and indicated strongly marked formations, with apparently but few characters in common; but we frequently got soundings which we could not exactly call either 'globigerina-ooze' or 'red clay;' and before we were fully aware of the nature of these we were in the habit of indicating them as 'grey ooze' (gr. oz.). We now recognise the 'grey ooze' as, in most cases, an intermediate stage between the globigerina-ooze and the red clay; we find that, on one side as it were of an ideal line, the red clay contains more and more of the material of the calcareous ooze, while, on the other, the ooze is mixed with an increasing proportion of 'red clay.'

When the section from Teneriffe to Sombrero was taken we had not fully recognised the importance of the transition stage, and the bottom was marked on the chart 'globigerina-ooze,' or 'red clay,' according as one or other gave a distinct and marked character to the sounding. The soundings at Stations 5 and 6, for example, might have been labelled 'grey ooze;' for although its nature has altered entirely from the 'globigerina-ooze,' the 'red clay' into which it is rapidly passing still contains a considerable admixture of carbonate of lime.

The depth goes on increasing, to a distance of 1150 miles from Teneriffe, when it reaches 3150 fathoms; there the clay is pure and smooth, and contains scarcely a trace of lime. From this great depth the bottom gradually rises, and, with decreasing depth, the grey colour and the calcareous composition of the ooze return. Three soundings in 2050, 1900, and 1950 fathoms on the 'Dolphin Rise,' gave highly characteristic examples of the *Globigerina* formation. Passing from the middle plateau of the Atlantic into the western trough, with depths a little over 3000 fathoms, the red clay returned in all its purity: and our last sounding, in 1420 fathoms, before reaching Sombrero, restored the 'globigerina-ooze' with its peculiar associated fauna.

This section shows also the wide extension and the vast geological importance of the red-clay formation. The total distance from Teneriffe to Sombrero is about 2700 miles. Proceeding from east to west, we have

About 80 miles of volcanic mud and sand,
,, 350 ,, 'globigerina-ooze,'
,, 1050 ,, 'red clay,'
,, 330 ,, 'globigerina-ooze,'
,, 850 ,, 'red clay,'
,, 40 ,, 'globigerina-ooze,'

giving a total of 1900 miles of 'red clay' to 720 miles of 'globigerina-ooze.'

The following Table, taken from the chart, gives a good general idea of the distribution of the two formations with regard to depth; it being understood, however, that while in all the soundings marked 'red clay' the characters of that formation greatly predominated, in several of the more shallow of these the change was by no means complete. The table gives an average of 1600 fathoms for our soundings in this section in the 'globigerina-ooze;' this is a datum of no value, for we sounded only once in shallow water (450 fathoms), and we know that this formation covers large areas at depths between 300 and 400 fathoms; but the mean maximum depth at which it occurs is important, and may be taken at about 2250 fathoms. The mean depth of the 'red clay' soundings is about 2750 fathoms. The general concurrence of many observations would go far to prove, what seems now, indeed, to stand in the position of an ascertained fact, that wherever the depth increases from about 2200 to 2600 fathoms, the modern chalk formation of the Atlantic passes into a clay.

No. of Station.	Nature of the Bottom.		No. of Station.	Nature of the Bottom.	
	Glob.-ooze.	Red Clay.		Glob.-ooze.	Red Clay.
1	1890	...	13	1900	...
2	1945	...	14	1950	...
4	2220	...	15	...	2325
5	...	2740	16	...	2435
6	...	2950	17	...	2385
7	...	2750	18	...	2675
8	...	2800	19	...	3000
9	...	3150	20	...	2975
10	...	2720	21	...	3025
11	...	2575	22	1420	...
12	2025	...	23	450	...

The nature and origin of this vast deposit of clay is a question of the very greatest interest; and although I think there can be no doubt that it is in the main solved, yet some matters of detail are still involved in difficulty. My first impression was, that it might be the most minutely divided material, the ultimate sediment, produced by the disintegration of the land, by rivers, and by the action of the sea on exposed coasts, and held in suspension and distributed by ocean currents, and only making itself manifest in places unoccupied by the 'globigerina-ooze.' Several circumstances seemed, however, to negative this mode of origin. The formation seemed too uniform; whenever we met with it, it had the same character, and it only varied in composition in containing less or more carbonate of lime.

Again, we were gradually becoming more and more convinced that all the important elements of the 'globigerina-ooze' lived on the surface; and it seemed evident that, so long as the conditions on the surface remained the same, no alteration of contour at the bottom could possibly prevent its accumulation; and the surface conditions in the Mid-Atlantic were very uniform, a moderate surface-current of a very equal temperature passing continuously over elevations and depressions, and everywhere yielding to the tow-net the ooze-forming foraminifera in the same proportion. The Mid-Atlantic swarms with pelagic Mollusca; and in moderate depths, the shells of these are constantly mixed with the 'globigerina-ooze,' sometimes in number sufficient to make up a considerable portion of its bulk. It is clear that these shells must fall in equal numbers upon the red clay; but scarcely a trace of one of them is ever brought up by the dredge on the red-clay area. It might be possible to explain the absence of shell-secreting animals *living on the bottom* by the supposition that the nature of the deposit was injurious to them; but the idea of a current sufficiently strong to sweep them away if falling from the surface, is negatived by the extreme fineness of the sediment which is being laid down; the absence of surface shells appears to be intelligible only on the supposition that they are in some way removed by chemical action.

We conclude, therefore, that the 'red clay' is not an additional substance introduced from without, and occupying certain depressed regions on account of some law regulating its deposition; but that it is produced by the removal, by some means or other, over these areas, of the carbonate of lime which forms probably about 98 per cent. of the material of the 'globigerina-ooze.' We can trace, indeed, every successive stage in the removal of the carbonate of lime in descending the slope of the ridge or plateau where the 'globigerina-ooze' is forming, to the region of the clay. We find, first, that the shells of pteropods and other surface mollusca, which are constantly falling on the bottom, are absent, or if a few remain, they are brittle and yellow, and evidently decaying rapidly. These shells of mollusca decompose more easily, and disappear sooner, than the smaller and apparently more delicate shells of rhizopods. The smaller foraminifera now give way and are found in lessening proportion to the larger; the coccoliths first lose their thin outer border and then disappear, and the clubs of the rhabdoliths get worn out of shape and are last seen, under a high power, as minute cylinders scattered over the field. The larger foraminifera are attacked, and instead of being vividly white and delicately sculptured, they become brown and worn, and finally they break up, each according to its fashion;—the chamber-walls of *Globigerina* fall into wedge-shaped pieces which quickly disappear, and a thick rough crust breaks away from the surface of *Orbulina*, leaving a thin inner sphere, at first beautifully transparent, but soon becoming opaque and crumbling away.

In the meantime, the proportion of the amorphous 'red clay' to the calcareous elements of all kinds, increases, until the latter disappear, with the exception of a few scattered shells of the larger foraminifera, which are still found, even in the most characteristic samples of the 'red clay.'

There seems to be little doubt that a considerable proportion of the fine molecular matter of which the 'red clay' is, to a great degree, composed; is the insoluble residue, the *ash*, as it were, of the calcareous organisms which form the 'globigerina-ooze,' after the calcareous matter has been by some means removed. I do not suppose that the material of the 'red clay' exists in the form of the silicate of alumina and iron in the living foraminifera or pteropods; but that inorganic salts other than salts of lime exist in all animal tissues, soft and hard, is undoubted; and I hazard the speculation that during the decomposition of these tissues in contact with sea-water and the sundry matters which it holds in solution or suspension, these salts may pass into more stable combinations.

Some careful observations which have been made by Mr. Murray have, however, shown that a great part of the 'red clay' is produced, as clays usually are, by the decomposition of felspathic minerals. The source of the felspar is singular; Mr. Murray has undoubtedly established the fact that over a large part of the bed of the ocean *pumice* occurs in quantity in different stages of decay, and that this is more specially evident in the 'red clay' area; and he traces, I have no doubt correctly, a great part of the material of the 'red clay' to this source. Nodules containing a large proportion of manganese peroxide are usually more or less abundant in the 'red clay;' and Mr. Murray believes that their materials are also derived from the decomposition of volcanic products.

Our dredgings in the Atlantic, and a subsequent careful examination of the soundings, certainly give us the impression that the siliceous bodies, including the spicules of sponges, the spicules and tests of radiolarians, and the frustules of diatoms, which occur in appreciable proportion in the 'globigerina-ooze,' diminish in number, and that the more delicate of them disappear in the transition from the calcareous ooze to the 'red clay;' and it is only by the light of subsequent observations that we are now aware that this is by no means necessarily the case. I think it may be well to anticipate here these later results in order to make the nature of the deep-sea deposits more clear.

On the 23rd of March, 1875, in the Pacific, in lat. 11° 24' N., long. 143° 16' E., between the Caroline and the Ladrone groups, we sounded in 4,575 fathoms. The bottom was such as would naturally have been marked on the chart from its general appearance 'red clay'; it was a fine deposit, reddish brown in colour, and it contained scarcely a trace of lime. It was somewhat different, however, from ordinary 'red clay'—more gritty; and the lower part of the contents of the sounding-tube seemed to have been compacted into a somewhat coherent cake, as if already a stage towards hardening into stone. When placed under the microscope, it was found to contain so large a proportion of the tests of radiolarians, that Mr. Murray proposed for it the name 'radiolarian-ooze.'

The RADIOLARIA, whose name recurs so frequently in these pages, and which play so important a part in supplying material for these new geological formations, are not very familiar to British naturalists. It seems that a very insignificant current of cold water passing southwards from the Arctic Sea divides against the north of Scotland, the main body of it flowing into the German Ocean, the temperature of which it lowers sensibly, and a very narrow belt passing down along the west coasts of the British Islands. It is in this belt that we usually work the dredge and tow-net; and, with the exception of some very curious compound forms which sometimes swarm in the West Highland Lochs, radiolarians are scarce. Whenever the belt of water of northern derivation is passed, which is only from 60 to 80 miles from the shore, these forms, which frequently occur in the Atlantic, the Mediterranean, the Pacific, and all moderately warm seas in sufficient numbers to discolour the water, become abundant. The RADIOLARIA form a class of the somewhat negative sub-kingdom PROTOZOA,—a sub-kingdom retained for the reception of all those animals of comparatively simple structure, such as the INFUSORIA, &c., whose relations we cannot yet fully make out. The radiolaria consist essentially of a little mass of sarcode, with no very definite bias as to form, but tending when irritated to assume more or less that of a sphere. The sarcode consists usually of rounded or oval granular masses of a brownish or yellowish colour, interspersed with very characteristic round oil-cells, bright yellow, and very refractive; the whole cemented together by soft transparent sarcode including fine granules. Near the centre of the body there is usually a very evident rounded mass of bioplasm which colours deeply with carmine; and the same dye brings out smaller bioplasts scattered through the general substance. When the animal is at rest and happy in a cell with abundance of fresh sea-water, soft sarcodic matter from the peripheral layer stretches itself out all round in a maze of straight radiating pseudopodia, only visible under a high magnifying power, and in these the peculiar flowing motion, which is so characteristic of sarcode feeding-filaments, is well marked. In many radiolarians, and especially in some very peculiar compound forms, a spherical internal chamber, called the 'central capsule,' whose function we do not fully understand, is very prominent. This capsule is however absent, or at all events, exists in a very modified form, in the more typical groups.

The body may be entirely naked, a mere sphere of sarcode giving off pseudopodia; or it may have a more or less fully-developed skeleton; sometimes in the form of separate horny or siliceous spicules, very like the spicules of sponges, disposed in an irregular net-work over the surface. In one interesting form (Fig. 51), which is especially abundant on the surface in some parts of the Pacific, minute echinated *calcareous* spheres, looking in outline like the rowels of spurs, are scattered irregularly in a fine gelatinous envelope which incloses the granular sarcode and oil-cells. We were familiar with these calcareous bodies in the soundings, but we had always taken them for the spicules of a Holothurian, which they much resemble. In the two groups which are of greatest importance in a geological point of view, the POLYCYSTINA and the ACANTHOMETRINA, the skeleton is much more regular and complete. In the former it consists of a delicate external shell of silica, minutely fenestrated, and often presenting

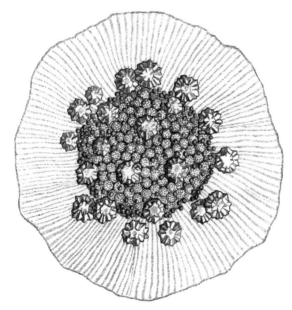

FIG 51.—*Calcaromma*[1] *calcarea*, WYVILLE THOMSON. With the pseudopodia contracted. From the surface. Two hundred times the natural size.

very remarkable and beautiful forms (Fig. 52); in the latter it is essentially internal, and is formed of a varying number of siliceous spicules, radiating from a centre round which the sarcode is accumulated (Fig. 53). The spicules are often most elegantly ornamented, and in an intermediate family,

[1] *Calcar*, a spur.

the Haliommatidæ, they give off a set of anastomosing branches, which form one or several concentric lacey shells which invest the sarcode nucleus (Fig. 54).

The observation of the great abundance of radiolarian tests at great depths, led to the reconsideration of the deposits from the deepest

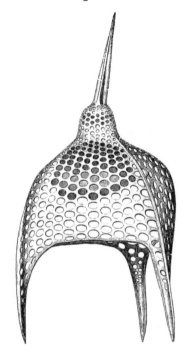

Fig. 52.—*Dictyopodium.* Sp. n. From the surface. Two hundred times the natural size.

soundings; and Mr. Murray now believes, and in this I entirely agree with him, that shortly after the 'red clay' has assumed its most characteristic form, by the total removal of the calcareous shells of the foraminifera, at a depth of say 3,000 fathoms, the deposit in many cases begins gradually to alter again, by the increasing proportion of the shells of radiolarians, until, at such extreme depths as that of the sounding of the 23rd of March, it has once more assumed the character of an almost purely organic formation—the shells of which it is chiefly

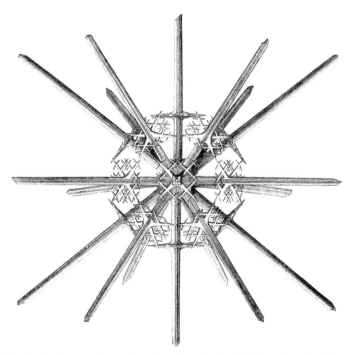

Fig. 53.—*Xiphacantha.* Sp. n. From the surface. One hundred times the natural size. The skeleton only.

composed being, however, in this case siliceous, while in the former they were calcareous. The 'radiolarian-ooze,' although consisting in great part of the tests of radiolarians, contains even in its purest condition a very considerable proportion of 'red clay.' I am certainly inclined to accept an explanation of this second change which was first suggested by Mr. Murray, and which is indeed almost a necessary sequel to his investigations.

We have every reason to believe, from a series of observations, as yet very incomplete, with the tow-net at different depths, that while foraminifera are

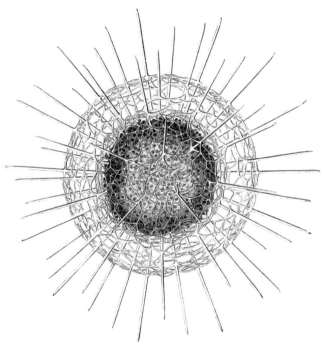

Fig 54.—*Haliomma*. Sp. n. From the surface. Two hundred times the natural size.

apparently confined to a comparatively superficial belt, radiolarians exist at all depths in the water of the ocean. At the surface and a little beneath it the tow-net yields certain species; when sunk to greater depths additional species are constantly found; and in the deposit at the bottom, species occur which have been detected neither on the surface nor at 1,000 fathoms, the greatest depth at which the tow-net has yet been systematically used; and specimens taken near the bottom of species which occur on or near the surface give us the impression of being generally larger and better developed. The results from the tow-net are not so directly satisfactory as those from the trawl or dredge, which usually bring up animals which we know from their nature must have lived on the bottom, and it requires a little consideration to arrive at their precise value. We have not yet contrived a tow-net which can be sent down closed, which will open its mouth when it reaches a certain depth and close it again after a certain lapse of time; although there is little doubt that such a net might be made easily enough. At present the tow-net, which consists simply of a conical bag of muslin or buntine attached to an iron ring, is constantly open, —descending, dragging along, and ascending. If worked on the surface there is of course neither difficulty nor question, but if it be brought up from 500 fathoms, at which depth it has been towing for some time, the net may be supposed to contain chiefly the species living at that depth; but mixed with these there must be a considerable number of more superficial forms, some taken when the net was going down with its open mouth downwards, and many more captured during its long ascent of half a mile through the upper layers. We cannot therefore as yet say with certainty whether the surface species live in the deeper belts or not, but we are justified in concluding that species which are absent on the surface, and present only when a certain depth has been gained, are special to that and probably to greater depths. If again species differing both from those procured on the surface and at intermediate depths are found in the bottom deposits, it is a legitimate inference that these live below the zone of our deepest tow-net observations.

Now, if it be the case that ooze-forming foraminifera are confined to an upper layer of say not more than 500 fathoms in thickness, the supply of their shells and consequently the supply of the 'red clay,' which, according to our view, is to a great extent the product of their decomposition, must be pretty constant over the area where foraminifera abound; while, on the other hand, if the radiolarians live at all depths in the sea, the number of their skeletons falling to the bottom at one place must increase with the increasing depth of the water, and it becomes quite intelligible that in a bed which is being formed at the prodigious depth of five and a half nautical miles the tests of the radiolarians should so preponderate over the 'red clay' as to entirely alter the character of the deposit. I must repeat, however, that it must not be supposed that these deep-sea formations which from their general appearance we put down on the chart as 'red clay' or 'grey ooze' are in all cases entirely or even chiefly, produced by the more or less complete decomposition of the shells of surface animals with calcareous shells, and the mixture of their ash in varying proportions with the tests of radiolarians. That they are so to a great extent I fully believe, but they always contain a certain amount of mineral matter; very usually, perhaps universally, small particles of a substance which has nearly the composition of 'wad,' and which often occurs, as I have already mentioned,

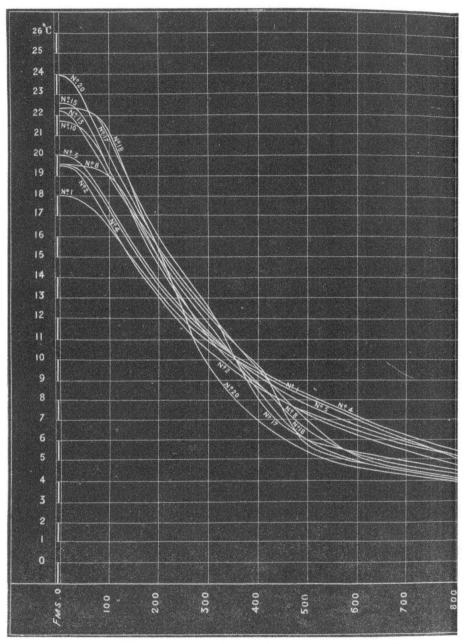

Fig. 55.—Curves constructed from Serial Temperature

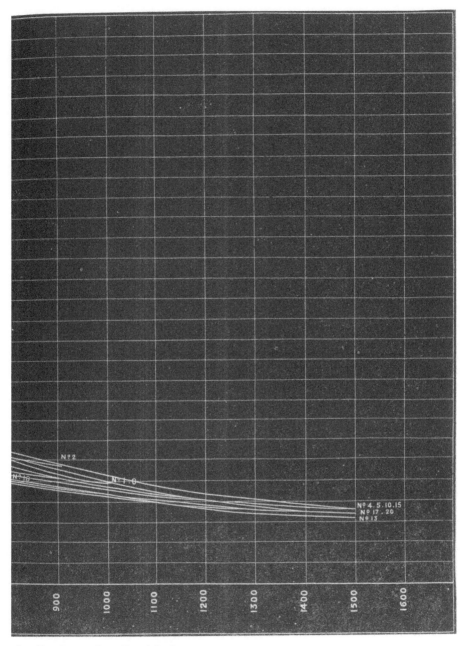

Soundings between Teneriffe and Sombrero.

in the form of large concretionary nodules and cakes; and very frequently crystals and groups of crystals of sulphate of lime. Everywhere near land, for example at Station 25, near St. Thomas, where the depth was 3,875 fathoms, the deposit is coloured greyish with foreign matter; and everywhere in volcanic regions, and notably over nearly the whole area of the Pacific, the red and grey clays owe a considerable portion of their material to the disintegration of pumice, which appears to be drifted about and distributed by currents until it becomes water-logged, when it falls to the bottom and undergoes slow decomposition.

The serial temperature soundings of this section (Appendix A.), taken with extreme care, bear internal evidence of accuracy in the way in which, in almost all cases, the curves or lines representing a number of observations indicate permanence or change in perfect harmony. Thus on Plate V. the isotherm of 11° C., although the combined result of twelve separate observations, is almost straight, while at every station in the section the aggregate of the isothermal lines either spread out consistently, or gather together as if by a common impulse.

The thermometers have been read with very great care, and the corrections for pressure required for each individual thermometer have been in all cases calculated; but it is difficult to read a thermometer of the ordinary construction to tenths of a Fahrenheit degree, and it is possible that when reduced to a diagrammatic form, as in Plate V. and Figs. 57 and 58, slight differences, almost within the limit of error of observation, may appear somewhat exaggerated.

One marked feature in this section is the comparative thinness of the surface layer of heated water and the enormous thickness of the mass of underlying cold water, varying little in temperature throughout its depth; and a second is the great regularity of the curve of fall of temperature from the surface to the bottom (Fig. 55). The bottom temperature of the western basin of the Atlantic is slightly but very decidedly lower than that of the eastern at corresponding depths.

One point connected with some of our earlier observations is of interest in its bearing upon the interpretation of the temperature observations of the 'Porcupine' cruises of 1869 and 1870. Curve A, Fig. 56, represents the vertical distribution of heat in the Bay of Biscay determined by the 'Porcupine' in 1869, and Curve B the distribution of heat derived chiefly from bottom temperatures, off the coast of Portugal. The marked peculiarity in these curves is a rise or 'hump' between 200 and 900 fathoms, indicating a temperature in that particular stratum of water from some cause abnormally high.

This has been accounted for by the 'banking down' against the coasts of Europe of the Gulf Stream, the north-eastern reflux of the equatorial current; and it is naturally most marked at that point where the impact, as it were, of the Gulf Stream is most direct, the coast of the Lusitanian peninsula.

It is indicated on all current charts that about the 40th parallel of north latitude a great part of the Gulf Stream bends southwards along the coast of Africa, a portion of it curving round and rejoining

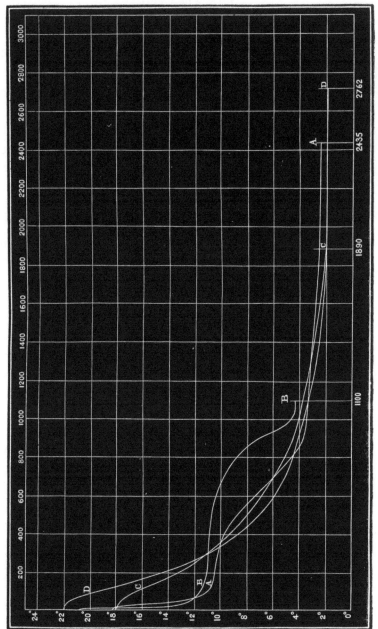

Fig. 56.—Curves constructed from Serial and Bottom Soundings. A, in the Bay of Biscay. B, off the coast of Portugal; and C and D, between Teneriffe and Sombrero.

the equatorial drift, and another portion joining the Guinea current. Some of the observations in the present section appear to confirm in a remarkable way the view that this irregularity in the curve is due to the cause to which it has been attributed. The southern branch of the Gulf Stream is not impeded by land, and there is little or no 'banking down.' Curve C is reduced from the serial sounding at Station 1 (Plate IV.), 70 miles from Santa Cruz. The 'hump' has now disappeared as such, but it is still represented by the space between C and D raising the portion of the curve corresponding with the 'hump' and representing the course of the southern branch of the Gulf Stream moving southwards, now without a special impediment. Curve D represents the middle station of the section in the course of the trade-wind drift where the distribution of temperature is very regular, the isotherms being closely pressed together and the heated water confined to a thin surface layer; and the diagram, Fig. 57, represents a vertical section at the same station, showing the proportion of water at different temperatures. Fig. 58 gives for comparison the proportion of water at different temperatures at Station 1.

Between Teneriffe and Sombrero three series of temperature observations were taken by Lieutenant Bethell with Mr. Siemens's resistance deep-sea thermometer, and there seems to be little doubt from the general correspondence of the results with those of the protector thermometers, that the instrument answers its purpose. In some of the observations, however, there were wide discrepancies, but these

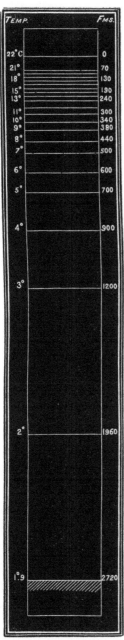

Fig. 57.—Diagram constructed from Serial Sounding No. 10.

Fig. 58.—Diagram constructed from Serial Sounding No. 1.

may have been due to want of practice in observing.

Should it turn out that Mr. Siemens's instrument can be thoroughly depended upon to give accurate results it will undoubtedly be invaluable for certain purposes. In our Antarctic cruise, for instance, on several occasions the surface-layer was colder than any layer beneath; our thermometers on the Miller-Casella construction registered accordingly their minimum on the surface, a warmer layer succeeded in which they registered their maximum, but after fixing these two points they were useless. A workable instrument on Mr. Siemens's principle would have given us the bottom temperature, which we most of all desired.

I must say, however, that I doubt if this resistance apparatus can be applied in general practice to its present purpose without great modification. The instrument is bulky, and, with the battery and galvanometer and thermometers and freezing mixture, troublesome to work on board a ship, and difficult to observe if there is any motion. The cable is bulky and tender, and would probably not stand the fatigue of being often used as a sounding line.

Throughout this section Mr. Buchanan has taken the specific gravity of the sea-water from the surface daily, and from the bottom and from intermediate depths as often as it was possible to obtain samples. A description of the instruments used in procuring the water from the bottom and from intermediate depths, and for determining the specific gravity, has been already given (p. 34 *et seq.*).

I need only mention here that on the voyage from Teneriffe to Sombrero Mr. Buchanan found the remarkable and unexpected result that the water has virtually the same specific gravity from the bottom to within 500 fathoms of the surface. From 500 fathoms the specific gravity rapidly rises, till it usually attains its maximum at the surface. Some minor variations in the specific gravity both of the upper and of the lower layers have manifested themselves from time to time, but to these we shall refer hereafter.

In the table of specific gravities appended, one or two instances occur (marked with an asterisk on the table) in which the specific gravity is as great at the bottom as it is at the surface. In these cases we are forced to believe that by some misadventure the cylinder of the water-bottle became disengaged just as it reached the water. The lowering of the slip water-bottle requires some care until it is fairly beneath the surface, after which there is no chance of the cylinder falling until the instrument reaches the bottom.

In this our first section across the Atlantic, the dredge was lowered thirteen times, and nine times brought up a sufficient sample of the bottom, containing animals living, or which had been living when they entered the dredge; for in most cases when brought to the surface they were perfectly dead. Of the nine successful hauls, four were in depths above 1500 fathoms and under 2000, four above 2000 and under 3000, and one in a depth of 3150 fathoms. In each case, what with the slack of the line and the movement of the ship, the dredge had to travel from eight to ten statute miles through the water, and necessarily each dredging operation occupied a whole day. Under these circumstances it was of course out of the question to attempt to make a perfect collection of the bottom fauna, particularly since all our investigations tend to show that animal life, represented by the higher groups, is scattered and by no means abundant at extreme depths. Our object was to get a fair representation of the deep-sea fauna of this region, and to settle finally the question whether abysses existed where the condition depending upon depth was so extreme as to place a limit to the distribution of living beings. In the former of these objects I believe we have succeeded fairly; in the latter completely.

The dredgings yielded at least 28 species, not including the foraminifera. Of these, 7 were mollusca (2 gasteropods and 5 lamellibranchs); 3 crustaceans; 4 annelids and 1 gephyrean; 4 molluscoids (1 brachiopod and 3 bryozoa); 2 alcyonarian and 2 zoantharian corals; 3 echinoderms and 3 sponges.

Four species of mollusca, referred to three genera which are abundant in water of moderate depth, *Arca*, *Limopsis*, and *Leda*, were dredged at a depth of 2740 fathoms, where they were associated with two species of bryozoa allied to well-known forms.

All the mollusca which we took were small, and several of them appeared to be identical with species procured at great depths in the 'Porcupine' dredgings; a fact in favour of an opinion expressed by Professor Lovén and myself, that many forms of the abyssal fauna may be found to have a very wide, perhaps even a universal, distribution.

It is unfortunate that in the deepest haul of all, 3150 fathoms, no living thing was brought up higher in the scale than a foraminifer; but this I attributed at the time to the nature of the bottom, and this opinion was afterwards borne out by the abundance, at scarcely a less depth, and on a bottom differing only in being somewhat less uniform and containing sand-grains and a few shells of foraminifera, of tube-building annelids of a very common shallow water type. The crustacea do not appear to suffer from the peculiarity of the circumstances under which they live, either in development or in colour. The singular fact of the suppression of the eyes in certain cases has already been referred to. The Echinoderms and sponges which enter so largely into the fauna of the zone ending at 1000 fathoms are not abundant at extreme depths.

APPENDIX A.

Table of Temperatures observed between Teneriffe and Sombrero Island, February and March, 1873.

Depth in Fathoms.	Station No. 1. Lat. 27° 24′ N. Long. 16° 55′ W.	Station No. 2. Lat. 25° 52′ N. Long. 19° 14′ W.	Station No. 3. Lat. 25° 45′ N. Long. 20° 12′ W.	Station No. 4. Lat. 25° 28′ N. Long. 20° 22′ W.	Station No. 5. Lat. 24° 20′ N. Long. 24° 28′ W.	Station No. 6. Lat. 23° 22′ N. Long. 27° 49′ W.	Station No. 7. Lat. 23° 15′ N. Long. 30° 56′ W.	Station No. 8. Lat. 23° 12′ N. Long. 32° 56′ W.
Surface.	18°·0 C.	19°·5 C.	20°·0 C.	19°·5 C.	20°·0 C.	20°·7 C.	20°·0 C.	19°·5 C.
100	16·5	16·6	...	17·2	17·0	...	...	19·0
200	13·2	13·1	...	13·7	13·9	...	...	14·9
300	10·9	11·1	...	11·0	11·2	...	...	11·1
400	9·3	8·6	...	9·4	9·4	...	...	9·3
500	8·0	7·3	...	7·6	7·7	...	...	6·7
600	7·2	6·5	...	6·5	6·8	...	...	6·3
700	...	5·8	...	6·1	6·0	...	...	5·2
800	5·1	5·2	...	5·6	5·2	...	...	4·3
900	3·8	4·8	...	4·7	4·7	...	...	4·0
1000	4·0	...	...	4·6	4·0	...	...	...
1100	...	...	...	3·8	3·5	...	...	...
1200	...	...	...	3·5	...	...	...	...
1300	...	...	...	3·1	2·9	...	...	...
1400	...	...	...	2·8	2·5	...	...	...
1500	...	...	...	2·6	2·7	...	...	...
Depth at Bottom	1890	1945	1525	2220	2740	2950	2750	2800
Temperature at Bottom.	2°·0	2°·0	2°·2	...	2°·0	2°·0	2°·0	2°·0

Depth in Fathoms.	Station No. 9. Lat. 23° 23' N. Long. 35° 10' W.	Station No. 10. Lat. 23° 10' N. Long. 38° 42' W.	Station No. 11. Lat. 22° 45' N. Long. 40° 37' W.	Station No. 12. Lat. 21° 57' N. Long. 43° 29' W.	Station No. 13. Lat. 21° 38' N. Long. 44° 39' W.	Station No. 14. Lat. 21° 1' N. Long. 46° 29' W.	Station No. 15. Lat. 20° 49' N. Long. 48° 45' W.
Surface.	20°·6 C.	22°·2 C.	22°·2 C.	22°·8 C.	22°·2 C.	23°·3 C.	22°·5 C.
100	...	19·5	19·9	...	18·9	...	19·4
200	...	14·3	14·7	...	15·5	...	14·9
300	...	11·5	...	...	12·4	...	11·9
400	...	8·7	...	...	9·6	...	8·5
500	...	6·7	...	...	6·7	...	5·9
600	...	5·2	...	...	5·2	...	5·0
700	...	4·8	...	...	4·7	...	4·6
800	...	4·6	...	...	4·1	...	4·3
900	...	4·0	...	...	3·6	...	4·0
1000	...	3·5	...	...	3·4	...	3·3
1100	...	3·3	...	...	3·0	...	3·2
1200	...	3·1	...	...	2·5	...	2·9
1300	...	2·6	...	...	2·5	...	2·7
1400	...	2·7	...	...	2·3	...	2·5
1500	...	2·6	...	...	2·2	...	2·6
Depth at Bottom.	3150	2720	2575	2025	1900	1950	2325
Temperature at Bottom.	1°·9	1°·9	2°·0	2°·2	1°·9	1°·8	1°·7

Depth in Fathoms.	Station No. 16. Lat. 20° 39′ N. Long. 50° 33′ W.	Station No. 17. Lat. 20° 7′ N. Long. 52° 32′ W.	Station No. 18. Lat. 19° 41′ N. Long. 55° 13′ W.	Station No. 19. Lat. 19° 15′ N. Long. 57° 47′ W.	Station No. 20. Lat. 18° 56′ N. Long. 59° 35′ W.	Station No. 21. Lat. 18° 54′ N. Long. 61° 28′ W.	Station No. 22. Lat. 18° 40′ N. Long. 62° 56′ W.
Surface.	23°·3 C.	23°·3 C.	23°·9 C.	23°·9 C.	23°·9 C.	24°·4 C.	24°·4 C.
100	...	21·0	21·1	...	20·0	...	...
200	...	16·0	15·3	...	15·6	...	...
300	...	12·5	12·3	...	9·7	...	...
400	...	8·1	8·6	...	7·0	...	...
500	...	7·4	5·8	...	5·3	...	...
600	...	5·6	5·0	...	5·0	...	...
700	...	5·0	4·6	...	4·6	...	...
800	...	4·7	4·2	...	4·0	...	...
900	...	4·2	...	...	3·5	...	...
1000	...	3·6	...	...	3·3	...	...
1100	...	3·4	...	...	2·9	...	...
1200	...	3·2	...	...	2·7	...	...
1300	...	...	...	...	2·8	...	...
1400	...	2·5	...	...	2·7	...	...
1500	...	2·5	...	...	2·5	...	...
Depth at Bottom.	2435	2385	2675	3000	2975	3025	1420
Temperature at Bottom.	1°·7	1°·9	1°·6	1°·3	1°·6	1°·3	3°·0

APPENDIX B.

Table of Specific Gravities observed between Teneriffe and Sombrero Island.

Date 1873:	Latitude N.	Longitude W.	Depth at the Station.	d. Depth at which the sample was taken.	t. Temperature at d.	t'. Temperature at which the Specific Gravity was observed.	Specific Gravity at t'. Water at 4° = 1.	Specific Gravity at 15°·5. Water at 4° = 1.	Specific Gravity at t. Water at 4° = 1.
Feb. 15	27° 22'	16° 57'	Fms. 1890	Fms. Bottom.	2°·8C.	17°·9C.	1·02594	1·02652	1·02850
,,	...	...		Surface.	18·1	18·5	1·02658	1·02731	1·02664
16	26 40	17 53		Surface.	18·8	17·8	1·02678	1·02734	1·02649
17	25 52	19 22	1945	Bottom.	2·7	18·8	1·02537	1·02604	1·02803
,,	...	...		Surface.	19·4	19·6	1·02639	1·02742	1·02642
18	25 45	20 12	1530	Surface.	19·2	19·6	1·02619	1·02722	1·02628
19	25 28	20 22	2220	Surface.	20·1	20·3	1·02601	1·02723	1·02606
20	24 56	21 30		Surface.	19·6	19·9	1·02619	1·02730	1·02624
21	24 22	24 4		Surface.	20·2	20·5	1·02629	1·02756	1·02634
,,	24 20	24 28	2740	Bottom.	2·0	20·0	1·02633*	1·02747	1·02944
22	24 15	24 59		Surface.	20·0	20·5	1·02626	1·02753	1·02638
23	23 22	27 49		Surface.	20·75	20·6	1·02633	1·02762	1·02622
,,	23 14	28 22	2950	Bottom.	2·0	19·6	1·02645*	1·02748	1·02945
24	23 23	31 31		Surface.	20·7	20·7	1·02633	1·02765	1·02630
,,	...	...	2750	Bottom.	2·0	20·3	1·02492	1·02614	1·02811
25	23 12	32 56	2800	Bottom.	2·0	19·6	1·02514	1·02617	1·02813
,,	...	...		Surface.	21·0	21·3	1·02627	1·02775	1·02632
26	23 23	35 11		Surface.	21·6	21·9	1·02615	1·02781	1·02621
,,	...	...	3150	Bottom.	2·0	20·2	1·02587	1·02657	1·02854
27	23 28	36 42		Surface.	21·6	22·0	1·02608	1·02777	1·02616
28	23 10	38 42	2720	Bottom.	2·0	22·1	1·02585	1·02757	1·02954
,,	...	...		Surface.	22·2	22·5	1·02595	1·02778	1·02601
March 1	22 45	40 37	2575	Bottom.	2·0	21·6	1·02469	1·02629	1·02825
,,	...	...		2100	2·0	21·3	1·02458	1·02610	1·02806
,,	...	...		500	8·0	21·2	1·02474	1·02606	1·02765
,,	...	...		Surface.	22·4	22·4	1·02591	1·02772	1·02529
,,	...	...		850	4·2	22·3	1·02448	1·02625	1·02814
2	22 30	42 6		Surface.	22·3	22·5	1·02595	1·02780	1·02600
3	21 57	43 29		980	3·3	21·9	1·02450	1·02616	1·02806
,,	...	...	2025	Bottom.	2·2	21·7	1·02484	1·02646	1·02843
,,	...	...		Surface.	22·5	22·6	1·02579	1·02765	1·02580
,,	...	...		400	8·6	21·4	1·02485	1·02636	1·02763
4	21 38	44 39	1900	Bottom.	1·9	22·1	1·02527	1·02699	1·02896
,,	...	...		300	12·5	20·0	1·02550	1·02664	1·02728
,,	...	...		Surface.	22·7	22·8	1·02589	1·02780	1·02587
,,	...	...		200	15·5	21·4	1·02626	1·02778	1·02778
5	21 1	46 29	1950	Surface.	23·5	23·7	1·02543	1·02758	1·02542
6	20 49	48 45	2325	Bottom.	1·7	21·0	1·02480	1·02621	1·02819
,,	...	...		Surface.	22·5	23·3	1·02566	1·02772	1·02582
,,	...	...		500	5·9	21·2	1·02579	1·02725	1·02892
,,	...	...		300	12·1	21·8	1·02488	1·02650	1·02725
7	20 39	50 33	2435	Bottom.	1·7	22·0	1·02586*	1·02754	1·02951
,,	...	...		200	15·5	21·6	1·02525	1·02682	1·02697
,,	...	...		500	6·6	20·5	1·02629	1·02756	1·02913
,,	...	...		400	8·3	20·9	1·02482	1·02620	1·02750
,,	...	...		300	12·2	21·1	1·02520	1·02665	1·02743
,,	...	...		Surface.	23·3	24·4	1·02536	1·02778	1·02578

Date 1873.	Latitude N.	Longitude W.	Depth at the Station.	d. Depth at which the sample was taken.	t. Temperature at d.	t'. Temperature at which the Specific Gravity was observed.	Specific Gravity at t'. Water at 4° = 1.	Specific Gravity at 15°·5. Water at 4° = 1.	Specific Gravity at t. Water at 4° = 1.
March 8	20 7	52 32	Fms.	Fms. Surface.	23 · 3C.	24 · 2C.	1·02538	1·02777	1·02567
,, 10	... 19 41	... 55 13	2385 2675	1370 Bottom.	2 · 5 1 · 6	22 · 6 23 · 8	1·02427 1·02402	1·02614 1·02626	1·02809 1·02823
,, 11	... 19 15	... 57 47		Surface. Surface.	23 · 3 23 · 9	24 · 4 24 · 9	1·02498 1 02480	1·02740 1 02739	1·02530 1·02508
,, 12	... 18 56	... 59 36	3000 2975	Bottom. Bottom.	2 · 0 2 · 2	22 · 9 22 · 6	1·02426 1·02545*	1·02623 1·02732	1·02820 1·02930
,, 13	... 18 54	... 61 28	3025	Surface. Bottom.	24 · 0 1 · 3	24 · 7 21 · 5	1·02485 1·02536*	1·02739 1·02690	1·02508 1·02888
,,	...	...		Surface	24 · 5	25 · 1	1·02431	1·02697	1·02455
,,	...	...		50	23 · 5	25 · 5	1·02446	1·02723	1 02518
,,	...	...		100	20 · 6	23 · 7	1·02527	1·02748	1·02634
,,	...	...		150	18 · 5	23 · 3	1·02550	1·02757	1·02692
,,	...	...		200	16 · 2	22 · 5	1·02503	1·02687	1·02687
,,	...	...		500	8 · 7	22 · 0	1·02450	1·02618	1·02788
,, 14	... 18 40	... 62 56	1420	Surface.	24 · 5	25 · 8	1·02423	1·02710	1·02430

CHAPTER IV.

ST. THOMAS TO BERMUDAS.

Dredging in moderate depths in the West Indian Seas.—New Blind Crustaceans.—Deep-sea Corals.—*Hyalonema toxeres.*—An accident.—A deep sounding.—The Miller Casella Thermometers.—Temperatures.—Arrival at Bermudas.—History of the Islands—Their general Appearance.—'Red' and 'blue' Birds. — The Corals which form the Reefs.—The Geology of Bermudas.—General Nelson's description. — Æolian Rocks. — Calcareous concretions simulating Fossils.—The Topography of the Islands—Their Products—Their Climate—Their Vegetation.

APPENDIX A.—Report from Professor Abel, F.R.S., to H. E. General Lefroy, C.B., F.R.S., on the Character and Composition of Samples of Soil from Bermudas.

APPENDIX B.—Abstract of Temperature—observations taken at Bermudas from the year 1855 to the year 1873.

ON Saturday the 15th of March, before going in to the harbour of St. Thomas, a sounding was taken in 450 fathoms off the Island of Sombrero. The bottom brought up by the sounding machine was globigerina mud largely mixed with broken shells, chiefly those of pteropods. The dredge was put over early and veered to 1000 fathoms. At noon it was hauled up half-filled with calcareous ooze. It was again sent down, and brought up in the afternoon with a like freight. These dredgings, which we did not regard as entering into the regular work of the sections, but which were only undertaken to give us a general idea of the deep-water fauna of the West Indian province, may be taken in connection with one or two hauls made with the same object and under the same circumstances, in water of nearly equal depths, on the 25th of March after leaving St. Thomas. The careful examination of the zone between 300 and 1200 fathoms among the West Indian Islands will undoubtedly add enormously to zoological knowledge. The objects of the present expedition do not of course include a detailed investigation of this kind, which must be done quietly in a small steamer by some one on the spot, and will require the patient work of several years. Even the few hauls of the dredge which we had it in our power to make brought to light a number of new and highly interesting animal forms representing nearly all the invertebrate groups. A thorough investigation of the belt must yield a wonderful harvest.

In dredging on the 15th, we got several sponges belonging to the hexactinellidæ, very closely allied to those which we had previously met with in moderately deep water off the coast of Portugal, showing that the distribution of this remarkable order in deep water is very wide. Several stony corals occurred, but of all of these, with the exception of a species of *Stylaster* which was very abundant at this station, we got better examples on a subsequent occasion. The *Stylaster* agrees very closely with the description and figure given by Pourtales of *S. complanatus*. The only marked difference is, that the primary and secondary septa do not unite to the same

PLATE VIII. *The Track of the Ship between St. Thomas, Bermudas, and Halifax.*

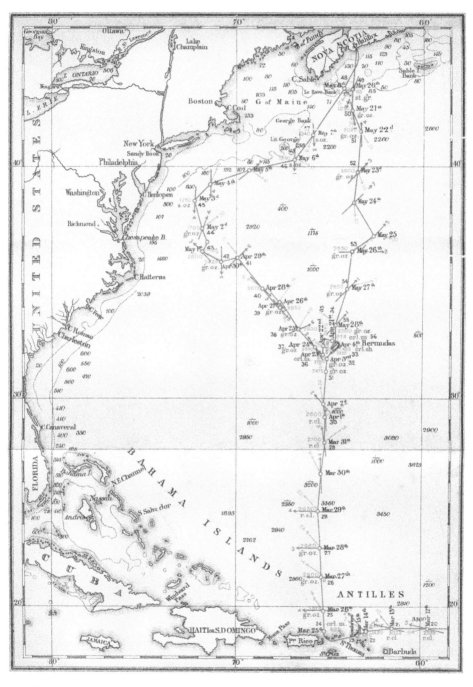

extent as shown in Count Pourtales's figure. The genus *Stylaster* is recent and widely distributed. One or two very elegant alcyonarian zoophytes, as yet undetermined, adhered to the tangles. The ECHINODERMATA were represented by the variety *abyssicola* of *Cidaris hystrix;* several fine star-fishes, among them a fine *Archaster*, a species of *Luidia*, and *Astrogonium longimanum*, a species described by Oerstedt from a specimen in the Hamburg Museum from an unknown locality; and some large ophiurids, including two species of *Ophiomusium*, and some undescribed forms.

In this dredging two very interesting crustaceans occurred, both belonging to the macrurous decapoda, and both participating in a singular deficiency—the total absence of eyes. One of these has been referred to the genus *Willemœsia* (GROTE). It agrees with the species described in the previous chapter in all its leading characters, although certain marked differences must lead to a slight modification of the characters of the genus as formerly defined. In *Willemœsia leptodactyla* all the five pairs of ambulatory legs bear chelæ, while it is a character of the typical astacidæ that chelæ are present on three pairs only. In the new species there are chelæ on four pairs of the ambulatory legs, the fifth pair ending in simple curved claws. The two species agree with one another, and with *Astacus*, in possessing a lamellar appendage at the base of the outer antennæ, and along with this they have the flattened carapace of *Palinurus*. These characters have not been hitherto observed in combination in any recent form, and their so occurring seems to be a more valuable generic character than the variable one of the form of the limbs. The character of the genus will now stand thus:—

Willemœsia, n. g.—Cephalothorax flattened, with a compressed, free, lateral margin. A lamellar appendage at the base of each of the outer antennæ. At least four pairs of ambulatory legs bear chelæ. No trace of eyes or of eye-stalks.

W. leptodactyla, v. W-S.—All the ambulatory feet bearing chelæ.

W. crucifer, v. W-S.—Four pairs of the ambulatory feet bearing chelæ.

The single example dredged of *Willemœsia crucifer* (Fig. 59) is a male. The total length is 42 mm. (cephalothorax 19 mm., and abdomen 23 mm.), and the extreme width of the cephalothorax 18 mm. The carapace is flattened and compressed laterally, and the flattened lateral portions are curved upwards like wings. The lateral margins are denticulated, and divided by two deeper incisions into three parts, the first bordered by seven, the second by four, and the third by seventeen teeth. The surface of the carapace is granulated, not spiny, as in *W. leptodactyla*. Two ridges crossing one another in the middle of the back divide the cephalothorax into four areas, the two anterior indicating the hepatic, and the posterior the branchial regions.

The abdomen consists of six segments and the telson. A ridge runs along the dorsal surface of each segment in the middle line, and rises on the first segment into one, and on the four succeeding segments into two spines, directed forwards. As in *W. leptodactyla*, not only are the eyes and eye-stalks absent, but there is no indication of a space for their accommodation in the position in which eyes are

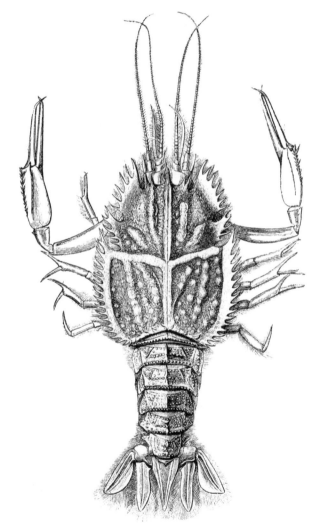

Fig. 59.—*Willemœsia cruciger*, v. W-S. x 2 (No. 23.)

normally developed. The antennæ are placed one

pair beneath the other. There is a lamellar appendage, which scarcely rises to the top of its basal segment, attached to the outer antenna. The flagella of the outer antennæ are 21 mm. in length, and the larger flagella of the inner antennæ nearly equal them. The parts of the mouth are normal. Four pairs of the ambulatory legs bear delicate chelæ sparsely spined along the upper edge, and the fifth pair end in a simple curved process. The first pair of ambulatory legs are not so long in proportion to the body or so slender as in *W. leptodactyla*. The first abdominal feet are style-like, and flattened at the end. The swimmerets have three joints, to the two first of which the palpi are attached. The telson and the caudal appendages are in no way remarkable. They, with the palpi of the swimmerets, the basal joints of the ambulatory legs, and the basal joints of the antennæ, are fringed with fine hairs.

Willemœsia crucifer certainly differs widely in general appearance from the recent Astacidæ, at the end of which family we should, however, be inclined to place it for the present. It has a very close resemblance to some fossil forms, particularly to the various species of the genus *Eryon*. It has been already remarked that *Willemœsia* in its flattened cephalothorax approaches the Palinuridæ; in all the living members of that family, however, the first pair of legs are monodactylous, while in *Willemœsia* they are didactylous. The fossil genus *Eryon* forms an exception in this particular among Palinurids, with which it has hitherto been arranged, and has the first pair of limbs didactylous, as in *Willemœsia*. It may also be noticed that *Eryon* is the only Palinurid which like *Willemœsia* has a lamellar appendage at the base of the outer antenna. It is very likely that when the recent deep-sea forms near the Astacidæ and Palinuridæ come to be carefully correlated with the cretaceous and jurassic species, it may be necessary to establish an additional family, the Eryonidæ.

The second crustacean, although having little of the facies of the typical *Astaci*, presents apparently no characters of sufficient value to warrant its separation from that genus.

Astacus zaleucus v. W-S. (Fig. 60), with its long, compressed cephalothorax, flattened abdomen, and unequal chelæ, has at first sight somewhat the appearance of a *Calianassa*.

The total length of the animal is 110 mm., the cephalothorax 50 mm., and the abdomen 60 mm. The carapace is hard and firm, though only slightly calcified. It is greatly compressed laterally, rising into a high arch. It terminates in front in a slender spiny rostrum 8 mm. in length. The rostrum is covered with a thick felting of hair, which extends backwards, forming two hairy triangles on the anterior part of the cephalothorax. In front of the carapace, between its anterior and upper edge and the insertions of the antennæ, in the position of the eyes in such forms as *Astacus fluviatilis*, there are two round vacant spaces, which look as if the eye-stalks and eyes had been carefully extirpated and the space they occupied closed with a chitinous membrane. The lamellar appendage of the outer antenna has teeth along its inner border. It extends to the middle of the second basal segment of the antenna, which is remarkably long. The flagella of

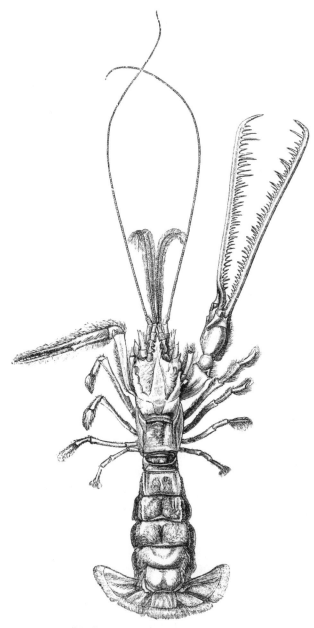

Fig. 60.—*Astacus zaleucus*, v. W·S. (No. 23.)

the outer antennæ are 130 mm in length. The inner antennæ originate on a line with the outer. The funiculus is shorter, and the flagella, which are equal in length, are much shorter than those of the outer antennæ.

The parts of the mouth are normal. The three first pairs of ambulatory legs are terminated by chelæ, the fourth pair bear recurved claws, and the fifth abortive stump-like claws. The chelæ of the first pair of legs are strangely developed, particularly the right chela, which is double the length of the left, and with its formidable ranges of long spines along the inner border of each claw reproduces on a small scale the jaws of the Gangetic gavial. The last segment of the pereion is not covered by the carapace, but is in movable connection with it. The first segment of the abdomen is very small, and the segments gradually increase up to the fourth, which the fifth and sixth equal in size. The abdominal segments are flattened from above downwards. The telson is quadrate, and combines with the two pairs of caudal appendages, which are widely expanded laterally to form the caudal fin. The dorsal surfaces of the second, third, and fourth abdominal segments, and the margins of the tail, are thickly covered with woolly hair. The individual being a male, the first pair of swimmerets consist of long, slender appendages, and the four succeeding pairs have one strong, round, basal joint, to which are attached two palpi fringed with hair. As has been already mentioned, there is some resemblance between this form and *Calianassa;* but in that genus the lamellar appendage to the outer antenna is absent, there are four pairs of limbs with chelæ instead of three, and the carapace is soft. To the genus *Astacus*, therefore, with which it has all characters in common except the great development of the right chela and the total absence of eyes—neither characters of generic value —the present species must be referred.

A. zaleucus v. W-S.

Rostrum spiny, elongated. Lamellar appendage of the outer antenna reaching to the middle of the second joint of the funiculus. Chelæ on three pairs of ambulatory feet, those on the first pair strongly but unequally developed. Cephalothorax very much compressed laterally. Eye-stalks and eyes entirely wanting.

On Sunday the 16th of March we anchored in the Gregaria channel at the entrance of the harbour of Charlotte Amalia. We spent a few very pleasant days at St. Thomas, some of the civilians of our party enjoying greatly their first experience of life and scenery within the tropics. M. Gardé, the Danish Governor, received us with the most friendly hospitality. He is a naval man, and was greatly interested in our investigations; and his aide-de-camp, Baron Eggers, had collected and worked out the plants of the island with care, and was otherwise well acquainted with its natural history. St. Thomas, like most of the West Indian Islands, has suffered greatly since the emancipation, from the difficulty of procuring free labour; and most of the sugar estates, formerly in high cultivation and a source of large revenue, are now waste and covered with a second jungle. The comparative prosperity of this island seems to be entirely due to the excellence of the harbour, which marks it out as the principal packet and coaling station for that part of the West Indies.

The natural history of the Island of St. Thomas is tolerably well known, and large collections of its fauna and flora have been sent home from time to time by many competent naturalists to the museum at Copenhagen. On the present occasion our time was much too limited to attempt to make collections, so the naturalists contented themselves with a little shallow-water dredging, and such a general survey of the island and its shores as might familiarize them with the more characteristic forms of animal and vegetable life; for while the Atlantic Islands, Madeira and the Canaries, although gradually assuming a more tropical character, maintain the most intimate relations in natural products with the south of Europe, in tropical America everything is changed, and it takes a little time to become familiar with new acquaintances, whom one has hitherto known, if he have known them at all, only from figures or descriptions, or at best mummied or pickled, or otherwise in inadequate effigy. Ophiurideans are particularly plentiful at St. Thomas, and we made large collections of these, particularly of the many large and characteristic West Indian species of the genus *Ophioderma*.

On the 24th of March we left the harbour of Charlotte Amalia, and proceeded with a light northeasterly breeze towards the Culebra passage. The next morning, the weather still continuing favourable, we sounded in 390 fathoms, the bottom globigerina mud, with fragments of coral and an unusually large proportion of the shells of pteropods of the genera *Cleodora*, *Diacria*, and *Styliola*. In the forenoon the large dredge was put over with an ample fringe of tangles. A fine specimen of *Brisinga endecacnemos*, and one or two very elegant alcyonarian zoophytes came up on the tangles. The dredge-bag was empty. Later in the day we sounded in 625 fathoms. The ooze was closer and more free from shells and coral than in the former haul, but otherwise much of the same character. This time the dredge came up about half-full, and on sifting its contents many interesting additions were made to our collections. There we met for the first time with the curious little crinoid *Rhizocrinus lofotensis*, for which we had been on the outlook since the beginning of the cruise, and *Salenia varispina*, which we now recognised as a very widely-distributed inhabitant of the deeper water.

A singular Gephyrean (Fig. 61), taken in this dredging, was referred by Dr. v. Willemœs-Suhm to Professor Lovén's genus *Chætoderma*. The length of the specimen was 52 mm., and its width 2 mm. The pharyngeal portion, with the proboscis, is smooth, and the remainder of the body is covered with delicate calcareous spines, directed backwards. The mouth is at the anterior extremity in a small, soft papilla, which, when the specimen was brought up, was inflated with a red liquid. The posterior extremity was abruptly truncated, and the two feathered appendages observed by Lovén were either

Fig. 61.—*Chætoderma nitidulum*, Lovén. (No. 24.)

absent or completely retracted. This species seems to be closely allied to if not identical with *C. nitidulum* (LOVÉN) from the coast of Sweden.

The corals, which were abundant in individuals, were all deep-water forms. They have been examined by Mr. Moseley, who refers the majority to species which have been described by M. de Pourtales [1] from the Strait of Florida. I abstract the following notice of the corals from Mr. Moseley's notes.

In the family Turbinolidæ we dredged two examples of a species of *Caryophyllia*, both dead, and one with a branch of *Stylaster* attached to the margin of the calicle. Five fine specimens of a solitary coral seem to agree with the figure and description of *Trochocyathus coronatus*, briefly described by Count Pourtales from a single imperfect specimen brought up on the lead by one of the U.S. coast survey parties from 460 fathoms, in lat. 30° 41′ N., long. 77° 3′ W. (Fig. 62). The corallum is circular in horizontal section, with a broad flat base. At its junction with the wall of the calicle the base is continued outwards into twelve stout, pointed tubercles, irregularly beset with small projections, the tubercles corresponding in position with the primary and secondary costæ. The base has thus, when viewed from beneath, an irregularly circular outline with a deeply-indented margin. In the centre of the base there is a conical projection, and at its summit a very small, oval, smooth space, the indication of a point of attachment. Twelve radiating ridges

---

[1] Illustrated Catalogue of the Museum of Comparative Zoology at Harvard College. No. IV. 'Deep-Sea Corals.' By L. F. de Pourtales. Cambridge (Mass.), 1871.

proceed from the base of the central projection, one to each of the tubercles, becoming more marked as they proceed outwards. The ridges are beset with small, pointed tubercles, which, with the ridges themselves, increase in size from the centre outwards. These small tubercles are arranged to some extent at regular intervals along the ridges, and there are traces of a series of concentric wavy lines, corresponding in position to the several rings of tubercles. These are evidently lines of growth, showing the

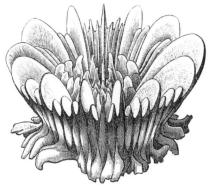

Fig. 62.—*Trochocyathus coronatus*, POURTALES, × 2. (No. 24.)

outline of the corallum at successive steps, the tubercules corresponding to these lines having been originally marginal, but having become nearly obliterated by fusion with other tubercles successively formed outside them, and sometimes entirely lost in the resulting ridge. In one specimen a second series of very delicate ridges are clearly marked, radiating outwards for a short distance from the base of the central cone, and corresponding with the secondary and tertiary costæ. The whole surface of the base of the corallum is covered with small, rounded, closely apposed granules. The large marginal tubercles of the base are from 4 to 5 mm. long, and about 4 mm. broad at their origin. Superiorly they are joined by slightly elevated, rounded ridges, the continuations into them of the primary and secondary costæ. The majority of the tubercles taper outwards, but some are obtuse. They terminate in three or four irregularly disposed spines. Two, or even three, marginal tubercles are sometimes fused together laterally into one mass.

From the region of origin of the marginal tubercles of the base, the wall of the calicle slopes upwards and outwards at an angle of 60° with the plane of the base, its height above the base being about 14 mm. The rounded ridges described as passing into the tubercles reach upwards to the margin of the calicle, and the external edges of the exsert primary and secondary septa are continued downwards along the middle of these costal ridges for about one-third the height of the wall of the calicle. The costæ of the third and fourth order are present as much smaller ridges separated by fine vertical grooves. The whole surface of the wall of the calicle is scattered over with small pointed granulations.

The arrangement of the septa is irregular. In the two smaller specimens there are six systems and four cycles. In the three larger specimens, in several of the large inter-septal spaces included between the primary and secondary septa, two septa are developed in addition to the usual three; there is thus a tendency in this species to form a fifth cycle of septa. The septa are complete, with the exception of those of the fourth and the partial fifth cycles. The primary and secondary septa are very prominently exsert, projecting vertically above the margin of the wall of the calicle 4 mm. Their edges are rounded, and they slope gradually down to the point where the pali take origin. The septa of the third and fourth orders are also exsert, but to a much less degree.

The primary, secondary, and tertiary septa are provided with pali, which form three circlets. The pali are very conspicuous and prominent, those of the third cycle being, as usual, the largest, and projecting to a height of 4 mm. above the level of the summit of the columella, while those of the primary cycle do not project more than 1·5 mm. The pali of the secondary septa are placed at a slightly greater distance from the centre of the columella than those of the primary; the pali thus form three rings or crowns.

The whole of the septa and pali are formed of thin but strong laminæ slightly thickened at the line of origin from the calicle, and terminating superiorly in sharp knife-edged margins.

The surfaces of the septa and pali are covered with very small granular projections, which, in the primary and secondary septa, are seen to be arranged with considerable regularity in radiating rows, which mark out a series of successive lines following the course of the margins of the septa, and representing their lines of growth.

The columella is about 4 mm. in diameter, and is composed of a number of contorted laminæ, finely granulated, and more or less fused together in the older individuals.

The extreme diameter of the largest specimen,

measured from the outer edges of the exsert septa is 32 mm. The height of the calicle is 16 mm.; total height to the top of the septa 20 mm.; diameter of the base of the calicle 21 mm.; of circlet of basal spines 28 mm. This is a very marked and handsome coral. Fresh specimens are of a pure white colour, and the crown-like shape is very elegant.

Thirteen specimens were procured of another pretty coral belonging to the Turbinolidæ. *Deltocyathus agassizii*, POURTALES (Fig. 63). This species has been also described and figured by Pourtales.[1]

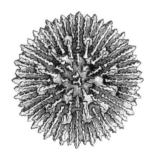

FIG. 63.—*Deltocyathus agassizii*, POURTALES, x 4 (No. 24).

Our specimens vary in diameter from 4 to 13 mm. Dr. Martin Duncan considers this species identical with *D. italicus*, and the distinctions between the two species seem very critical. M. de Pourtales considers that the recent differs from the fossil species in the point of junction in the pali of the second and third cycles not being exsert, and in the V, or Δ, not being so prominent; and, further, in the costæ being covered with fine, sharp granulations, while in *D. italicus* they are " composed of series of

[1] Loc. cit. p. 14.

very regular granules." The columella is also less developed in the recent than it is in the fossil form. All these characters are minute, and may depend, to a certain extent, upon the condition of the specimens; but it seems to be best on the whole to retain the name given by M. de Pourtales for the present.

A surveying party, sounding from a boat in 200 fathoms off Bermudas, brought up in the cup-lead a very beautiful specimen of a variety of this species described by Count Pourtales, in which the primary costæ are large, and prolonged beyond the margin of the calicle. Pourtales's specimen was imperfect; ours was finely preserved, and the horn-like appendages were developed to a remarkable degree. The diameter of the calicle is 9 mm.; the length of the horns 3·5 mm., or more than one-third the diameter of the calicle; they are slender and rounded, and they taper to a fine point. A smaller horn is developed in relation with one of the secondary costæ, but the remaining secondary costæ show no tendency to elongation. The horned or stellate variety of *D. agassizii* appeared at first to Pourtales to present good specific characters, but the examination of intermediate forms showed that it could not be regarded as specifically distinct. Our Bermudas specimen, which is in most excellent preservation, shows some other points of interest; the pali as a whole project more prominently above the general elevation of the septa than they do in the unarmed variety, and the Δ formed by the junction of the secondary with the tertiary pali is prominent and conspicuous, in this respect approaching the fossil species, *D. italicus*.

Among the Oculinidæ, we obtained from both localities abundant examples of *Lophohelia carolina*, POURTALES. A large number of these were living; they were of various sizes and ages; some of the specimens were 18 centimetres in length and some of the older stems 1 centimetre in diameter.

Some dead but very fresh and perfect specimens of *Fungia symmetrica*, POURTALES, came up at this station. These resemble *Deltocyathus* greatly at first sight in size and general appearance, but they may be readily distinguished by a certain difference in form, and by the very evident synapticulæ uniting the septa.

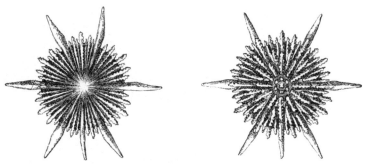

FIG. 64.—*Deltocyathus agassizii*, POURTALES. Stellate variety. From a depth of 200 fathoms near Bermudas.

Several beautiful specimens were procured in the haul of a coral which we had taken before at Station 3 at a depth of 1525 fathoms, *Cryptohelia pudica*, MILNE-EDWARDS. The genus *Cryptohelia* was established by Milne-Edwards and Haime for a stylasteracean obtained from New Guinea. The Stylasteraceans are remarkable in having their septa all equal, and as has lately been observed by Sars in the case of *Allopora*, the only member of the group in which the soft parts are known, the tentacles lie between the (spurious?) septa. Mr. Moseley observed the same fact in a stylasteracean nearly allied to *Cryptohelia* dredged in 500 fathoms off the Meangis Islands. The genus

Fig. 65.—*Cryptohelia pudica*, Milne-Edwards. Twice the natural size. (No. 24.)

Cryptohelia has its branches disposed in a uniform vertical plane, with the calicles all directed towards one face of the plane. The coral tissue is unusually dense and white; a thin lamina of coral tissue produced from the margin of each calicle on one side projects in front of the mouth of the calicle and forms

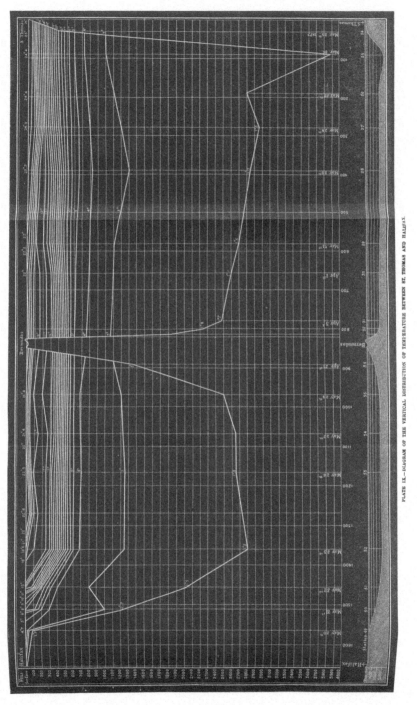

PLATE IX.—DIAGRAM OF THE VERTICAL DISTRIBUTION OF TEMPERATURE BETWEEN ST. THOMAS AND HALIFAX.

The material originally positioned here is too large for reproduction in this reissue. A PDF can be downloaded from the web address given on page iv of this book, by clicking on 'Resources Available'.

a kind of shield or operculum. M. de Pourtales has described a series of specimens of *Cryptohelia* obtained in 270 fathoms water off Bahia, under the name of *C. piercei*, but after an examination of M. de Pourtales' specimens and of the more abundant material obtained by the 'Challenger' Mr. Moseley entertains little doubt that his and our specimens belong to the same species, again identical with *C. pudica* of Milne-Edwards. The specimens vary greatly in size and in the amount of development of the characteristic opercular fold; the specimen figured is the largest and most fully developed which we obtained, and the one in which the operculum was most marked. The whole group of Stylasteraceans requires careful revision, when very likely it may be found necessary to merge *Cryptohelia* and *Endohelia* in *Stylaster*.

Two examples of the sponge-body of a very handsome *Hyalonema* were sifted out of the coral-mud. Unfortunately, in both cases the sponge had been torn from the central coil, and the absence of the coil might have thrown some little doubt upon the form and mode of finish of the complete animal; so that it was extremely fortunate that a young specimen of the same species, about 40 mm. in length, was caught in the tangles quite perfect.

Hyalonema toxeres, Wy. T., resembles closely the other known species, *H. lusitanicum* and *H. sieboldi*, in general appearance and in the arrangement of its parts. A more or less funnel-shaped sponge presents two surfaces, covered with a network of different patterns formed by varying arrangements of large, five-rayed spicules. The upper concave surface shows a number of oscular openings, irregularly arranged;

T

and the lower surface a more uniform network of pores, some of which seem to be inhalent and others exhalent. The central axis of this sponge is closely warped into the upper part of a coil of long and strong glassy spicules, which, as in the other species, serve to anchor the sponge in the soft mud. Both of the specimens dredged have the sponge more flattened and expanded than it is in *H. lusitanicum*. In

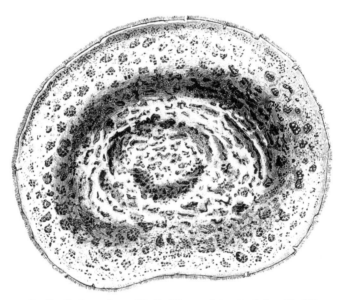

FIG. 66.—*Hyalonema toxeres*, WY. T. Upper surface, natural size. (No. 24.)

one of them it is nearly flat (Fig. 66), forming a uniform cake-like expansion, 80 mm. in length, by 70 mm. in width, and about 8 mm. in thickness. The upper or oscular surface is covered by an exceedingly close network, with groups of large openings at nearly equal intervals. It is slightly raised in the centre. The central elevation is followed by a slight depression, and the upper wall then passes out nearly horizontally to a sharp peripheral edge, fringed with long, delicate spicules, each consisting of a slender central shaft, with a cross of four short transverse processes in the centre. The outer half of the central axis is delicately feathered. The lower surface of the sponge (Fig. 67) is protected by a singularly elegant network of sarcode, with wide oval

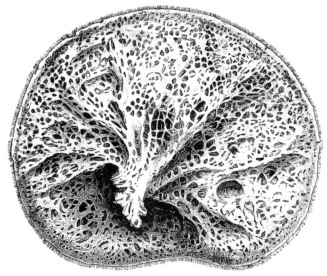

Fig. 67.—*Hyalonema toxeres*, Wy. T. Lower surface of the sponge, natural size. (No. 24.)

and round meshes radiating irregularly from a central point. The membrane is traversed by irregularly radiating ridges of firmer substance, which unite in the centre in a projecting boss at the point where in this specimen the 'glass-rope' has unfortunately been torn out.

In minute structure, *Hyalonema toxeres* corresponds in all essential respects with *H. sieboldi* and

T 2

H. lusitanicum. All the spicules are of the same ground-forms, with some little differences in detail, with the exception of one remarkable spicule which enters largely into the structure of *H. toxeres*, and serves to distinguish even the smallest portion of it. This is a large spicule, the largest above a centimetre in length, and more than half a millimetre in width in the centre, shaped like a bow or boomerang. These spicules are distributed in all parts of the sponge, and are particularly abundant near the insertion of the coil. No analogous form occurs in the other species of *Hyalonema*.

The large *Amphidisci* are much larger than in any other known sponge. They are upwards of half a millimetre in length, and visible to the naked eye— twice as large as in *H. lusitanicum*. The feathered shafts of the five-rayed spicules which fringe the openings are longer than in the other species, and the rays of the cross are much shorter (Fig. 68).

The second specimen of the sponge-body agreed with the one described in all essential points of structure, but was more conical in form. The young specimen (Fig. 69) differed from the young of *H. lusitanicum* of the same age in being wider and more cylindrical, but the external wall, which afterwards becomes that of the lower surface, showed the same arrangement in squares which we find in the young of the other species, so that apparently the graceful, round-meshed, wide netting of the under surface does not appear in the early stages.

The coil is developed much in the same proportion and in the same way as in *H. lusitanicum*, the fibres spreading out and incorporating with the sponge substance. The characteristic bow-like spicules are abundant in the young sponge, and these, with the large *Amphidisci*, place it beyond a doubt that it is the young of *H. toxeres*.

A quantity of loose spicules brought up in the dredge at the same time were referred to this species.

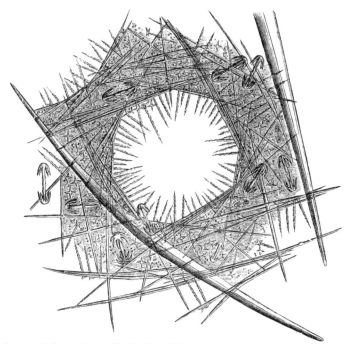

Fig. 68.—*Hyalonema toxeres*, Wy. T. Part of the membrane from the upper surface, x. 40. (No. 24.)

They were somewhat stouter than those of *H. lusitanicum*, and less regular in outline. There was one coil nearly complete, involved in a calcareous expansion of a branch of *Diplohelia profunda*. Two very young polyps, apparently of *Palythoa fatua*, were commencing the formation of their investing crust at the top of the coil of the young specimen, just below the sponge-body.

A sad accident occurred during this dredging, which threw a temporary gloom over our party, and brought

Fig. 69.—*Hyalonema toxeres*, Wy. T. A young specimen, x. 2. (No. 24.

home to us very forcibly the critical nature of our work. The large iron dredge which we were using in preference to the trawl, the ground being rather rough, caught upon a rock or a mass of coral, and brought a sudden strain upon the dredge-rope; and before the rope could be veered, or any other steps taken to relieve the strain, the hook of the foremost span carried away, and the leading block which was hooked to it flew back and struck William Stokes, one of the sailor lads, with such violence that he was driven against the ship's side. His thigh was broken in two places, and he was so seriously injured otherwise that he never recovered consciousness, and died a few hours afterwards. He was buried the following day, and singularly enough, just before joining in the solemn service which " committed his body to the deep," we had ascertained that his grave was in the very deepest spot which had ever been fathomed in the ocean. His death is recorded on a cross in the crowded little burial-ground in Ireland Island, Bermudas, with the fitting legend,—" In the midst of life we are in death."

On Wednesday, the 26th of March, we sounded (Station 25) in lat. 19° 41′ N., long. 65° 7′ W., nearly 90 miles north of St. Thomas, in 3,875 fathoms. The bottom brought up in the 'hydra' tube was reddish mud, containing, however, a considerable quantity of carbonate of lime. The colour and composition of the mud was not uniform. The upper layer—that which had been forced farthest into the tube—was much redder than that which was nearest the mouth of the tube, and which had consequently come from a greater depth.

This is a phenomenon with which we are now very familiar, particularly among the red clays from great depths. It seems to be due to some deoxidizing process taking place slowly, and consequently visibly affecting only those deeper layers which have been deposited for a considerable length of time.

As to the determination of depth, this sounding was perfectly satisfactory. Being still near the Islands, we did not expect the depth to be nearly so great as it was, and the 'hydra' tube was weighted with 3 cwt. only; but the sea was smooth and the weather fine, and it was easy to keep the ship nearly stationary. The following table gives the time-intervals in the running out of the line.

Fathoms.	Time.			Interval.		Fathoms.	Time.			Interval.	
	H.	M.	S.	M.	S.		H.	M.	S.	M.	S.
500	6	36	17	...		2400	7	9	38	2	14
600	6	37	16	0	59	2500	7	11	53	2	15
700	6	38	29	1	13	2600	7	14	11	2	18
800	6	39	48	1	19	2700	7	16	26	2	15
900	6	41	12	1	24	2800	7	18	37	2	11
1000	6	42	48	1	36	2900	7	20	59	2	22
1100	6	44	25	1	37	3000	7	23	26	2	27
1200	6	46	3	1	38	3100	7	25	45	2	19
1300	6	47	45	1	42	3200	7	28	11	2	26
1400	6	49	32	1	47	3300	7	30	38	2	27
1500	6	51	20	1	48	*3400	7	33	5	2	27
1600	6	53	12	1	52	3500	7	41	40	—	
1700	6	55	7	1	55	3600	7	44	20	2	40
1800	6	57	5	1	58	*3700	7	47	12	2	52
1900	6	59	2	1	57	3800	7	57	22	—	
2000	7	1	6	2	4	3900	8	0	19	2	57
2100	7	3	8	2	2	...		...		{ At the rate of	
2200	7	5	13	2	5	3925	8	1	18	3	56
2300	7	7	24	2	11	3950	8	2	23	4	20

It will be seen from the table that the intervals between 3,200, 3,300, and 3,400 fathoms are the same, while they ought to have been regularly increasing. Captain Nares thought that the sounding machine had probably reached the bottom, and that a uniform under-current might be dragging out the line. He accordingly commenced heaving in, but the strain on the accumulators at once showed that the weights on the 'hydra' had not detached. At 3,700 fathoms another attempt was made to heave in, but the weights were still there. Close to the 3,900 fathom mark the line suddenly came almost to a stop—50 fathoms more were let out, and the time taken at intervals of 25 fathoms, and the complete change of rate at once showed that the instrument was on the bottom. On reeling up it was evident from the decreased strain that the 'hydra' tube had been relieved of the weights, and was coming up with the instruments attached to it alone.

Two thermometers were sent down in this sounding, and a slip water-bottle. The thermometers were broken, and as the mode in which the fracture occurred is in itself curious, and has an important bearing upon the use of these instruments at extreme depths, I will briefly describe the condition of the thermometers when they came to the surface.

No. 39, a valuable instrument with a small and constant error, which we had used for some time whenever for any reason we required extreme accuracy, was shattered to pieces (Fig. 70 A).

In No. 42 the instrument was externally complete, with the exception of a crack in the small unprotected bulb on the right limb of the U-tube. The inner shell of the protected bulb was broken to pieces (Fig. 70 B).

In both of these cases there seems little doubt that the damage occurred through the giving way of the unprotected bulb. In No. 39 the upper part of that bulb was ground into coarse powder, and the fragments packed into the lower part of the bulb and the top of the tube. The large bulb and its covering shell were also broken, but into larger pieces, disposed as if the injury had been produced by some force acting from within. The thermometer tube was broken through in three places; at one of these, close to the bend, it was shattered into very small fragments. The creosote, the mercury, and bubbles of air were irregularly scattered through the tube, and it is singular that each of the steel indices had one of the discs broken off. The whole took place no doubt instantaneously by the implosion of the small bulb, which at the same time burst the large bulb and shattered the tube.

In No. 42 a crack only occurred in the small bulb, either through some pre-existing imperfection in the glass or from the pressure. When the pressure became extreme the crack yielded a little and the sea-water was gradually forced in, driving the contents

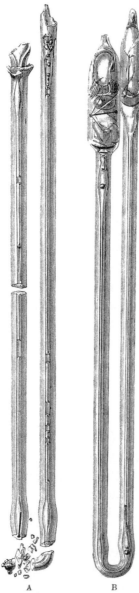

Fig. 70 —Thermometer tubes broken by pressure at Station 25. A, Thermometer No. 39; B, No. 42.

of the thermometer before it, and, taking it at a disadvantage from within, breaking the shell of the large bulb, which was unsupported on account of the belt of rarefied vapour between it and its outer-shell. The pressure was now equalized within and without the instrument, and the injury went no farther. Alcohol, creosote, mercury, and sea-water were mixed up in the outer case of the large bulb with the débris of the inner bulb, and one of the steel indices lay uninjured across the centre of it.

It now becomes an important question why the thermometer should give way at that particular point, and one still more important, how the defect is to be remedied. At first sight it is difficult to imagine why the small bulb should give way rather than the outer shell of the large one. The surface exposed to pressure is smaller, the glass is thicker, and it is somewhat better supported from within, as the tube is nearly filled with fluid under the pressure of an atmosphere. I believe the cause must be that the end of the small bulb is the last point of the instrument heated and sealed after the tube is filled with liquid, and that, consequently, the annealing is imperfect at that point. It is evidently of no use to protect the small bulb in the same way in which the large bulb is protected. The outer shell is merely a precaution to prevent the indications being vitiated by the action of pressure on the elastic bulb. Against crushing, it is no protection; it is rather a source of weakness, from its greatly increasing the exposed surface. The only plan which seems to be feasible is to thicken the small bulb itself, and, if possible, to improve its temper. It is only fair to say that these thermometers were tested and guaranteed to only three tons on the square inch, and that the pressure to which they were subjected was equal to four tons. The water-bottle appeared to have answered its purpose, although the wooden plug closing an air-vent in the tube had succumbed to the pressure, and had been washed in its contracted state out of its place.

Mr. Buchanan finds that the bottom water has a specific gravity slightly greater than usual at great depths (see Appendix C to Chapter V.), but not materially so. The amount of carbonic acid is somewhat in excess.

As this was the deepest sounding which we had taken, we were anxious to try whether the dredge would still prove serviceable. The small dredge was accordingly lowered with the usual bar and tangles, and from the centre of the bar a 'hydra' sounding-tube, weighted with 4 cwt., was suspended about two fathoms behind the dredge. A 2-inch rope was veered to 4,400 fathoms; a toggle was stopped on the rope 500 fathoms from the dredge, and when the dredge was well down, two weights of 1 cwt. each were slipped down the rope to the toggle. We commenced heaving in about 1.30, and the dredge came up at 5 P.M., with a considerable quantity of reddish-grey ooze, mottled like the contents of the sounding-tube. The bluer portion effervesced slightly with acids, the redder scarcely at all. The mud was carefully examined, but no animals were detected, except a few small foraminifera with calcareous tests, and some considerably larger of the arenaceous type. This dredging, therefore, only confirms our previous conviction, that very extreme depths, while not inconsistent with the existence of animal life, are not favourable to its development.

In the afternoon a series of temperatures were taken at intervals of 100 fathoms from the surface to 1,500 :—

Surface	. . . .	24°·5 C.		800 fathoms	. .	3°·6 C.
100 fathoms	. .	20·3		900 ,,	. . .	——
200 ,,	. . .	16·7		1000 ,,	. . .	3·1
300 ,,	. . .	12·1		1100 ,,	. . .	2·9
400 ,,	. . .	8·4		1200 ,,	. . .	2·9
500 ,,	. . .	6·4		1300 ,,	. . .	2·8
600 ,,	. . .	5·0		1400 ,,	. . .	2·5
700 ,,	. . .	——		1500 ,,	. . .	2·4

The curve constructed from this series indicates a very rapid and uniform fall of about 20° C. during the first 600 fathoms, and generally a distribution of temperature almost identical with that of some of the later stations on the section from Santa Cruz to Sombrero. (Curve No. 25, Fig. 71.)

In the evening we pursued our course northwards under all plain sail.

On the following day we sounded in much shallower water—2,800 fathoms; the bottom was much of the same character: and on the 28th in 2,960 fathoms, with a like result: but at our next sounding in 2,850 fathoms on the 29th, the calcareous element in the mud had almost entirely disappeared; and the contents of the tube seemed to be identical with the 'red clay' which occupied so large a part of our first section.

A temperature-sounding was taken at this station at every 100 fathoms from the surface to 1,500 (curve No. 28, Fig. 71), and in the afternoon the operation

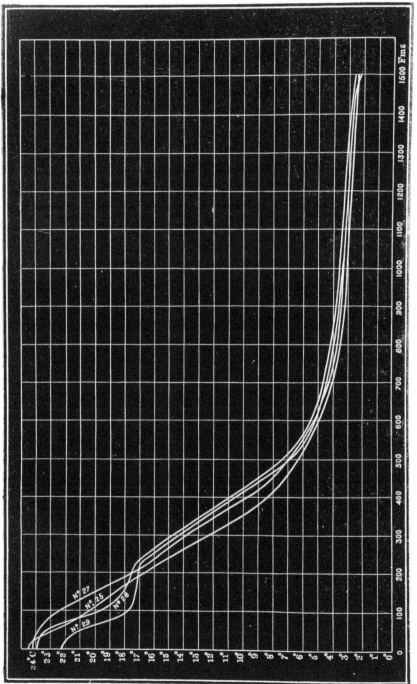

Fig. 71.—Curves constructed from Serial Temperature Soundings between St. Thomas and Bermudas.

was repeated at intervals of 50 fathoms, with the result of verifying the temperatures taken earlier in the day, which seemed to indicate the presence of a mass of water nearly 200 fathoms thick, at a temperature ranging from 16° C. to 18° C., bringing down the isothermal line of 16° C. on Pl. IX. from the 200 to the 300 fathom line and producing a hump on the curve deduced from the sounding somewhat resembling that which we had traced along the west coasts of Europe.

On Monday the 31st, the weather still continuing very pleasant, with a light south-easterly wind, we sounded and dredged in 2,700 fathoms. The dredge brought up a quantity of reddish-grey mud of the usual character. Carbonate of lime was in some quantity, and the clay contained a number of the fresh shells of calcareous miliolines. The only higher animal sifted out was a scarlet caridid shrimp. A serial sounding (curve No. 29, Fig. 71) gave a still more decided indication of a stratum of warm water between the 1-and 300 fathom lines.

Surface	. . . .	22°·2 C.
100 fathoms	. . . .	17·7
200 ,,	. . . .	17·0
300 ,,	. . . .	14·3
400 ,,	. . . .	10·6
500 ,,	. . . .	6·8
600 ,,	. . . .	—
700 ,,	. . . .	4·7
800 ,,	. . . .	4·1
900 fathoms	. . . .	3°·5 C.
1000 ,,	. . . .	3·4
1100 ,,	. . . .	3·1
1200 ,,	. . . .	—
1300 ,,	. . . .	2·8
1400 ,,	. . . .	2·8
1500 ,,	. . . .	2·4
Bottom	. . . .	1·6

Two additional series of temperature were taken at this station, one at intervals of ten fathoms down to one hundred, and a second at the odd fifties down to four hundred and fifty. These soundings completed and entirely corroborated the first series.

10 fathoms	. . .	22°·8 C.	60 fathoms	. .	20°·3 C.
20 ,,	. . .	22·1	70 ,,	. . .	19·6
30 ,,	. . .	21·9	80 ,,	. . .	19·2
40 ,,	. . .	21·2	90 ,,	. . .	18·0
50 ,,	. . .	20·9	100 ,,	. . .	18·3

50 fathoms	. .	21°·0 C.	350 fathoms	. .	12·4 C.
150 ,,	. . .	17·6	450 ,,	. . .	8·2
250 ,,	. . .	16·1			

Slip water-bottles were used to bring up samples of water from different depths for the determination of the specific gravity and for analysis.

On the 1st of April the mud brought up from 2,600 fathoms was the same as that dredged the day before, and it seemed to give so little promise of a successful haul that we did not repeat the operation. The second gig was sent out to bring in some of the patches of gulf-weed which were passing the ship in large numbers, and to collect surface-animals.

On the 2nd of April, at a distance of 134 miles from Bermudas, a series of temperature-soundings was taken at intervals of 20 fathoms from the surface to 300 fathoms, with the following result:—

Surface	. . . .	20°·9 C.	160 fathoms	. .	17°·9 C.
20 fathoms	. . .	19·5	180 ,,	. . .	17·7
40 ,,	. . .	19·4	200 ,,	. . .	16·7
60 ,,	. . .	19·3	225 ,,	. . .	17·4
80 ,,	. . .	18·9	250 ,,	. . .	16·8
100 ,,	. . .	19·0	275 ,,	. . .	16·5
120 ,,	. . .	18·6	300 ,,	. . .	15·9
140 ,,	. . .	17·7			

It will be seen by reference to Plate IX. that not only does the layer of super-heated water maintain its position, but its temperature actually rises, notwithstanding the more northern latitude. The lines representing 18°, 17°, and 16° C. are depressed, each nearly 50 fathoms; and a subsequent observation (Station 57), about fifteen miles from Bermuda, shows that the same conditions of temperature are maintained right up to the islands.

On the 3rd of April soundings were taken successively in 2,475, 2,250, 1,820, and 950 fathoms, gradually passing up the slope of the reef. The bottom in all cases had a basis of soft, white, calcareous mud, evidently produced by the disintegration of the Bermuda reef and of the multitude of pteropod shells which sink down from the surface. At a distance of ten miles or so from the reef the soundings are sometimes actually composed of the fragments of surface-shells to the almost entire exclusion of the homogeneous detritus of the coral.

The next day we sounded at various depths from 780 to 120 fathoms. The fauna seemed to be on the whole scanty, the finely-divided calcareous mud being probably unfavourable to the existence of most of the higher forms of animal life. Among the few interesting species which we met with at this station were a fine specimen of the Euryalid *Ophionereis lumbricus* (LYMAN), attached to a gorgonia; a large, handsome *Spatangus*, allied to *S. purpureus;* and some fragments of *Cœlopleurus floridanus* (A. AGASSIZ), a very singular urchin, near Cidaris in some respects, but with the spines enormously long in proportion to the body.

A pilot came on board off St. George's, and we

passed slowly through the intricate and dangerous 'narrows' between the reefs, the natural defences of the northern coast of Bermudas which make any artificial fortifications almost unnecessary, and anchored in Grassy Bay in the evening.

Bermudas or 'Somers,' or, by corruption, 'The Summer Islands,' seems to have been discovered about the year 1503, by Juan Bermudez, a Spaniard, in the vessel 'La Garza,' having on board Oviédo, the well-known author of the history of the West Indies. Oviédo says, addressing the emperor Charles V., "In the year 1515, when I first came to inform your Majesty of the state of things in India, I observed that in my voyage when to windward of the Island of Bermudas, otherwise called 'Gorza,' being the most remote of all the islands yet found in the world, I determined to send some of the people ashore, both to search for what might be there and to leave certain hogs upon it to propagate. But on account of a contrary wind I could not bring the ship nearer than cannon-shot."

The first English printed account of Bermudas is by Henry May, a sailor, who was wrecked there in 1593, in a French ship commanded by M. de la Barbotier. May states that he and the French crew found on the island many hogs, but these so lean as to be unfit for food, and abundance of birds, fish, and turtle. By good luck the chests of carpenter's tools were saved from the wreck, along with some sails and rigging, and May and his companions contrived to build a vessel of considerable size of the native cedar, in which, after remaining about five months on the islands, they stood for the banks of Newfoundland.

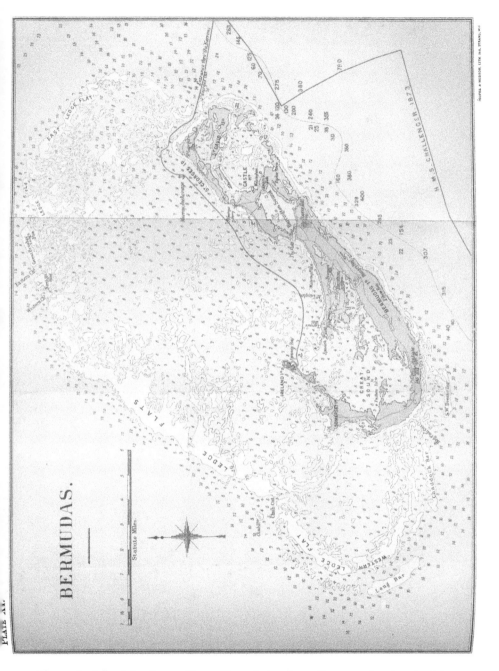

The material originally positioned here is too large for reproduction in this reissue. A PDF can be downloaded from the web address given on page iv of this book, by clicking on 'Resources Available'.

"Here they met with many ships, but none of them charitably inclined towards them, when it pleased God they fell in with 'the honest English barque "Fawmouth,"' which received them on board. While in this vessel they 'tooke' a French ship, into which Captain de la Barbotier and his seamen were transferred; May himself remaining with the English vessel, which arrived at Falmouth in August, 1594." [1]

The next we hear of Bermudas is from an account by one of her crew of the wreck of the 'Sea Adventure,' in the year 1609.

The 'Sea Adventure' was one of a small fleet dispatched from London to convey the newly-appointed Governor, Sir Thomas Gates, Admiral Sir George Somers, and some other officials, to the young colony of Virginia.

On Monday, July 24th, St. James's day, when they reckoned themselves within seven or eight days' sail of Cape Henry, "the clouds began to thicken around, and a dreadful storm commenced from the northeast, which, swelling and roaring as it were by fits, at length seemed to extinguish all the light of heaven and leave utter darkness. The blackness of the sky and the howling of the winds were such as to inspire the boldest of our men with terror, for the dread of death is always more terrible at sea, as no situation is so entirely destitute of comfort or relief as one of danger there."

After seeing St. Elmo's fires on the rigging, springing a leak, and undergoing every possible trial, moral

[1] 'The Naturalist in Bermuda,' by John Matthew Jones, Esq., of the Middle Temple. London, 1859.

and physical, for five days, Sir George Somers at length sighted land, and the wind lulling a little, they ran their ship ashore, where she became a complete wreck.

"We now found that we had reached a dangerous and dreaded island, or rather islands, called the Bermudas, considered terrible by all who have touched at them; and from the dreadful tempests, thunders, and other alarming events prevailing, are commonly named the Devil's Islands." Sir Thomas Gates, Sir George Somers, and their companions, found the islands totally uninhabited, but capable of yielding abundance of food. Hogs which had been set adrift by some earlier visitors, whose names have been lost, were so numerous that thirty-two were brought in by a party after one day's hunting. Fish abounded on the coasts, and were caught with the greatest ease; and turtles added daily to the luxury of their fare. "It is such a kind of meat as a man can neither absolutely call fish nor flesh; the animal keeps chiefly in water, feeding on sea-grass like a heifer, in the bottom of the coves and bays; and the females lay their eggs, of which we found five hundred at a time on opening a she-turtle, in the sand of the shore. They are covered close up, and left to be hatched by the sun."

The party found the islands so pleasant and so productive, and their ideas about Virginia were so vague, that there was a very general disposition to remain where they were; and for nearly a year, during which they were constructing vessels to continue their voyage, the governor and the admiral had great difficulty in keeping their party together, and in suppressing conspiracies to obstruct their work, and to prevent their further progress.

Early in May, 1610, however, Sir George Somers had completed his pinnace, the 'Patience,' and on the 10th of May the little party set out, and about midnight of the 18th they " were sensible of a charming odour from the land resembling that from the coast of Spain, near the Straits of Gibraltar."[1] They reached Fort Algernon in safety, but they found the Virginian colony so badly off for provisions that Sir George Somers volunteered to return to Bermudas for supplies; and during that trip he died near the site of the present town, St. George's, where there is a monument erected to his memory. His nephew, Captain Matthew Somers, carried his uncle's body to England in the old cedar pinnace. Owing to Captain Somers' representations, a company was formed in England to colonise Bermudas, and in 1612 the first party of settlers arrived, under the charge of Governor Richard More.

Since that time Bermudas has been a British colony, though perhaps not a very successful one; and latterly an important naval and military station. During the earlier part of its history, Bermudas was intimately connected with Virginia, and the account given of it in Smith's 'History of Virginia' is at once so quaint and so generally correct that I cannot refrain from giving a somewhat lengthy extract :—

"Before we present you the matters of fact, it is fit to offer to your view the Stage whereon they were acted, for as Geography without History seemeth a carkasse without motion, so History without Geography

[1] 'Shipwrecks and Disasters at Sea.' London, 1846.

wandreth as a Vagrant without certaine habitation. Those Islands lie in the huge maine Ocean, and two hundred leagues from any continent, situated in 32 degrees and 25 minutes of Northerly latitude, and distant from *England*, West-South-West, about 3,300 miles, some twenty miles in length, and not past two miles and a halfe in breadth, enuironed with Rocks, which to the North-ward, West-ward, and south-East extend further than they have been yet well discouered: by reason of those Rocks the Country is naturally very strong, for there is but two places, and scarce two vnlesse to them who know them well, where shipping may safely come in, and those now are exceeding well fortified, but within is roome to entertaine a Royall Fleet: the Rocks in most places appeare at a low water, neither are they much couered at a high, for it ebbs and flowes not past fiue foot; the shore for most part is a Rocke, so hardened with the sunne, wind and sea, that it is not apt to be worne away with the waues, whose violence is also broke by the Rocks before they can come to the shore: it is very vneuen, distributed into hills and dales, the mold is of diuers colours, neither clay nor sand, but a meane between; the red which resembleth clay is the worst, the whitest resembling sand and the blackest is good, but the browne betwixt them both, which they call white, because there is mingled with it a white meale, is the best; vnder the mold two or three foot deep, and sometimes lesse, is a kinde of white hard substance which they call the Rocke: the trees vsually fasten their roots in it, neither is it indeed Rocke or stone, or so hard, though for most part more harder than chalke; nor so white, but pumish-like and spungy, easily recieuing and containing much water. In some places Clay is found under it, it seemes to be engendered with raine water draining through the earth, and drawing with it of his substance vnto a certaine depth where it congeales; the hardest kinde of it lies under the red ground like quarries, as it were thicke slates one upon another, through which the water hath his passage, so that in such places there is scarce found any fresh water, for all or the most part of the fresh water commeth out of the sea draining through the sand, or that substance called the Rocke leaving the salt behinde, it becomes fresh."

Representative government was introduced in Bermudas so early as the year 1620, and in 1621 the Bermudas Company of London, in whom the government was at that time vested, issued a liberal charter. That charter remained in force only till 1685, when, probably on account of the importance of the islands as a military station, it was annulled by the Home Government; and since then the governors have been appointed by the Crown, and the laws of the colony have been enacted by a legislature consisting of the governor and nine members of council appointed by the Crown, and thirty-six members of assembly elected by the nine parishes into which the islands are divided. Slavery appears to have existed in Bermudas from the first in a mitigated and patriarchal form. The legislative bodies of Bermudas and of Antigua were the only two among our colonies which abolished slavery without the intervention of apprenticeship. The proportion received by Bermudas of the compensation voted by Parliament was 50,584*l*.—27*l*. 4*s*. 11*d*. for each of 4,203 slaves. The number of the civil population in 1871 was 12,426, of whom 5,030 were white, and 7,396 coloured. The coloured element in Bermudas is by no means entirely African. In the earlier days of the settlement many labourers and slaves were brought from Virginia and other parts of North America; and one may often recognise the aquiline nose and characteristic features of the North American Indian, now, however, except in one or two families, very much masked by negro intermixture.

Approaching the islands from the southward, their general effect is somewhat sombre. The land is low, rising nowhere to a height greater than two hundred and sixty feet, and by far the greater part forming gentle undulations at a height of from twenty to sixty feet above the sea level.

Although very valuable crops are raised, it is by a system of market-gardening in isolated patches rather than by agriculture, and the islands cannot be said to be generally or uniformly cultivated. A great part of the higher land is covered with a natural pasture of inferior grasses, mixed with a low scrub of what they call wild sage, a species of *Lantana*, which has been introduced in comparatively late times, and has spread in a wonderful way, so that it is now a perfect nuisance. The whole area of the islands is not more than 12,000 acres, and of these only about 1,200 are under cultivation.

The principal islands are well wooded, but the great preponderance of the Bermudian cedar (*Juniperus bermudiana*), with a close and rigid foliage of the darkest green, gives a gloomy character to the woods;

as we got a little nearer, however, and the white houses of St. George's and the white tents of the encampment on Prospect Hill came into view, and the long fringing beach of bright coral sand with its outer border of intensely blue water breaking into dazzling white surf, the gravity of the scene was greatly relieved.

As we shall see hereafter, there is a total want of springs and wells of fresh water on the island, and it has become an almost universal custom to roof the houses with thin slabs of white limestone, and, further, to whitewash both roof and walls; the rain-water collected on the roof, and kept clean and fresh by the constantly renewed whitewash, is carefully led into a tank, and forms the only supply of pure water. Every house of any pretension is provided with such a tank, also covered with a sloping whitewashed roof, which, while it checks evaporation, adds to the contents of the tank by its own rain-catch. The white roofs are altogether peculiar, and as the houses and cottages of the rural population are scattered over the whole island, so as almost to run into one continuous straggling village, the white squares gleaming among the dark trees produce rather a pleasing effect, and one which is certainly very characteristic of Bermudas.

Saturday, April 5th.—A lovely clear morning; the sea perfectly smooth, and the sky almost cloudless. It was so early in the season that the temperature (68° F. in the shade) was not oppressive. The view from the ship—the 'Camber,' the government basin with the floating dock, the largest in the world, and the substantial buildings of the dockyard, and the 'Royal Alfred,' the 'Terror,' the 'Irresistible,' and a number of gun-boats, and the life and music and colour inseparable from a military station, to the west of us; the tortuous channels with which we were soon to be very familiar between the reefs, marked out by divers-coloured buoys and leading among the many islands of the 'great sound,' to Hamilton, to the south; the north coast of the main island stretching in a succession of shallow bays and wooded knolls and low cliffs, from Spanish Point to the high grounds at the entrance of Castle Harbour, to the east; and the wonderful variety and brilliancy of colour of the sea all around us — was very beautiful.

Captain Nares and I went in the forenoon to pay our respects to Admiral Fanshawe, commanding on the North American station, and to the governor, General Lefroy. We rowed across the glassy sea, clearly mapped out into patches of bright purple and stripes of the most vivid green by the reefs and the sandy spaces between them. Over the reefs in some places the water was only a fathom deep or less, and we could see the great round masses of brain coral beneath us, and the groves of purple *gorgoniæ* and all kinds of feathery zoophytes, interspersed with yellow sponges and bunches of sea-weed in all shades of olive and bright green and red. Clarence Cove, the landing place for Clarence Hill, the admiral's official residence, is an inclosed little bay, with the dark cedar woods coming close down on all sides to the water's edge. A garden rich with the luxuriant foliage-vegetation which suits the climate and the sheltered situation so well, runs along one side of the cove under the hill; and in the garden a little mound marks the grave of a middy, who died, poor boy, on the station, and who was buried in that quiet place in accordance with a not unnatural wish.

A winding path leads through the wood up to the house. Outside the drawing-room window there is a verandah looking down upon another small terraced garden, and commanding a very lovely view over the islands. While we were standing at the window enjoying it, a little flock of birds, some of them bright scarlet, and others of a splendid metallic blue, more like the fancy birds on an old brocade than real pirates of a kitchen garden, alighted on one of the trees below us. The 'red bird' (*Guarica cardinalis*), and the 'blue bird'—the blue-robin of the States (*Sialia wilsoni*), are probably commoner and more tame at Clarence Hill than elsewhere, because Mrs. Fanshawe especially protects and encourages them; but they are frequent all over the islands, and they are so very ornamental that various laws have been enacted by the legislature to prevent their extermination.

From Clarence Hill we went on to Mount Langton, the residence of the governor. We were unfortunate in not finding the governor at home on this occasion, but we had the pleasure of seeing him very frequently afterwards, and the kindness with which he did everything in his power to make our visit pleasant and profitable will long be remembered by all of us. We were indebted to Admiral Fanshawe and to Captain Aplin, captain-superintendent of the dockyard, for every possible accommodation and assistance in carrying on our work, in addition to the greatest personal kindness; but General Lefroy's time was less occupied with official duty. Himself a trained observer and deeply interested in the welfare of the colony, he was thoroughly acquainted with its physical conditions, and it was chiefly under his friendly guidance that we gathered what information we could during the short period of our stay.

Mount Langton has perhaps the best situation on the island. It stands high near the north shore, and only a mile or so from Hamilton, the principal town. Successive governors have done a great deal in laying out and improving the grounds, and in introducing ornamental and economic plants suitable to the climate; and General Lefroy especially has almost converted it into a *jardin d'acclimatisation*.

We had an excellent view of a great part of the islands from the signal station at Mount Langton. Bermudas is practically an 'atoll,' or annular coral reef. The reef is about twenty-four miles in length by twelve in width. Its long axis extends from N.E. to S.W. It is situated in a region of variables, but the most prevalent and by far the most violent winds are from the S.W. The portion above the level of the sea stretches along the southern or weather side, and consists of a chain of five narrow islands and a multitude of islets and detached rocks, which raise the number of the elements of the archipelago to over three hundred. The edge of the reef on the lee side is under water, with here and there a ledge showing above it at low tide, and with a single rock, the 'North Rock,' rising to the height of sixteen feet. There are only two or three channels through which vessels can come in through the reefs on the north

Fig. 72.—Group of gru-gru palms on the croquet-lawn, Mount Langton.

side, and all of these are difficult. The best passage is that round St. Catherine's Point and past Murray's Anchorage, and so on to Grassy Bay, the usual rendezvous of her Majesty's ships. An unbroken reef stretches along the south shore about a quarter of a mile from the land, from one end of the islands to the other. The central portion of the reef forms an imperfect lagoon, with an average depth of seven to eight fathoms.

The general character of this atoll is much the same as that of like reefs in the Pacific, with certain peculiarities depending upon the circumstance that it is the coral island farthest from the equator, almost on the limit of the region of reef-building corals. Accordingly, some of the great reef-building genera, such as *Madrepora*, *Cladocora*, and *Astrangia*, which are common even in the West Indies where the coral fauna is scanty, are absent.

The water over the reefs is extremely clear, and by using a water-glass—a square bucket with the bottom of plate-glass, just lowered so far as to get rid of the ripple and reflections on the surface, every detail can be made out of the economy of the reefs, and that of their inhabitants. The reefs and ledges are of all sizes, and they are separated from one another by channels from a yard to a quarter of a mile in width, floored with white coral sand, the débris of the coral worn down by the action of the waves, mixed with dead shells. The reefs project abruptly above the level of the floors of these passages to the height of six to ten feet. The surface of the reef is covered with massive, branching, and feathery things of very many kinds, occupying it as closely and as irregularly as the various weeds do a fallow-field. First we have the growing corals themselves, which may represent the dockens and the thistles, or rather a mass of beautiful marigolds and carnations, and daisies and gazanias, which have been thrown over the hedge in clearing a luxuriant garden and have taken root and gone on flowering. Most of the Bermudas corals, such as *Oculina diffusa*, *Symphyllia dipsacea*, *Astræa radians*, &c., are like sea-anemones or groups of sea-anemones in every shade of purple, orange, or green. The base or stock of the coral is dead and forms part of the reef; but each of the living branches is tipped with its sea-anemone, and the stars of plates by which its cups are supported are the earthy skeleton of the mesenteric plates which hang the stomach of the sea-anemone in its body cavity. In most cases the bodies of these sea-anemones, with their ranges of tentacles and their high colouring, are so prominent that they entirely mask the coral; but in a few, as for example in a brain-coral, *Diploria cerebriformis*, which seems to thrive at Bermudas better than almost anywhere else, forming domes six or eight feet in diameter, the animal matter is in comparatively small quantity, and covers the coral with what appears little more than a coating of greyish or yellow mucus. The *Gorgoniæ*, the *Bryozoa*, and the *Hydroid* zoophytes are like the other more prominent weeds in the field, as abundant and as irregularly distributed, growing in the spaces between the clumps of the different kinds of coral. One form, *Millepora*, which has been latterly classed with the hydroids, but which would seem to be more nearly related to the lost order *Anthozoa rugosa*, represented by two species, *M. alcicornis* and *M. ramosa*, is extremely abundant at Bermudas, where it acts in every way the part of a coral, forming massive additions to the reef of carbonate of lime abstracted from the sea. Beneath these large things there is usually a close felting of an undergrowth, consisting partly of sponges and smaller zoophytes, but chiefly of what are sometimes called lithophytes, sea-weeds of such genera as *Corallina*, *Melobesia*, and *Nullipora*, which like corals take carbonate of lime from the sea-water and incorporate it with their tissues.

All these things living and dying are constantly yielding a fine powder of lime, which sinks down and compacts in the spaces among their roots, and every breaker of the eternal surf grinds down more material and packs it into every hollow and crevice capable of receiving and retaining it. A great order of worms including the genus *serpula* secrete carbonate of lime and form thick and large calcareous tubes, or make galleries through the partially consolidated calcareous mud and harden it and bind it together. So great a part do these worms play in the construction of the reefs at Bermudas that General Nelson, in an admirable paper on the Geology of the Island, published by him in the Transactions of the Geological Society in the year 1834, calls some small circular reefs found everywhere round the shores in the wash of the breakers, and which appear to be due to their agency alone, 'Serpuline reefs.'

As I have already said, the Bermudas Islands, in common with most other coral islands, are formed by the raising of the weather edge of the reef above the level of the sea. This appears to be accomplished, in the first place, by the agency of the waves alone. Fragments, many of them with the inherent power of increasing themselves and cementing themselves together through the growth of the living things which invest them, are piled up on one another until they reach the highest point accessible to the sea in storms.

The moment the ridge appears above water a beach of coral-sand is formed against it. The top of the beach dries at low water, and the sand is blown on, first among the crevices of the breakwater already formed, which it widens and strengthens, and then over the breakwater to the ledges and reefs beyond, which it tends to raise to the surface. In this way in all coral seas islands have a tendency to form along the windward edges of annular reefs. The windward island then forms a shelter to the leeward portion of the ring, depriving it of the main source of its elevation, the piling up of fragments by the waves, so that on the leeward side we usually have more or less of the reef remaining submerged, and any passages of communication between the central lagoon and the outer sea.

I have little to add to the excellent account of the geology of Bermudas given by General Nelson. The Bermudas of the present day is simply a bank of blown sand in various stages of consolidation. The depth of water increases round the island with extreme rapidity. Seven miles to the north there is a sounding of 1,375 fathoms, and about two miles further off one of 1,775 fathoms. To the north-east there is water of 1,500 fathoms at a distance of ten

x

miles, to the north-west of 2,100 fathoms at a distance of seven miles, and to the southward of 2,250 fathoms at ten miles. The only direction in which there would seem to be a series of banks is along an extension of the axis of the reef to the south-west. We anchored for a night in 30 fathoms water on this line about twenty miles from the edge of the reef, and a shoal is mentioned at a still greater distance in the same direction. About three hundred miles farther on, however, a sounding is given of 2,950 fathoms, and there seems little probability that there is any connection between the Bermudas reef and the Bahamas. What the basis on which the Bermudas reef rests may be we have no means of telling, in fact its having the form of an atoll precludes the possibility of our doing so. There seems to be little doubt from Darwin's beautiful generalisation, which has been fully endorsed by Dana and other competent observers, that the atoll form is due to the entire disappearance by subsidence, of the island round which the reef was originally formed. The abruptness and isolation of this peak, which runs up a solitary cone to a height about equal to that of Mont Blanc, is certainly unusual; probably the most reasonable hypothesis may be that the kernel is a volcanic mountain comparable in character with Pico in the Azores or the Peak of Teneriffe.

There is only one kind of rock in Bermudas. The islands consist from end to end of a white granular limestone, here and there becoming grey or slightly pink, usually soft and in some places quite friable, so that it can be broken down with the ferrule of an umbrella; but in some places, as on the shore at

Hungry Bay, at Painter's Vale, and along the ridge between Harrington Sound and Castle Harbour very hard and compact, almost crystalline, and capable of taking a fair polish. This hard limestone is called on the islands the 'base rock,' and is supposed to be older than the softer varieties and to lie under them, which is certainly not always the case. It makes an excellent building stone, and is quarried in various places by the Engineers for military works. The softer limestones are more frequently used for ordinary buildings. The stone is cut out of the quarry in rectangular blocks by means of a peculiarly constructed saw, and the blocks, at first soft, harden rapidly, like some of the white limestones of the Paris basin, on being exposed to the air.

As I have already indicated, this limestone is entirely what General Nelson aptly calls an 'Æolian formation.' The fine coral-sand, which surrounds the islands to a distance of about twenty miles, is washed in by the sea; it is then caught at certain exposed points by the prevailing winds and blown into sand-hills often forty or fifty feet in height. The sand is spread over the surface in a certain sense uniformly, but that uniformity is liable to be interfered with by anything which for a moment affects the direction or force of the wind; for instance, the sand is blown up and heaped round any obstacle, or it may be swept out by irregular gusts into hollows which are afterwards filled up by a secondary series of layers; or a total change may be made on the whole arrangement of the surface by a sub-tropical rain-flood. All the appearances produced with great rapidity by such causes are of course perpetuated in the rock which is formed by the consolidation of the sand, so that we have often repeated again and again in the distance of a quarter of a mile all the phenomena,—denudation, unconformability, curving, folding, synclinal and anticlinal axes, &c., which are produced in real rocks, if I may use the expression, by combined aqueous and metamorphic action, extending over incalculable periods of time. The principal roads, which are extremely good as they are laid out and maintained partly with a view to military operations, frequently pass through deep cuttings and give excellent geological sections, exhibiting an amount of confusion sufficient to perplex the most experienced geologist, if he did not hold the key. The general tendency of the layers of sand to wrap themselves round opposing objects, using the vortices into which the sand itself is thrown by swirls of the wind, as nuclei, if they encounter nothing more prominent or resisting, sufficiently accounts for the prevalence of saddle-back sections.

One phenomenon of these limestones especially gives a false idea of their age and permanence. Every here and there in all parts of the islands we have long stretches of limestone rock dipping in different directions, but very often towards some northern point, at a constant angle between 28° and 32°. Such beds are often overlaid unconformably by nearly horizontal layers, and they certainly give the idea of having assumed their present position by upheaval. This is not so, however. The sand-hills terminate landwards in a more or less regular glacis, and as the sand advances, layer after layer is added uniformly to the face of the glacis, producing a very regular stratification at the angle of rest of dry sand of this particular kind, an angle of about 30°, entirely

Fig. 73.—Stratified 'Æolian' rocks, Bermudas.

corresponding with that of the limestone. Although I examined the greater part of the islands very carefully, I never met with an instance of a dip in the limestone at an angle higher than 32°, usually it is several degrees lower.

There is a wonderful 'sand-glacier' at Elbow Bay on the southern shore of the main island. The sand has entirely filled up a valley, and is steadily progressing inland in a mass about five-and-twenty feet

FIG. 74.—'Sand-glacier' overwhelming a garden, Elbow Bay, Bermudas. (From a Photograph.

thick. The day we examined it under the guidance of the Hon. Mr. Gosling there was a light breeze blowing from the southward, and a light haze or dimness lay just over the surface of the sand, and on holding up a sheet of paper perpendicular to the surface and transverse to the direction of the wind, the travelling sand rapidly fell from the windward surface of the paper and banked up before it. The glacis is very regular. It has partially overwhelmed a garden, and is moving slowly on (Fig. 74). When our photograph was being taken the owner of the garden was standing with his hands in his pockets, as is too much the habit of his race, contemplating the approach of the inexorable intruder. He had made some attempt to stay its progress by planting a line of oleanders and small cedars along the top of the slope. A neighbour, a little more energetic or more seriously menaced, managed to turn the flank of the sand by this means just as it was on the point of engulfing his house; but another was either too late in adopting precautionary measures, or perhaps submitted helplessly to his fate, for all that now remains of his cottage is the top of one of the chimneys projecting above the white sand like a tombstone, with a great bush of oleander drooping over it (Fig. 75). On its path upwards from the beach, this 'glacier' has overwhelmed a wood of cedars (Fig. 76). Firewood is valuable in Bermudas, and it is probable that in this case the trees were cut down when their fate seemed inevitable. This is always an unwise step, for sometimes an apparently very slight obstacle will stay the movement of the sand in a particular direction. The only way of stopping it artificially seems to be to cover it with vegetation. If planted in large numbers, and tended and watered for a time, it seems that oleanders and the native juniper will grow in the pure sand, and if they once take root the motion of the sand ceases. Some native plants, which form a peculiar vegetation, sending out enormously long runners or roots—such as *Ipomœa pes-capræ* and *Coccoloba uvifera*, and the crab-grass, *Agrostis virginica*, which is said to have been introduced, but which is now among the most valuable pasture-grasses on the islands—then take hold of it, and it becomes permanently fixed. The outer aspect of the sand-hill of course slopes downwards towards the sea, and whenever its progress landward—its growth—has been arrested, the tendency of the incoherent mass is to travel back again

FIG. 75.—Chimney of a Cottage which has been buried by a sand-glacier, Elbow Bay. (From a Photograph.)

by gravitation, and the action of rain; accordingly it is not unusual to be told that one of these *coulées* is gradually disappearing.

The process by which the free coral-sand is converted into limestone is sufficiently simple, and involves no great lapse of time. The sand consists almost entirely of carbonate of lime in a state of fine ultimate subdivision, owing to its having entered into the structure of the skeletons of animals; it is therefore easily soluble in water containing carbonic acid. When rain, which always contains a considerable quantity of carbonic acid derived from the atmosphere, falls upon the surface of the sand, it takes up a little

Fig 76.—'Æolian' Limestone Beds in process of formation, showing stratification, and the remains of a grove of Cedars which has been overwhelmed. Elbow Bay, Bermudas. (From a Photograph.)

lime in the form of bi-carbonate, and then, as it sinks in, it loses the carbonic acid and itself evaporates, and it leaves the previously dissolved carbonate of lime as a thin layer of cement, coating and uniting together the grains of sand. A crust is thus formed, and such successive crusts form lines of demarcation between successive layers of sand, and give the character of stratification and lamination which these wind-rocks always possess. Usually harder and softer layers alternate, indicating the greater or less degree in which the previous layer had been cemented and hardened before receiving the next addition of dry sand. The rocks remain permeable to water and soluble, so that this process of solution and the deposition of cement in the interstices of the stone goes on constantly. The extreme result is a compact marble-like limestone, in which the grains of sand are combined in a continuous magma with stalagmite or travertine.

This dissolving and hardening process takes place irregularly, the water apparently following certain courses in its percolations, which it keeps open and the walls of which it hardens; and in consequence of this the rock weathers most unequally, leaving extraordinary rugged fissures and pinnacles, and piling up boulders, the cores of masses which have been eaten away, more like slags or cinders than blocks of limestone. The ridge between Harrington Sound and Castle Harbour is a good example of this. It is like a rockery of the most irregular and fantastic style, and there seems to be something specially productive in the soil, for every crack and crevice is filled with the most luxuriant vegetation, massing over the stones and straining up as tier upon tier of climbers, clinging to the trees and rocks. Frequently the percolation of hardening matter from some cause or other only affects certain parts of a mass of rock, leaving spaces occupied by free sand. There seems to be little doubt that it is by the clearing out of the sand from such spaces, either by the action of running fresh water or by that of the sea, that those remarkable caves are formed which add so much to the interest of Bermudas.

Wherever, throughout the islands, a section of the limestone is exposed of any depth, it is intersected by one or two horizontal beds of an ochre-like substance, called locally 'red earth;' and the same substance is met with in greater purity in cracks and pockets all through the limestone. This 'red earth,' mixed with varying proportions of decayed vegetable matter and coral-sand, forms the surface layer of vegetable soil. As Smith says, when this red earth is pure the soil is inferior; when it is black—that is to say, when it contains much decomposed vegetable matter—it is better, and the best soil of all is probably that which consists of red earth, humus, and coral-sand.

The origin of this 'red earth' is a matter of great interest, as it seems to afford a singular illustration and confirmation of our view as to the organic origin of the 'red clay' of the Atlantic sea-bed. There is ample evidence all over the islands that there has been an enormous amount of denudation,—that while in some places coral-sand has been encroaching and new rocks have been thus formed, in other places masses of rock of great thickness have been removed by the disintegrating effect of rain-water.

During the disintegration of the stone, the softer parts are first removed, leaving a kind of skeleton, consisting of the harder parts, and as rough as a mass of scoriæ. The ridge between Harrington Sound and Castle Harbour, which rises to the height of about 150 feet, is, as we have already said, entirely composed of limestone in that condition; passing where it is not eroded, into a very compact hard stone, worked in a large quarry at Painter's Vale for building purposes. The height of these ridges and crests, taken in connection with their structure and distribution, gives a very good idea of the amount of denudation which has taken place,—at all events during this last episode in the geological history of the islands.

I am indebted to his Excellency General Lefroy for the report on the analyses of soils from Bermudas given in Appendix A to this chapter, and I believe it holds the clue to the mode of formation of the 'red earth.' The coral-sand, like the mass of skeletons of surface animals accumulated at the bottom of the ocean, does not consist of carbonate of lime alone. It contains about one per cent. of other inorganic substances, chiefly peroxide of iron and alumina, silica, and some earthy phosphates. Now these substances are to a very small degree soluble in water charged with carbonic acid; consequently, after the gradual removal of the lime, a certain sediment, a certain ash as it were, is left behind. One per cent. seems a very small proportion, but we must remember that it represents one ton in every hundred tons of material removed by the action of water and of the atmosphere, and the evidences of denudation on a large scale are everywhere so marked, that even were some portion of this one per cent. residue further altered and washed away, enough might still be left to account fully for the whole of the 'red earth.' The vegetable soils containing a large proportion of red earth, are accumulated, usually to no very great depth, in hollows, the fine ultimate sediment naturally finding its way down to the lowest point. Fig. 74 gives an excellent illustration of the formation of one of the intercalated beds. The soil of the garden, which consists of red earth mixed with decayed vegetable matter, rests upon limestone. The sand glacier creeps over it, and it is covered by a series of beds, twenty to thirty feet thick, dipping at an angle of 30°. The water percolating through the sandstone gradually removes the organic matter, and the inorganic residue is left.

Wandering about among the pretty hill-and-dale scenery of Bermudas one is not at first conscious of a singular omission, until all at once it bursts upon him that there is not a drop of water to be seen anywhere —no river, stream, or lake, not even a ditch or a duck-pond. The heavy rain falls upon the porous sand-heap, and runs through it as if it were a sieve. After a heavy shower, it may remain for a little collected in pools along the beaten road, or it may rush down a steep incline, but an hour after the rain is over every trace of it has disappeared. From the whole of the islands about low-water mark being composed of the same porous rock, the sea-water passes through it horizontally as freely as the rain-water passes through it vertically, so that up to high-water mark, and probably considerably above it on account of capillarity, the rock is completely saturated. There are some marshes and ponds on the main island, the marshes covered with a luxuriant vegetation, but in all of them the water is brackish, and they are all more or less affected by the tide, though the rise and fall is almost imperceptible in those at a distance from the sea.

If a well be sunk in almost any part of the island, it is filled with water at once, but it is only the upper layer which is fresh. The water at the bottom of the well is brackish, and is affected by the tide, and the fresh-water, which is merely the rain-catch of the surrounding ground, lies on its surface. As there is always a certain amount of mixture, the wells do not yield good drinking-water, and the people trust greatly to their rain-water tanks.

The direct evidences of subsidence are everywhere very palpable. The rocks exposed between tide-marks, and now being subjected to denudation, are not reef-rocks formed under water, but are, in most cases, stratified Æolian rocks.

The little pinnacle off the shore of Ireland Island, figured in the vignette at the end of this chapter, has its base composed of the ordinary blown sand of the sand-hills; the middle part is a shred of an old glacis; and the top is again horizontally stratified sand which has been laid down unconformably on the cut edges of its laminæ, after it had been greatly 'denuded' by rain and wind. The North Rock has almost exactly the same structure, so that we can scarcely doubt that the dry land of Bermudas at one time occupied a space considerably larger than it does at present. Tradition and the accounts of some of the earlier voyagers would seem to corroborate this; but soft though the rocks may be, and rapid the changes which take place in them in a geological sense, it seems difficult to believe that after they were consolidated, any great change could have taken place in their distribution in the short period during which they have been the subject of tradition. A very careful survey was made in the year 1843, and up to the period of our visit there did not seem to have been the least alteration, even in the depth and extent of the passages among the living reefs; a matter of jealous interest where there are only a few inches to come and go upon, in the question of the entrance of a vessel of a certain draught.

Perhaps even a more satisfactory proof of subsidence was given a few years ago. In preparing a bed for the great floating dock it was necessary to make an excavation in the Camber extending to a depth of fifty feet

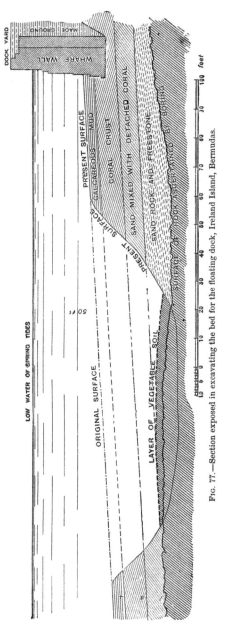

Fig. 77.—Section exposed in excavating the bed for the floating dock, Ireland Island, Bermudas.

below low water. First they came in the cutting, at a depth of twenty-five feet below the surface, to a bed of calcareous mud, five feet thick, forming the floor of the basin; next to loose beds, twenty feet thick, of what has been called 'coral crust'—coral-sand mixed with detached masses of *Diploria* and isolated examples of smaller corals and of many shells, and passing into 'freestone,' the coral-sand cemented together but somewhat loosely coherent. Beneath this, at a depth of about forty-five feet, there is a bed of a kind of peat, and vegetable soil containing stumps of cedar in a vertical position, and the remnants of other land vegetation, with the remains of *Helix bermudiensis*, and of several birds; the bed of peat was ascertained by boring to lie upon the ordinary hard 'base-rock.' (Fig. 77.)

Fossils, or semi-fossils, are very common throughout the islands, and they are generally such as we find associated with sand-hills. One or two species of *Helix*, showing many varieties, are by far the most abundant; the shells can be picked out of the soft rock in some places in thousands, as at Mount Langton and along the road-side in Somerset Island.

It is difficult to imagine, where there has been so much change of level, that elevations, at all events of a local character, should not have occurred from time to time; and yet we could not satisfy ourselves that we had detected any absolute proof of this. The only case in which we felt an approach to confidence was a rock about ten feet above high-water mark on Boaz Island, which seemed to contain serpula borings in position. In various places, however, a peculiar kind of calcareous tubing which forms round roots of bushes and grass presents itself under such various aspects, that we may have mistaken this for *serpulæ*. In many spots the rock far above high water contains marine shells, but these seem to be all such as might have been blown along with the sand into their elevated position, or have been carried there by other means. *Turbo pica*, for example, one of the commonest, is too heavy to be carried by the wind, but it is constantly transported far inland and to any height by the 'soldier crab.'

In the collection of Mr. Bartram, a very enthusiastic amateur naturalist living near St. George's, I found a small worn and rounded fossil, which seemed to be the cup of a crinoid allied to *Holopus*. Mr. Bartram kindly gave me the specimen, and we looked most carefully in all subsequent dredgings near Bermudas for some further indication of the presence of the species, but in vain.

On the 21st of April we left Bermudas and crossed the Gulf Stream to Halifax, Nova Scotia. The observations made during that trip will be recorded in the next chapter. Towards the end of May we returned southwards, and on the 31st we had the pleasure of finding ourselves once more lashed alongside the dockyard at Bermudas.

The weather was now much warmer, the thermometer in the shade ranging from 21°·5 to 25° C., and with the advancing season the appearance of the islands had changed considerably. Even from a distance many additional shades of green might be seen brightening and softening the sombre uniformity of the cedar woods, and, conspicuous above all others, brilliant masses of the splendid foliage of the 'Pride

Y

of India' (*Melia azedarach*), which, during our former visit, showed its delicate trusses of lilac flowers only. I have already referred to the use of the common oleander (*Nerium oleander*) for arresting the progress of moving sand. The plant was introduced into Bermudas in modern times, and appears to have been encouraged partly on account of its value for that purpose, and partly doubtless for its showy flowers. The climate and soil seem to suit it wonderfully, and it has regularly taken possession of the islands. Large bushes, twenty feet high, are everywhere—round the cottages, along the road-sides and in the woods; and thick hedges of great height, planted partly as bounding fences and partly for shelter, intersect the cultivated ground in every direction. Nothing can be imagined more ornamental. There are all varieties of colour, from white through pale rose and lilac to nearly crimson; the flowers tend to come double or semi-double, and they bloom most profusely. The country round Hamilton, and Somerset Island, were a perfect blaze of colour in June, and as the flowers come in succession and stand a long time, they would remain so during the greater part of the season. The oleander is not now so popular as it was, and although it is still planted in large numbers in shifting sand, it is suspected that high thickets and hedges of it near dwellings are not healthy, and latterly they have been cleared away in many places.

June 9th.—A party started early in the morning in the galley and the steam pinnace for the Walsingham Caves. We called at Mount Langton for the Governor, who had arranged to join us, and then went on to 'The Flats,' the entrance to Harrington

Sound. The strait, which is the only communication between this beautiful land-locked sheet of water and the sea, is very narrow, and spanned by a low bridge. The rising tide rushes in through it with great force, and when we arrived the ebbing tide was rushing out with the velocity of a mill-race. We sent on the steam pinnace with the photographer and some of our party by way of Castle Harbour to Walsingham, to try to photograph the interior of one of the caves with the magnesium light; and we warped the galley against the rapids, and she was soon in the still, clear water of the Sound. Harrington Sound is a most peculiar basin, and certainly it is extremely beautiful. It is nearly rectangular, a mile wide by about two miles long in the direction of the axis of the island. To the south a low narrow band separates it from the sea, and on the other three sides the land rises in irregular, richly wooded ridges, forming nearly the highest ground in Bermudas. The sheet of water is thus completely land-locked, and as it is of considerable depth, if there were any good access to it, it would make one of the finest harbours conceivable. It was at one time proposed to cut a canal, opening a communication to the southward, and to make it the Government harbour; but the project was abandoned in favour of the present arrangement at Ireland Island.

We had taken along with us one of the native fishermen who was in the habit of collecting corals and knew the localities of the different species, and under his guidance we explored some low caves under the cliffs along the northern shore. The roofs of the caves were covered with green algæ, and below the scarcely perceptible line where the air met the still, clear water, there was a complete incrustation of the delicate coral *Mycedium fragile,* standing out in undulating purple crescents like some luxuriant lichen. Some of the sailors stripped and dived for it, and soon there was a pile of beautiful specimens in the boat. A colony of tropic birds (*Phaëton æthereus*) were building in the cliffs above, and as they sailed over us, their two long, white tail-feathers gleaming in the sun, their white breasts reflected the colour of the water, and they looked as if they had been moulded in pale green glass. We rowed slowly round, in full enjoyment, to the corner of the Sound nearest Walsingham, and a few minutes' walk brought us to the caves.

As in all limestone districts, the caves at Bermudas consist of large vaulted chambers hollowed out in the rock by the removal of its material by running fresh water or by the action of the sea. The process is probably more rapid in a coral island than it is where the rock belongs to one of the older formations. Dana observed similar caverns in Metia or Aurora Island, one of the western Paumotus, in which the geological structure may greatly resemble that of Bermudas, and he quotes from the Rev. Mr. Williams an account of a cavern in the coral rock of Atiu, one of the Hervey group, in which he " wandered two hours without finding a termination to its windings, passing through chambers with fretwork ceilings of stalagmite, and stalactite columns which sparkled brilliantly with the reflected torch-light." [1] The

---

[1] 'United States Exploring Expedition,' vol. x., *Geology,* by James D. Dana, A.M., Geologist to the Expedition, p. 67.

CHAP. IV.] ST. THOMAS TO BERMUDAS. 325

entrances are usually small crevices in the rock, often

FIG. 78.—Entrance to the 'Convolvulus Cave,' Walsingham, Bermudas. (From a Photograph.)

almost masked by vegetation. One, which we passed on our way, called the Convolvulus Cave, is covered with a glorious mantle of *Ipomœa nil*, its ephemeral flowers changing during the day from a brilliant azure to rich purple. (Fig. 78.)

A curious circumstance had given me a particular interest in one of the Walsingham caves. In the year 1819 the late Sir David Milne, at that time commanding in chief on the North American and West Indian station, had a very fine stalagmite upwards of eleven feet in length, averaging two feet in diameter, and weighing three-and-a-half tons, removed from the cave and placed in the Museum of the University of Edinburgh, where the course of circumstances has now placed it in my custody. The stalagmite was sawn over near the floor of the cave, and in the year 1863 Sir Alexander Milne, then commanding in chief on the same station, visited the cave and examined carefully the stump of the column which had been removed forty-four years before by his father. It had made some attempt at reparation, and in the year 1864 Mr. David Milne Home gave the results of his brother's observations in a notice to the Royal Society of Edinburgh. He observed five drops of water falling on the stump, two at the rate of three or four drops in the minute, the others much less frequently. At the spot where the two drops were falling two small knobs of calcareous matter had been formed. On the part of the stump where the three drops were falling, the deposit consisted of only a thin crust. The total estimated bulk of the stalagmite which had accumulated during forty-four years was about five cubic inches. Mr. Milne Home calculated that at that rate it would have taken 600,000 years to form the original stalagmite; but he points out very truly, that it is highly improbable that the supply has been uniform, and that in all likelihood it was very much greater at an earlier period, and has been steadily decreasing owing to the consolidation of the rock forming the roof of the cavern.

When we examined the stump, which was about ten years later, the two drops were still falling, but apparently somewhat more slowly, one not quite three times in a minute, the other twice; this must depend, however, in some measure upon the previous weather. The three drops were still falling, and adding silently to their crust. We could not determine that the bulk of the new accumulation was perceptibly greater than when it was measured by Sir Alexander Milne. We were very anxious to carry away with us a permanent record of the present condition of the stump of the stalagmite, and we twice tried to photograph it with the magnesium light. On the first occasion the picture came out fairly, but most unfortunately in the darkness and the difficulty of conducting such operations it was spoiled. When we tried it again there was something wrong with the bath, and it was a complete failure.

It then occurred to us that it might be possible to take another slice from the column, showing the amount of reparation during half a century, as an accessory and complement to the Edinburgh specimen. Our time was too short to allow us to do this ourselves, but Captain Aplin most kindly undertook to make the attempt after our departure, and I have just heard that he succeeded in his difficult task. The roof of the cave at the point whence the stalagmite was removed is at a height of about fifteen feet, and facing the stump there are two majestic columns uniting the roof and the floor, one of them upwards of sixty feet in circumference. They are beautifully fluted and fretted with stalactite, and shone out with a pure white-frosted surface in the magnesium light.

Sending the galley on before us we walked along the isthmus which forms the western boundary of Harrington Sound to Painter's Vale, where there was a small detachment of Engineers living under canvas, and another cave.

I think the Painter's Vale cave is the prettiest of the whole. The opening is not very large. It is an arch over a great mass of débris forming a steep slope into the cave, as if part of the roof of the vault had suddenly fallen in. At the foot of the bank of débris one can barely see in the dim light the deep clear water lying perfectly still and reflecting the roof and margin like a mirror. We clambered down the slope, and as the eye became more accustomed to the obscurity the lake stretched further back. There was a crazy little punt moored to the shore, and after lighting candles Captain Nares rowed the Governor back into the darkness, the candles throwing a dim light for a time—while the voices became more hollow and distant—upon the surface of the water and the vault of stalactite, and finally passing back as mere specks into the silence.

After landing the Governor on the opposite side Captain Nares returned for me, and we rowed round the weird little lake. It was certainly very curious and beautiful; evidently a huge cavity out of which the calcareous sand had been washed or dissolved, and whose walls, still to a certain extent permeable, had been hardened and petrified by the constant percolation of water charged with carbonate of lime. From the roof innumerable stalactites, perfectly white, often several yards long and coming down to the delicacy of knitting-needles, hung in clusters; and wherever there was any continuous crack in the roof or wall, a graceful, soft-looking curtain of white stalactite fell, and often ended, much to our surprise, deep in the water. Stalagmites also rose up in pinnacles and fringes through the water, which was so exquisitely still and clear that it was something difficult to tell where the solid marble tracery ended and its reflected image began. In this cave, which is a considerable distance from the sea, there is a slight change of level with the tide sufficient to keep the water perfectly pure. The mouth of the cave is overgrown with foliage, and every tree is draped and festooned with the fragrant *Jasminum gracile*, mingled not unfrequently with the 'poison ivy,' *Rhus toxicodendron*. The Bermudians, especially the dark people, have a most exaggerated horror of this bush. They imagine that if one touch it or rub against it he becomes feverish, and is covered with an eruption. This is no doubt entirely mythical. The plant is very poisonous, but the perfume of the flower is rather agreeable, and we constantly plucked and smelt it without its producing any unpleasant effect. The tide was with us when we regained the Flats Bridge, and the galley shot down the rapid like an arrow, the beds of scarlet sponges and the great lazy trepangs showing perfectly clearly on the bottom at a fathom depth.

Every here and there throughout the islands there are groups of bodies of very peculiar form projecting from the surface of the limestone where it has been weathered. These have usually been regarded as fossil palmetto stumps, the roots of trees which have been overwhelmed with sand and whose organic matter has been entirely removed and replaced by carbonate of lime. Fig. 79 represents one of the most characteristic of these from a group on the side of the road in Boaz Island. It is a cylinder, a foot in diameter and six inches or so high; the upper surface forms a shallow depression an inch deep surrounded by a raised border; the bottom of the cup is even,

FIG. 79.—Calcareous concretion simulating a fossil palm-stem, Boaz Island, Bermudas.

and pitted over with small depressions like the marks of rain-drops on sand; the walls of the cylinder are rough with transverse ridges and grooves singularly like the lines of insertion of endogenous leaves. The cylinder seems to end a few inches below the surface of the limestone in a rounded boss, and all over this there are round markings or little cylindrical projections like the origins of rootlets. The object certainly appears to agree even in every detail with a fossil palm-root, and as the palmetto is abundant on the islands and is constantly liable to be destroyed by and ultimately enveloped in a mass of moving sand, it seemed almost unreasonable to question its being one. Still something about the look of these things made me doubt, with General Nelson, whether they were fossil palms, or indeed whether they were of organic origin at all; and after carefully examining and pondering over several groups of them, at Boaz

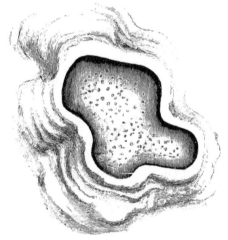

FIG. 80.—Calcareous concretion in Æolian limestone, Bermudas.

Island, on the shore at Mount Langton, and elsewhere, I finally came to the conclusion that they were not fossils, but something totally different.

The form given in Fig. 79 is the most characteristic, and probably by far the most common; but very frequently one of a group of these, one which is evidently essentially the same as the rest and formed in the same way, has an oval or an irregular shape (Figs. 80, 81, and 82). In these we have the same raised border, the same scars on the outside, the same origins of root-like fibres, and the same pitting of the bottom of the shallow cup; but their form precludes the possibility of their being tree-roots. In some cases (Fig. 83), a group of so-called

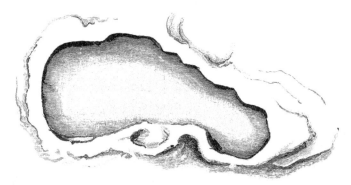

Fig. 81.—Calcareous concretion in Æolian limestone, Bermudas.

'palm-stems' is inclosed in a space surrounded by a ridge, and on examining it closely this outer ridge is found to show the same leaf-scars and traces of rootlets as the 'palm-stems' themselves. In some

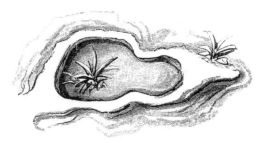

Fig 82.—Calcareous concretion, Bermudas.

cases very irregular honey-combed figures are produced which the examination of a long series of intermediate forms shows to belong to the same category. (Fig. 84.)

In the caves in the limestone, owing to a thread of water having found its way in a particular direction through the porous stone of the roof, a drop falls age after age on one spot on the cave-floor accurately directed by the stalactite which it is all the time creating. The water contains a certain proportion of carbonate of lime, which is deposited as stalagmite as

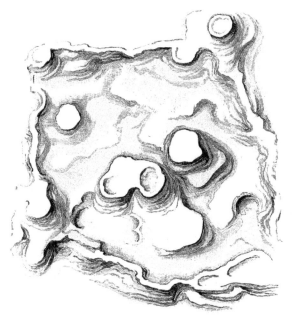

Fig. 33.—Calcareous concretions in Æolian rock, Bermudas.

the water evaporates, and thus a ring-like crust is produced at a little distance from the spot where the drop falls. When a ring is once formed, it limits the spread of the drop, and determines the position of the wall bounding the little pool made by the drop. The floor of the cave gradually rises by the accumulation of sand and travertine, and with it rise the walls and floor of the cup by the deposit of successive layers of stalagmite; and the stalagmite produced by the drop percolating into the limestone of the floor hardens it still further, but in this peculiar symmetrical way. From the floor and sides of the cup the water oozes into the softer limestone around and beneath; but, as in all these limestones, it does not ooze indiscriminately, but follows certain more free paths. These become soon lined and finally blocked with stalagmite, and it is these tubes and threads of

Fig. 84.—Concretions in Æolian rock, Bermudas.

stalagmite which afterwards in the pseudo-fossil represent the diverging rootlets.

Sometimes when two or more drops fall from stalactites close to one another the cups coalesce (Figs. 80, 81, and 82); sometimes one drop of two is more frequent than the other, and then we have the form shown in Figs. 81 and 82; sometimes many drops irregularly scattered form a large pool with its raised border, and a few drops more frequent and more constant than the rest grow their 'palmetto stems' within its limits (Fig. 83); and sometimes a number of drops near one another make a curious regular pattern, with the partitions between the recesses quite straight (Fig. 84).

I have already referred to the rapid denudation which is going on in these islands, and to the extent to which they have been denuded within comparatively recent times. The floors of caves, from their being cemented into a nearly homogeneous mass by stalagmitic matter, are much harder than the ordinary porous blown limestone; and it seems that in many cases, after the rocks forming the walls and roof have been removed, disintegration has been at all events temporarily arrested by the floor. Where there is a flat surface of rock exposed anywhere on the islands it very generally bears traces of having been at one time the floor of a cave, and as the weather-wearing of the surface goes on the old concretionary structures are gradually brought out again, the parts specially hardened by a localized slow infiltration of lime resist disintegration longest and project above the general surface. Often a surface of weathered rock is so studded with these symmetrical concretions, that it is hard to believe that one is not looking at the calcified stumps of a close-growing grove of palms.

All the figures are portraits, and are taken from a single group on Boaz Island.

Ireland Island, the extreme island of the chain to the westward, contains the dockyard with the Government basin and the wonderful iron floating-dock, which was made in England, and towed across the Atlantic with so much labour and risk a few years ago. It is covered with Government buildings, and is under strict naval discipline—an appanage and extension, in fact, of the guardship, H.M.S. 'Terror.' Boaz Island succeeds; it is united to Ireland Island by a bridge, and is the site of a military hospital and barrack. A short ferry then leads to Somerset Island, the richest and best cultivated and perhaps the prettiest part of Bermudas. The houses here are numerous and good, and the market-garden style of culture is fully carried out. The soil is excellent; 'red earth' with decayed vegetable matter, and 'mixed with it a white meale'—that is to say, the large-grained free coral-sand. When we arrived in the beginning of April—very early in the season—they were already despatching to the New York market weekly shiploads of delicious early potatoes, for which they were getting 17*l*. a ton, onions at 7*s*. 6*d*. a box, and the earliest tomatoes—just beginning to ripen—at 3*s*. a small box. In Somerset the fields and gardens are small, separated and intersected by tall oleander hedges, and all the rugged ground is covered with cedar woods.

Long Island, or Bermudas, extends as a narrow strip from Somerset Island round the great Sound. At the head of the bay, where Hamilton, the principal town, is situated, it widens out, being about two miles in width; it then forms the singular quadrangular frame round Harrington Sound, and ends on the shore of Castle Harbour. The road from Somerset to Hamilton passes Gibb's Hill, 245 feet high, with a lighthouse 120 feet, showing a revolving light, brightening up every minute into a strong glare which continues for six seconds. As the height of the lantern is 360 feet above the level of the sea, the light can be seen for a distance of about thirty miles. Cottages and patches of cultivation are scattered along the side of the road for the whole distance.

Hamilton is a quaint, rather pretty little town of about 2,000 inhabitants. A half-street of irregular houses and stores with green verandas faces the harbour, with a commodious line of wharves and sheds along the shore. The better houses are scattered and a little inland, and some very pretty villas, the residences of the leading merchants and some of the Government officials, occupy pleasant situations round the head of the bay. The suburbs of Hamilton show well the peculiarity of the contrast between the white-roofed houses and the dark junipers. Nearly the best examples of these so-called 'cedars' form a fine avenue just behind the town. (Fig. 85.)

St. George's Island, running along the northern shore of Castle Harbour, terminates the series to the north-east. The town of St. George's is the second on the islands, even more prettily situated than Hamilton, and having likewise a good harbour for vessels of moderate size, and a good line of wharves. The streets of the town are, however, close and narrow, and the drainage is bad; so that St. George's, one of the principal military stations, is by no means healthy. St. David Island, Cooper Island, Castle Island, and many smaller islets, form a broken barrier, closing in Castle Harbour to the southward.

General Lefroy has paid great attention to all questions bearing upon the health and the material prosperity of Bermudas, and his reports to the Home Government and the reports submitted to him by

Fig. 85.—Cedar, Avenue Hamilton, Bermudas. (From photograph.)

the local authorities are of high interest. Very charming in many ways, it seems doubtful whether, with the scarcity of labour which followed slave emancipation, Bermudas would be a colony of any value except as a military station. In this point of view, however, it is of an importance perhaps second only to Malta. Judging from General Lefroy's careful statistics, Bermudas cannot be regarded as an essentially unhealthy place, or as one possessing a climate unfavourable to the life of Europeans. The death-rate among the white population is about twenty-two in a thousand—nearly the same as the general death-rate of England. The mean of the death-rate among the troops in Bengal for the five years up to 1869 was 23·8 in a thousand, in Bermudas 16·1, in Malta 15·9, and in Canada—an exceptionally healthy station—6·9. During the last thirty years Bermudas has suffered from four epidemics of yellow fever, and these, with one exception—that of 1856—told severely on the troops; but setting aside that terrible scourge, it must be regarded as one of the healthiest of our less healthy stations. A good deal can be done, doubtless, to improve the sanitary condition of the towns and of the military establishments; but the root of the evil is in the porous nature of the rock, preventing a full and purifying supply of running water, and rendering anything like effective sewerage extremely difficult.

There is little live stock on the islands; cattle and sheep do not thrive well, probably mainly owing to the want of a plentiful supply of good water. Butcher-meat is almost all imported from America. Horses are not very numerous, and only tolerably good. There are a few mules and asses.

The greatest extent and diversity of land is in

Long Island, between Hamilton and the shore of Harrington Sound. The country is undulating and well wooded, with here and there extensive brackish water marshes cumbered with a luxuriant vegetation of palmettos, mangroves, junipers, and ferns, including the common bracken *Pteris aquilina*, *Osmunda regalis* and *O. cinnamomea*, and *Acrostichum aureum* often seven or eight feet high (Fig. 86). It is intersected by good roads, and dotted with white-roofed houses, churches, and school-houses.

The principal crops raised are potatoes, tomatoes, and onions, for the New York market; these are of the very best quality, but they are raised in comparatively small quantity, owing partly to the scarcity of labour and partly to the patchy distribution of fertile soil and the want of a sufficient supply of manure. Arrowroot—the starch of *Maranta arundinacea*—was at one time a principal article of export; but the quantity produced has been steadily decreasing of late years. What is made is certainly excellent, and fetches a much higher price than the West Indian arrowroot which is driving it out of the market. The starch is contained in a long jointed tuber, or rather rhizome or under-ground stem which springs from the crown of the root of the *Maranta*. This, when it is ripe, is a foot or so in length, slightly flattened, and about an inch in diameter. When fresh it is covered with a brownish skin; but this separates in drying from the tuber, which is white and semi-transparent and little else than a mass of starch. Fragments of the rhizome, or small shoots which are sent off along with the rhizomes, are planted about the month of May, and

CHAP. IV.]   ST. THOMAS TO BERMUDAS.   341

Fig. 86.—Natural Swamp-vegetation, Bermudas. (From a photograph.)

they send up a stem three feet high, with handsome iris-like leaves. In about ten months each plant yields ten or twelve, sometimes as many as eighteen or twenty, tubers. These are partially dried and the skin removed, and then, after being carefully washed, they are ground in a mill worked by horses or oxen, into a coarse pulp. The pulp is passed, with an abundance of cold water, through sieves gradually diminishing in mesh, until the starch passes through free from fibre or other impurity. The greater part of the water is then poured off, as much of the remainder as possible forced out by a hand-press, and the cake of starch broken up and dried in shallow wooden trays exposed to a current of air, when it falls into the well-known snowy, glistening powder. The best arrowroot costs at the manufactory from 1*s.* 4*d.* to 1*s.* 6*d.* a pound; but the crop is a troublesome one, the labour connected with it extending over about a year, and the Bermudians find the culture of the potato, which lies only about one hundred days in the ground, gives much less trouble, and commands a certain and immediate market in America, a more profitable speculation. Maize is grown in small quantity, chiefly as a vegetable for the sake of the green heads. Indian corn is greatly used as an article of food, but it is imported from America. Cassava (*Jatropha manihot*) is common in gardens, and thrives well. It is much used about Christmas-time for making a very favourite dish of the season—'cassava pudding.' All the European vegetables grow in Bermudas, and, with care, seem to come to considerable perfection. It is singular that seed potatoes, and garden seeds of all kinds, including those of the onion and tomato, are imported every year, usually from America or Madeira. It is generally understood that seed grown on the islands will yield a deteriorated crop.

Some years ago Bermudas was famous for its oranges; several of the best varieties were cultivated in gardens, and the fruit arrived at wonderful perfection; while the lemon, the lime, and the bitter orange were self-sown, and sprang up everywhere, so that the country lanes and hedge-rows were redolent of the delicious perfume of orange blossom, and the fruit fell off and rotted on the ground.

About the year 1854 a minute insect of the family Coccidæ appeared on the orange-trees, and multiplied infinitely. The leaves, covered with scales, and glutinous with a viscid excretion from the animal, became yellow and fell off, the fruit dropped before ripening, and finally many of the trees died. Bermudas has never recovered from this plague, and now there is scarcely an orange grown on the islands. The wild lemons and limes still flourish, and perfume the air in the thickets about Walsingham and Painter's Vale; but the cultivated varieties have disappeared from the gardens. Fruit is by no means abundant. The smaller English fruits—gooseberry, raspberry, currant, &c., run to wood, and do not bear. Strawberries fruit fairly. Bananas are very generally cultivated and are good, although the varieties are not so carefully selected as they are in Madeira. There are some fine trees of the Avocada pear (*Persea gratissima*), which bear abundantly. The Mango has been introduced into some gardens, but the crop cannot be depended upon. The singular-looking

Papaw-trees (*Carica papaya*) are seen everywhere,

Fig. 87.—Papaw-trees (*Carica papaya*), in the garden at Clarence Hill. (From a photograph.)

male and female, round the cottages (Fig. 87); but the fruit is not much esteemed.

The climate of Bermudas is very genial; the mean annual temperature is 21° C., while that of Madeira, in almost exactly the same latitude, is 18° C. This difference of 3° C. is due partly to the prevalence at Bermudas of south-west winds blowing directly over the super-heated reflux of the equatorial current, and partly to the position of the islands within the region of the banked-down warm water of the Gulf Stream. The temperature of the coldest month (17°·2 C.) is, however, somewhat lower at Bermudas than at Madeira (17°·8 C.), while that of the warmest month is considerably higher (22° to 26° C.). This greater summer-heat, telling upon the flowering and the ripening of the seeds of plants, gives the flora of Bermudas a more tropical character than that of Madeira, and this is undoubtedly increased by the circumstance that while the vegetation of Madeira and the other 'Atlantic islands,' the Açores and the Canaries, appears to be to a great degree an extension of that of Southern Europe, that of Bermudas, if we except a large number of introduced plants, is in the main derived from the West Indies and the south of North America.

The fauna of Bermudas is singularly poor. There are no wild mammals except the rats and mice which have been imported with foreign produce. Only about half a dozen land-birds breed on the islands, and all of these are common North American species; probably the most abundant and most widely distributed are the American crow (*Corvus americanus*), and a pretty quaker-coloured little ground dove (*Chamœpelia passerina*). Many American birds are annual visitors; we saw during our stay two at least of these, a blue-and-white kingfisher (*Ceryle alcyon*), and the well-known 'rice bird' (*Dolichonyx oryzivorus*); several other birds, including a species of *Gallinula* and one or two of the smaller passeres, all common American species, alighted on the ship while we were cruising in this neighbourhood. Only one reptile is known, a lizard common in Carolina; and a small fish allied to the mullet occurs, not in great abundance, in the brackish-water marshes.

During our two visits Mr. Moseley collected the plants vigorously, and by a kind arrangement of General Lefroy's he spent a few days at the camp of the Engineers in Painter's Vale in the middle of the best botanical district. He dried about a hundred and fifty species of flowering plants, which were sent to Kew, and Dr. Hooker, in returning the rough list, expresses his surprise at finding the flora to possess so tropical a character.

It is pleasant to ride of an evening along the green roads in Bermudas. Things are so much alike all over the world, the exigencies of cultivation and traffic requiring everywhere the same palings or hedge-rows or low walls, and nature everywhere encumbering and ameliorating the road-sides with green weeds with blue or white or red or yellow flowers, that one might almost fancy oneself among the green lanes of the middle counties of England. The exotic character of the vegetation of Bermudas is not obtrusive. The universal cedar might be a yew or a dark-foliaged pine, and only here and there a graceful group of tall palmettos rises over a mangrove swamp. *Chamærops palmetto* is the only indigenous palm; the cabbage palm (*Oreodoxa oleracea*), the date palm (*Phœnix dactylifera*), the cocoa-nut (*Cocos nucifera*), and the grugru palm (*Astrocaryum cureum*), have been introduced, and grow well; but they do not ripen their fruit. The bananas round the cottages look tropical, and so do the stars of scarlet bracts of *Poinsettia* and the stars of crimson flowers of *Erythrina*; but the far more general tamarisks and oleanders are familiar. An exotic cast is given to the undergrowth by the prickly pear, the Mexican yellow poppy (*Argemone Mexicana*), the scarlet sage (*Lantana coccinea*), and the wild ipecacuanha (*Asclepias curassavica*), the food of the caterpillar of the finest Bermudian butterfly (*Danais archippus*); but nettles, chickweed, sow-thistles, rapes, clovers, and other cosmopolitan weeds, hold their accustomed place.

ÆOLIAN ROCK, BERMUDAS.

APPENDIX A.

Report from Prof. Abel, F.R.S., to H. E. General Lefroy, C.B., F.R.S., on the Character and Composition of Samples of Soil from Bermudas.

LABORATORY, 18, BILLITER STREET, E.C.,
January 4, 1873.

To PROFESSOR ABEL, F.R.S., &c.

LIEBIG, in his report to Lawes and Gilbert on the composition of the soil of the various plots under experimental cultivation at Rothamsted, has suggested the adoption of a uniform method by all chemists in the examination of soils, so as to lead to comparative results. The method he has adopted is the following, viz. :—

1. Solution in dilute acetic acid (one part of strong acid to four parts of water). This he considers the measure of plant food in the soil at immediate disposal.

2. Solution in dilute hydrochloric acid (also 1—4.) This yields the quantity which by the action of weather gradually disintegrates and becomes proximately available.

Of the remainder, that which is soluble in strong hydrochloric acid shows the readily decomposable part of the soil, though requiring a longer time for disintegration than that dissolved by weak acids, whilst the part which is rendered soluble only by fusion (when the examination is carried so far) represents the ultimate capability of the soil to renew its fertility after the lapse of time.

It is obvious, if a uniform method of analysis such as this

were adopted by all chemists, it would materially enhance the value of their work, by giving comparable results.

In the analysis of the Bermuda soils, which I have now to report, I have adopted as far as possible this plan of analysis, omitting the solution in weak hydrochloric acid, and using instead strong hydrochloric acid. The quantity dissolved by this re-agent represents the readily decomposable part, and that which would have been soluble in dilute hydrochloric acid after the acetic solution.

The considerable quantities of alumina dissolved both by weak and strong acids shows all the soils to be readily acted on by weather, and consequently having a proportionate facility for renewing their fertility. This is especially the case with No. 3 soil. There is no marked deficiency of any normal constituent in any of the soils, with the exception of chlorine and sulphuric acid. In No. 1 soil both are nearly absent; in soil No. 2 the chlorine is a mere trace, and in soil No. 5 the sulphuric acid is very minute. The actual weight of these minute quantities, however, when multiplied up to that actually existing in an acre of soil one foot deep, is, as has been shown by Professor Way, not at all insignificant.

Phosphoric acid is present in soil Nos. 1, 2, 3, in unusually large quantities, the average for a fertile soil being 0·10 per cent. With certain limitations it is scarcely fair when all the constituents of a soil are essential to the proper development of plants, to attach a higher importance to one than to another; some, however, are invariably present in large quantities, and seem to serve more as a medium of growth than by any chemical activity; some constituents are more easily and cheaply supplied than others; the direct application of some to an appropriate crop producing a corresponding increase in yield, when a supply of others would produce no increase.

The large amount of phosphoric acid in the three soils (Nos. 1, 2, 3), should, by disintegration and proper cultivation, maintain their fertility for ages to come, and the coral and chalk (Nos. 4-5) will furnish excellent dressings for clay sands where there is a deficiency of lime.

To maintain the fertility of a soil under proper cultivation it is necessary to supply in the manure the minerals which are taken off in the crop. It by no means follows that the constituents of a crop form the best fertilizers for that crop; in fact, the contrary statement seems to hold good—the special manure for some crops is just that constituent in which they are deficient. The wheat crop, for instance, which is starchy rather than nitrogenous, is greatly increased by the direct application of nitrogenous manures. The same may be said of the sugar-cane, and, I have little doubt, of arrowroot; in the case of wheat, the increase in the crop by the nitrogenous manures may be calculated within narrow limits; but, if supplied to a bean crop, not only no benefit is found, but in some cases positive injury. Green crops of all kinds, which are non-nitrogenous, are largely increased by nitrogenous manures.

The direct supply of phosphate in an assimilable form largely increases the turnip crop, whilst the same supply to the wheat crop, which is especially phosphatic, scarcely increases the yield.

The best natural fertilizer is doubtless farm-yard manure, or good stable dung. When this cannot be procured the artificial manures, such as guano-sulphate of ammonia, are valuable, especially on soils rich in the mineral constituents of plants, such as the Bermuda soils. The application of these nitrogenous manures would doubtless considerably increase the non-nitrogenous arrowroot crop, and probably the Banana. I regret I have not an analysis of the ash of the Banana to which I can refer.

These soils, were they in England, would doubtless produce large crops of wheat when manured with nitrogenous manures; and there is little doubt they would in a climate such as Bermuda.

Much interest would attach to experiments showing the power of these soils for the absorption and retention of ammonia and water. However, time precludes this at present.

FREDERICK A. MANNING.

SOIL I.—RED EARTH.

Soluble in dilute acetic acid. (1—4.)	Hygroscopic water	16·231
	Organic substance	11·210
	Silicic acid	0·126
	Sesquioxide of iron	0·040
	Alumina	0·146
	Lime	3·144
	Carbonic acid	2·251
	Sulphuric acid	Trace
	Chlorine	Do.
	Magnesia	0·015
Soluble in strong hydrochloric acid.	Sand	40·001
	Alumina	13·604
	Sesquioxide of iron	12·310
	Lime	0·364
	Magnesia	0·464
	Potash	0·118
	Soda	0·006
	Phosphoric acid (estimated in nitric acid solution)	0·626
		100·656

SOIL II.—RED EARTH.

Soluble in dilute acetic acid. (1—4.)	Hygroscopic water	23·200
	Organic substance	13·000
	Silicic acid	0·124
	Sesquioxide of iron	0·036
	Alumina	0·082
	Lime	7 840
	Magnesia	0·155
	Carbonic acid	6·750
	Sulphuric acid	0·031
	Chlorine	—
Soluble in strong hydrochloric acid.	Sand	31·640
	Alumina	7·368
	Sesquioxide of iron	9·964
	Lime	—
	Magnesia	0·169
	Potash	0·088
	Soda	0·026
	Phosphoric acid (estimated in nitric acid solution)	0·530
		100·003

SOIL III.

Soluble in dilute acetic acid.	Hygroscopic water	6·930
	Organic substance	14·910
	Silicic acid	0·034
	Sesquioxide of iron	0·332
	Alumina	0·110
	Lime	1·146
	Magnesia	0·091
	Carbonic acid	0·767
	Sulphuric acid	0·060
	Chlorine	0·042
Soluble in strong hydrochloric acid.	Sand	20·100
	Alumina	22·790
	Sesquioxide of iron	28·318
	Lime	2·188
	Magnesia	—
	Potash	0·155
	Soda	0·055
	Phosphoric acid (estimated in nitric acid solution)	0·620
		98·648

SOIL IV.—CORAL.

Completely soluble in dilute acetic acid. (1—4.)	Hygroscopic water	0·316
	Organic substance	3·806
	Sand	0·050
	Sesquioxide of iron	} 0·520
	Alumina	
	Lime	52·470
	Magnesia	1·686
	Potash	0·064
	Soda	0·243
	Carbonic acid	42·866
	Sulphuric acid	0·206
	Chlorine	0·020
	Phosphoric acid (estimated in nitric acid solution)	0·077
		102·324

SOIL V.—CHALK.

Completely soluble in dilute acetic acid. (1—4.)	Hygroscopic water	18·134
	Organic substance	3·961
	Sand	0·040
	Sesquioxide of iron } Alumina	0·180
	Lime	43·355
	Magnesia	0·638
	Potash	0·074
	Soda	0·059
	Carbonic acid	35·912
	Sulphuric acid	Trace.
	Chlorine	0·010
	Phosphoric acid (estimated in nitric acid solution)	0·105
		102·478

EXHAUSTED AND FERTILE SOILS.

Analyses of Soil from Plot 3 and Plot 2 of the Experimental Farm of Lawes and Gilbert, at Rothamsted. By HERMAN and LIEBIG.

PLOT 3.—An exhausted soil, having grown wheat thirty successive years without manure, corn and straw being carried off. (The permanently unmanured plot.)

PLOT 2.—A fertile soil. This has also grown wheat thirty successive years. The corn and straw have been carted off, and it has been highly manured each year with farm-yard manure.

SURFACE SOIL (Top nine inches).

	Constituents Estimated.	PLOT 3. Exhausted Soil Per cent.	PLOT 2. Fertile Soil Per cent.
Soluble in dilute acetic acid.	Hygroscopic water	1·825	1·810
	Organic substance	5·363	6·212
	Silicic acid	0·065	0·084
	Oxide of iron and Alumina	0·100	0·116
	Lime	2·065	1·785
	Magnesia	0·028	0·025
	Potash	0·015	0·041
	Soda	0·012	0·019
	Sulphuric acid	Traces.	0·008
	Phosphoric acid (in nitric acid)	0·075	0·093
Soluble in dilute hydrochloric acid after the solution in acetic acid.	Silicic acid	0·369	
	Oxide of iron and Alumina	4·363	
	Lime	0·233	
	Magnesia	0·064	
	Potash	0·070	
	Soda	0·054	
	Sulphuric acid	0·015	

APPENDIX B.

Abstract of Temperature Observations taken at Bermudas from the Year 1855 to the Year 1873.

YEAR.	JANUARY.				FEBRUARY.				MARCH.			
	9 A.M.		3 P.M.		9 A.M.		3 P.M.		9 A.M.		3 P.M.	
	Dry.	Wet.	Dry.	Wet.	Dry.	Wet.	Dry.	Wet.	Dry.	Wet.	Dry.	Wet.
1855	16°·4 C.	...	17°·6 C.	15°·3 C.	15°·8 C.	14°·2 C.	17°·1 C.	14°·9 C.	15°·4 C.	13°·6 C.	16°·3 C.	14°·4 C.
1856	...	14°·8 C.	...	...	16·9	15·3	18·0	16·1	16·9	15·4	18·6	16·5
1857	14·1	12·0	15·4	13·1	17·4	14·8	18·6	15·7	16·2	13·8	17·6	14·7
1858	17·1	15·2	17·8	15·4	16·8	15·3	18·1	15·9	16·7	14·4	18·1	15·4
1859	16·6	15·0	18·0	15·4	16·9	15·1	17·2	15·5	17·3	15·2	18·8	15·8
1860	17·5	16·0	19·4	17·1	17·1	15·3	18·7	16·4	16·9	14·4	18·9	13·4
1861	16·2	14·1	17·5	14·1	17·3	15·3	18·5	16·9	16·4	14·2	17·2	14·6
1862	17·4	15·5	18·4	16·4	18·1	16·1	18·4	16·6	17·5	15·3	18·4	15·5
1863	18·3	16·6	18·3	15·9	18·4	15·7	19·0	16·0	...	...	...	...
1864	19·6	17·5	19·6	16·9	18·1	15·4	18·2	13·3	20·7	16·7	20·9	16·4
1865	...	...	...	...	18·3	16·1	18·9	16·4	17·3	14·2	17·9	14·7
1866	18·8	15·8	19·0	16·1	18·3	15·9	19·0	15·8	18·9	16·3	19·1	16·4
1867	16·8	14·1	17·6	14·3	17·6	14·8	17·6	15·1	18·4	16·0	19·3	16·4
1868	18·2	15·6	18·7	15·6	17·8	14·8	18·4	15·3	18·3	15·1	18·7	15·0
1869	18·6	16·3	19·2	16·4	...	...	...	...	...	...	...	...
1870	...	...	...	...	18·5	15·1	19·1	15·3	19·6	15·8	20·4	16·4
1871	18·8	15·7	19·1	16·0	17·9	13·2	18·8	15·3	17·1	13·6	17·9	14·5
1872	17·8	15·2	18·9	15·9	...	...	...	...	...	...	...	...
1873	19·9	17·7	20·8	18·1								

YEAR.	APRIL.				MAY.				JUNE.			
	9 A.M.		3 P.M.		9 A.M.		3 P.M.		9 A.M.		3 P.M.	
	Dry.	Wet.	Dry.	Wet.	Dry.	Wet.	Dry.	Wet.	Dry.	Wet.	Dry.	Wet.
1855	18°·6 C.	17°·2 C.	19°·6 C.	17°·7 C.	20°·3 C.	19°·2 C.	21°·1 C.	19°·6 C.	24°·4 C.	22°·7 C.	25°·8 C.	23°·3 C.
1856	17·6	15·7	18·8	16·4	21·6	20·4	22·3	20·7	24·4	22·7	25·9	23·6
1857	18·3	16·8	18·6	17·4	19·9	18·3	21·4	19·2	23·3	21·7	24·9	22·4
1858	18·2	16·3	19·7	17·1	22·2	21·5	23·4	21·9	23·3	21·9	24·8	23·2
1859	18·6	16·3	19·8	17·1	20·6	17·9	21·9	18·9	23·2	20·9	23·6	21·7
1860	17·8	12·3	19·3	14·2	20·4	19·1	21·9	19·2	23·2	21·4	24·9	22·2
1861	19·9	16·7	21·4	18·1	22·4	20·1	23·3	21·2	25·2	32·1	25·6	23·2
1862	19·4	16·8	19·9	16·7	22·1	18·9	21·9	18·8	24·6	22·3	24·7	22·3
1863	...	...	...	...	...	...	...	...	...	...	...	...
1864	22·7	17·5	21·1	17·2	22·4	19·8	23·0	19·9	26·2	20·9	26·5	22·9
1865	20·5	16·9	20·4	16·7	22·9	20·6	23·4	20·8	23·3	21·2	23·7	20·1
1866	19·4	16·9	19·2	16·4	24·0	21·7	24·2	21·8	26·9	23·4	26·6	23·3
1867	20·7	17·5	21·3	17·9	22·8	17·8	23·3	19·9	26·3	22·8	27·7	22·8
1868	19·5	16·4	19·6	16·4	23·7	21·3	23·7	21·3	27·8	24·1	27·7	24·3
1869	20·9	17·1	21·7	17·4	...	...	...	...	...	...	28·6	24·6
1870	21·6	18·2	22·2	18·4	23·9	20·1	26·1	20·9	26·3	23·3	27·0	24·0
1871	19·6	16·0	20·4	16·7	...	...	...	...	...	...	...	...
1872	...	...	...	...	...	...	...	...	...	...	...	...
1873	...	...	...	...	...	...	...	...	...	...	...	...

	JULY.				AUGUST.				SEPTEMBER.			
	9 A.M.		3 P.M.		9 A.M.		3 P.M.		9 A.M.		3 P.M.	
YEAR.	Dry.	Wet.	Dry.	Wet.	Dry.	Wet.	Dry.	Wet.	Dry.	Wet.	Dry.	Wet.
1855	...	...	...	...	28°·3 C.	25°·6 C.	29°·0 C.	25°·9 C.	26°·4 C.	22°·8 C.	27°·5 C.	23°·2 C.
1856	27°·4 C.	24°·8 C.	28°·9 C.	25°·4 C.	27·2	24·8	29·0	25·4	25·7	23·2	26·6	23·7
1857	26·5	24·4	27·7	24·8	27·0	24·5	28·7	25·1	25·6	23·6	27·0	24·3
1858	24·5	22·9	26·3	23·9	26·3	24·2	28·4	25·8	25·5	23·4	27·3	23·7
1859	25·4	22·5	26·4	24·1	26·2	24·1	27·5	24·8	25·8	23·9	26·7	24·6
1860	25·9	24·1	26·8	24·6	27·1	24·3	27·7	25·1	26·7	22·8	25·8	23·3
1861	26·2	23·9	27·9	24·4	27·1	24·1	29·2	24·9	26·1	23·4	28·1	24·2
1862	25·7	23·0	25·6	22·9	28·7	23·1	29·2	23·3	27·3	21·3	28·3	21·6
1863	27·0	23·9	27·3	23·9	28·5	27·4	28·4	26·7	26·9	25·1	26·9	24·4
1864	...	...	...	...	...	...	...	...	...	...	...	...
1865	28·9	25·1	29·0	24·9	28·2	24·1	28·3	24·2	27·9	23·2	28·0	23·1
1866	27·8	24·4	27·9	24·5	27·9	24·2	27·9	24·0	28·4	24·7	28·7	23·3
1867	27·0	24·1	27·4	24·2	27·8	24·8	27·9	24·7	28·0	24·6	28·4	24·3
1868	27·9	24·1	28·1	24·3	29·6	24·6	27·6	24·1	28·2	23·9	28·3	23·8
1869	28·9	25·3	28·5	24·9	29·5	25·3	29·6	25·3	26·6	23·1	27·3	22·9
1870	29·9	25·7	31·2	26·1	29·4	24·7	30·2	25·0	26·6	23·0	26·8	23·2
1871	29·2	25·3	30·4	25·8	29·7	25·4	25·8	25·6	28·3	24·7	29·3	25·3
1872	28·1	24·7	29·6	25·5	29·2	25·3	25·5	26·2	26·9	22·7	27·6	23·1
1873	...	...	...	...	...	...	...	...	...	...	...	...

ST. THOMAS TO BERMUDAS.

YEAR	OCTOBER					NOVEMBER					DECEMBER			
	9 A.M.		3 P.M.		9 A.M.		3 P.M.		9 A.M.		3 P.M.	
	Dry.	Wet.	Dry.	Wet.	Dry.	Wet.	Dry.	Wet.	Dry.	Wet.	Dry.	Wet.
1855	22°·9 C.	21°·1 C.	24°·0 C.	21°·5 C.	21°·7 C.	19°·7 C.	21°·9 C.	19°·7 C.	18°·1 C.	16°·2 C.	19°·2 C.	16°·8 C.
1856	23·0	20·7	24·2	21·4	19·3	16·6	20·1	16·9	16·9	14·6	17·6	15·3
1857	22·1	20·4	23·1	20·9	20·5	19·2	20·9	19·2	19·2	17·1	19·8	17·3
1858	22·7	19·7	23·9	20·3	20·2	18·9	21·1	18·4	17·2	15·7	18·3	16·3
1859	22·4	20·7	23·9	20·8	20·4	18·9	20·9	19·1	18·4	16·9	19·6	17·3
1860	23·4	21·7	24·6	22·3	19·9	17·7	20·8	18·1	16·9	17·2	18·1	15·3
1861	23·9	21·9	25·7	22·7	19·6	16·6	20·7	17·1	15·9	14·2	17·6	14·6
1862	25·8	25·4	26·6	19·9	22·2	18·1	23·2	18·7	18·8	14·5	20·4	14·9
1863	24·7	22·0	24·7	22·2	22·2	19·3	22·4	19·2	18·6	15·6	18·7	15·3
1864	...	...	...	...	...	...	...	...	...	...	...	...
1865	24·0	20·3	24·5	20·4	21·8	18·2	21·3	18·1	19·9	17·2	20·2	17·2
1866	24·7	20·6	24·4	20·4	21·1	17·8	21·3	17·8	19·2	15·6	18·7	15·8
1867	24·1	19·9	23·9	19·4	21·1	16·9	20·9	16·9	18·7	15·7	18·8	16·1
1868	25·9	22·4	25·8	22·1	20·2	17·4	20·5	16·9	18·7	15·1	18·7	15·6
1869	24·6	21·2	24·8	21·1	21·4	16·7	21·3	16·7	...	...	...	...
1870	25·4	21·7	25·4	21·7	21·4	17·8	21·3	17·7	18·7	15·3	18·8	15·6
1871	24·9	20·7	24·8	20·6	20·9	16·7	21·0	17·3	20·1	17·3	19·8	17·3
1872	24·4	20·7	25·7	21·5	22·2	19·1	22·3	19·6	19·7	16·7	19·7	16·6
1873	...	...	...	...	...	...	...	...	...	...	...	...

The temperature observations from August, 1855, up to February, 1869, are taken from the Abstract published by Sir H. James in 1862; those from February, 1859, up to April, 1862, are from manuscript sheets furnished by Sir H. James in 1873; and the observations from April, 1862, up to January, 1873, are from the manuscript sheets of the Army Medical Department.

CHAPTER V.

THE GULF STREAM.

Departure from Bermudas.—Sounding and Dredging near the Islands. —*Madracis asperula.*—The determination of Surface and Deep Currents.—Difficulty and uncertainty of our present method of Observation.—The Current-drag.—Sounding in the Gulf Stream in rough weather.—The Temperature of the Stream.—*Aceste bellidifera.*—*Porcellanaster ceruleus.*—*Aërope rostrata.*—Dredged a huge Syenite Boulder.—Le Have Bank.—Mirage.—Halifax.— Ice-markings.—Recross the Gulf Stream.—General considerations. —Comparison between the Gulf Stream and the Japan Current. —*Calymne relicta.*—*Ophioglypha bullata.*—*Lefroyella decora.*— Return to Bermudas.

APPENDIX A.—Table of Serial Temperature-Soundings taken between St. Thomas, Bermudas, and Halifax

APPENDIX B.—Table of the Bottom Temperatures taken between St. Thomas, Bermudas, and Halifax.

APPENDIX C.—Specific Gravity Observations taken between St. Thomas, Bermudas, and Halifax.

APPENDIX D.—Table of Meteorological Observations made in crossing and recrossing the Gulf Stream.

As I have already mentioned, towards the end of April we left Bermudas for a time, crossing the Gulf Stream to the neighbourhood of Sandy Hook. We coursed along the outer edge of the fishing banks to Halifax, where we remained ten days ; and then we ran another section across the Gulf Stream in a southwardly direction and returned to Bermudas on the

31st of May. Besides the very important one of getting braced up by a little cold weather before our long cruise in the Tropics, our chief object in taking this trip was to see for ourselves the wonderful ocean-river which had excited so much admiration and interest, and given rise to so much controversy; we did not anticipate being able in the short time at our disposal to aid in throwing any additional light upon a phenomenon which has already been so carefully and admirably investigated by the United States Coast Surveyors, except possibly by tracing the relation of the deeper layers of water beneath the stream with the layers at similar depths in other parts of the ocean.

We steamed out of the Camber on the morning of the 21st of April, and, going through the narrows, took a series of soundings following as nearly as possible the hundred fathom line. The bottom falls off very suddenly along the eastern and northern coasts; we got a sounding in the afternoon, Gibbs'-Hill Lighthouses 2° E. $18\frac{3}{4}$ miles distant, in 1,375 fathoms; the bottom temperature was 3° C., and the bottom a fine grey mud with a few foraminifera. On the following day we proceeded outside the reefs to the westward and southward, and sounded successively in 2,450, 2,100, and 1,950 fathoms, finding in each case a bottom of grey mud, chiefly the detritus of coral with a scanty sprinkling of foraminifera, and a bottom temperature of 1°·6 C. In the evening we sounded in 32 fathoms about 13 miles to the south-west of Bermudas; this is a bank well known to the Bermudas fishermen, and is said to have been discovered from the large number of fish swimming near the surface.

We anchored on the bank and the fishing-lines were soon out, but we were very unfortunate, for only one or two 'snappers' were taken. Early on the morning of the 23rd the surveying boats left the ship to sound out the bank; it was cold, blustering, unpleasant weather with a falling barometer and rising wind. During the day we sent the jolly-boat away

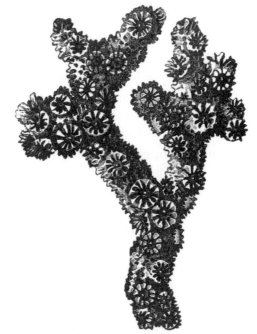

Fig. 88.—*Madracis asperula*.

to lower the small dredge a couple of hundred yards or so from the ship; the dredge was then slowly dragged to the ship by the donkey-engine. The bank, which seems to be about five miles across, consists mainly of large rounded pebbles of the substance of the Bermudas 'Serpuline reef.' There is an abundant growth all over the pebbles of the pretty little

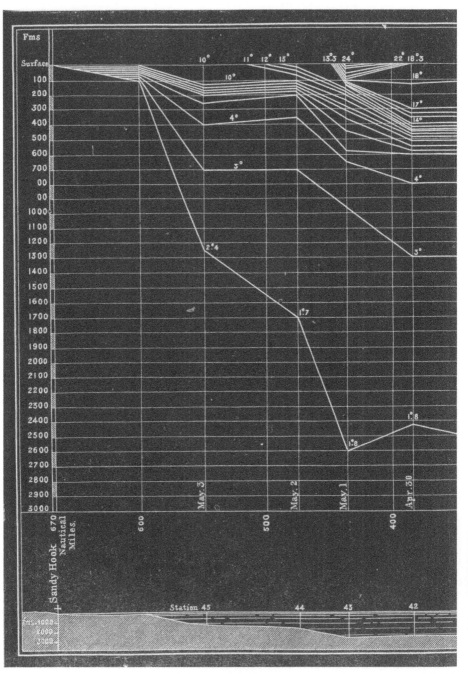

PLATE XI.—DIAGRAM OF THE VERTICAL DISTRIBUTION OF

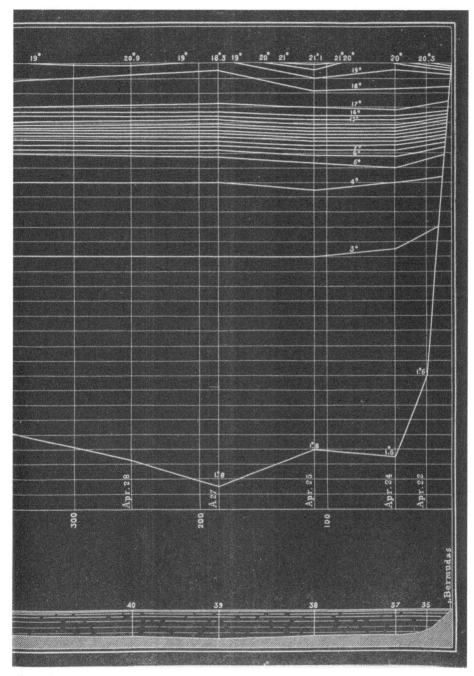

TEMPERATURE BETWEEN BERMUDAS AND SANDY HOOK.

branching corals *Madracis asperula* (Fig. 88), and *M. hellana;* and other invertebrates were abundant, particularly one or two large star-fishes. The fauna seemed however to be of the ordinary character, such as we found in shallow water everywhere round the Islands. As the weather did not look by any means promising we weighed anchor as soon as the boats returned and proceeded on our course to the north-west.

On the 24th we sounded and dredged in 2,650 fathoms 43 miles in a north-westerly direction from Gibbs'-Hill Lighthouse; the dredge brought up a little grey calcareous mud still evidently chiefly derived from the Bermudas reef; with *Globigerinæ, Orbulinæ,* some larger foraminifera, some otolites of fishes, and some shells of *Clio* and *Diacria.* It was evident, from the mud in the dredge-bag containing many more foraminifera than some which had been caught in bulk on the tangle-bar, that a large quantity of fine débris had been washed through the 'bread-bag' during the progress of the dredge to the surface; and this made it all the more remarkable that no animal higher in the scale than a rhizopod was contained in the dredge.

A series of temperature-soundings was taken at intervals of 50 fathoms down to 700 fathoms. (See Table—Appendix A to this chapter), and a second series to a depth of 150 fathoms at closer intervals gave the following results :—

Surface	. . . .	20°·0 C.	100 fathoms	. .	18°·5 C.
20 fathoms	. . .	19·6	120 ,,	. . .	18·3
40 ,,	. . .	19·0	140 ,,	. . .	18·3
60 ,,	. . .	18·5	150 ,,	. . .	18·2
80 ,,	. . .	18·6			

While the dredge was down observations were made on the direction and force of the currents at the surface and at different depths below it. The surface current can usually be determined without any great difficulty; indeed we get at all events a rough approximation to its determination, in the difference at the end of a given time between the position of the ship by observation and her position by dead reckoning. In fine weather, however, the surface current may be determined much more exactly. When the dredge is well on the bottom one of the quarter-boats is lowered and anchored to the dredge-line, the line between the boat and the ship being kept slack, and the ship drifting away. The boat thus becomes a fixed point, and from it a current-log is run out, the log-ship consisting of a triangular piece of wood weighted at the apex, and kept at the surface by an oar lashed across its base. The log-line is marked to fathoms, and is allowed to run for a given time, say six or twelve minutes; the line is then checked and the bearing of the log-ship taken from the boat, which gives the direction of the current; while the number of fathoms run out multiplied by the proportion which the time of running bears to an hour gives its hourly rate.

Various means have been devised for ascertaining the direction and rate of currents of water at different depths below the surface, but none of these can as yet be considered satisfactory. The difficulties may be best explained by describing the method employed in the 'Challenger.' It is necessary in order to investigate a deep current, to sink to the required depth some object which will be taken a good hold of by the current and carried along with it; and to measure the movement of this object it must be attached to a float on the surface which shall be affected as little as possible by the surface drift, by a line which shall be affected as little as possible by intermediate movements. For this purpose we use an instrument composed of three parts—the 'current-drag,' the 'line,' and the 'watch-buoy.' The first of these consists of a light frame of hammered iron four feet high and four feet across (Fig. 89), supporting four vertical fans of canvas. The iron cross-pieces are fastened by a bolt in the centre so that they can be folded away when not in use; and they are kept at right angles by a lanyard attached to their ends.

Fig. 89.—The 'Current-drag' and 'Watch-buoy.'

To the bottom of the frame a ½ cwt. lead is slung to sink it rapidly in the water, and the 'current-line,' a sufficient length of ordinary service 'cod-line,' is fastened to the top. When the drag has been let down to the desired depth, the line is attached on the surface to the 'watch-buoy,' a spindle-shaped iron buoy five feet long by one wide in the centre, so fashioned as to expose as little surface as possible to the drift or the wind, while it has sufficient buoyancy to sustain a weight of 70 lbs. in the water.

In using this instrument, the direction and force of the surface-current is first ascertained in the manner already described, and the boat then frees itself by letting go the dredge-rope, which is hauled in by the ship. The current-drag is lowered to say 50 fathoms from the boat, and the 'watch-buoy' attached. The boat with the observer then follows the 'watch-buoy' closely, without interfering with its movements; and the surface log is again dropped from the boat, allowed to run for a given fraction of an hour and checked, when its bearing and the length of line run out give the direction and rate of the surface-current from the boat. But the boat is no longer a fixed point, it is keeping with the 'watch-buoy;' that is to say it is moving in the direction and with the rate of a current at a depth of 50 fathoms. As the log-ship is free to move with the surface-current, all divergence whether in rate or direction between it and the watch-buoy must be due to influences acting upon the latter, for they would otherwise be drifted along together, and the rate and direction of the surface-current being already known, the deep-water movement can be readily calculated

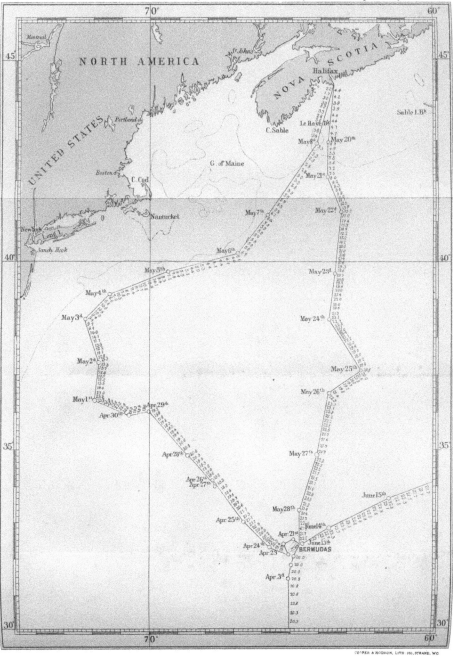

PLATE XII. *Diagram showing the Surface Temperature observed in crossing and re-crossing the Gulf Stream.*

The material originally positioned here is too large for reproduction in this reissue. A PDF can be downloaded from the web address given on page iv of this book, by clicking on 'Resources Available'.

from the relative positions of the watch-buoy and the log-ship; the actual movement of the watch-buoy with reference to a fixed point will be the resultant of the movement of the surface-water and the movement of the watch-buoy through the surface-water. This would be sufficiently simple if we could suppose that the surface-drift has no influence upon the watch-buoy and that the movements of intermediate water

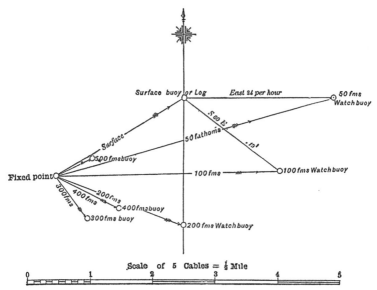

FIG. 90.—Result of Current Observations made on the 24th of April reduced to a diagrammatic form.

do not affect the 'current-line;' but we are well aware that this cannot be the case, and as we have no data for estimating the amount of error arising from these sources, our current determinations can as yet be regarded only as approximations. The following table gives the results for the 24th of April.

LAT. 32° 18′ N.   LONG. 65° 38′ W.

Surface current	. .	N. 60 E.,	0·24	of a mile an hour.
Current at 50 fathoms		N. 75 E.,	0·46	,, ,,
,, 100	,,	N. 87 E.,	0·36	,, ,,
,, 200	,,	S. 70 E.,	0·22	,, ,,
,, 300	,,	S. 40 E.,	0·08	,, ,,
,, 400	,,	S. 65 E.,	0·11	,, ,,
,, 500	,,	N. 65 E.,	0·06	,, ,,
,, 600	,,	No current detected.		

The results are shown in a diagrammatic form in Fig. 90.

April 25th.—We sounded in 2,600 fathoms, with a bottom temperature of 1°·8 C. The mud from the sounding was reddish and very smooth, containing scarcely any foraminifera and only a very small proportion of carbonate of lime. Serial temperatures were taken to 1,500 fathoms at intervals of 50 fathoms (Appendix A.), and for the first 200 at intervals of 20 fathoms:—

Surface	21°·1 C.	120 fathoms	18°·8 C.
20 fathoms	20·5	140 ,,	18·5
40 ,,	20·0	160 ,,	18·2
60 ,,	19·4	180 ,,	17·3
80 ,,	19·0	200 ,,	17·7
100 ,,	18·9		

During the day the force of the wind gradually increased, and towards night-fall it was blowing half a gale from the south-west, with a steadily falling barometer and vivid sheet-lightning on the northern horizon. The gale continued during the night and for the greater part of the following day, the wind veering round to the north-west, and the sea running so high as to render sounding or dredging impossible; towards evening, however, the

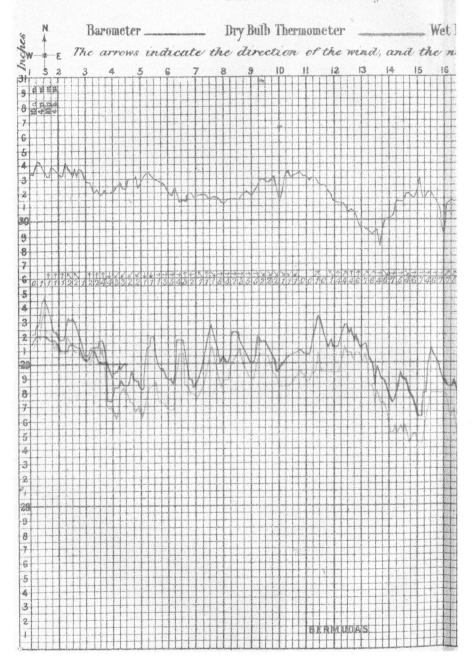

PLATE XIII. Meteorological Observations

Barometer ——— Dry Bulb Thermometer ——— Wet Bulb

The arrows indicate the direction of the wind, and the n...

BERMUDAS

rvations for the month of April, 1873.

Wet Bulb Thermometer —————   Temperature of Sea Surface —————

e numbers beneath its force according to Beaufort's scale

weather moderated and the barometer rose. Owing probably to the change of the direction of the wind the temperature of the air was 7° C., below that of the day before.

On the 27th, at a distance of 450 miles from Sandy Hook the depth was 2,850 fathoms, with a bottom of reddish-grey ooze containing some *Globigerinæ*, and a bottom temperature of 1°·8 C. As we were now approaching the southern border of the Gulf Stream two sets of serial temperatures were taken at intervals of 25 fathoms from the surface:—

First Series.		Second Series.	
Surface	18°·3 C.	Surface	—
25 fathoms	17·4	25 fathoms	18°·5 C.
50 ,,	17·9	50 ,,	18·9
75 ,,	—	75 ,,	18·6
100 ,,	17·6	100 ,,	18·3
125 ,,	—	125 ,,	17·9
150 ,,	17·6	150 ,,	18·2
170 ,,	17·8	175 ,,	17·8
200 ,,	17·8	200 ,,	17·7
225 ,,	17·1	225 ,,	17·6
250 ,,	17·2		

The small dredge was put over early on the following morning with a bread-bag lining to the net and a 28 lb. weight slung to the tangle-bar. In the forenoon a sounding was taken in 2,675 fathoms with a 'hydra' machine weighted with 3 cwt. having a water-bottle and two thermometers attached. It seems that the weights did not detach, for there was a great strain in heaving in and finally the line parted, and the observations on temperature and on the specific gravity of the water were lost. The dredge came up with a small quantity of mud containing many shells of foraminifera; the finest part of the mud having been, as is usual at these great depths, almost entirely washed through the bread-bag. The dredge contained no animals higher in the scale than foraminifera which we could say with any certainty had come from the bottom.

On the 29th it was blowing hard with a heavy swell, and our first attempt to take a series of temperatures was unsuccessful. In the morning the starboard wheel-rope was carried away, and the necessary repairs caused some delay. Later in the day the swell moderated somewhat, and we took a temperature sounding down to 650 fathoms at intervals of 50 fathoms (Appendix A), and of 25 fathoms to 250 :—

Surface	18°·3 C.	150 fathoms . . 17°·7 C.
25 fathoms . . .	18·3	175 ,, . . . 17·6
50 ,, . . .	18·2	200 ,, . . . 17·4
75 ,, . . .	18·2	225 ,, . . . 17·2
100 ,, . . .	18·1	250 ,, . . . 17·1
125 ,, . . .	18·0	

There was as yet no rise in the surface temperature sufficiently marked to indicate that we were in any way affected by the stream.

On the 20th of April, at 8 A.M., the temperature of the sea-surface was 18°·3 C., much as before; we took a sounding in 2,425 fathoms with a bottom of grey mud and a bottom temperature of 1°·8 C. The swell was still heavy, and sounding operations were carried on with some difficulty; we were able, however, to get a set of temperature soundings from 800 to 1,500 fathoms to complete the series of the previous evening. During the day we made about five knots an hour in a westerly direction, and at 2 P.M. the temperature of the sea-surface suddenly rose from 18°·6 C. to 22° C., showing that we had slipped over the southern edge of the Gulf Stream; owing to the cloudy and somewhat boisterous weather little difference was seen in the appearance of the water. At 8 P.M. the surface temperature of the water had reached 22°·8 C., while the thermometer in the air stood at 17°·8 C. The position of the ship at noon was lat. 35°58′ N., long. 70°39′ W., and the distance from Sandy Hook 308 miles.

We began to sound early on the 1st of May. The surface temperature was 24° C., and the stream was very manifest, running past the ship and surging up against the 'Burt's nippers' of the sounding-line with the rapidity and force of a mill-race. The sea was still running high; two attempts were made to obtain soundings; during the first of these the wind was from the eastward blowing with a force = 4, and the current dead against it, the ship lying broadside on with her head to the northward. To reach the bottom as quickly as possible 4 cwt. of sinkers were piled on the 'hydra' tube, and a new No. 1 sounding-line was run freely from the lee side of the ship. After a short time the line went clear of the ship's side, showing that the weight and the instruments attached had got beneath the rapidly-moving water and that the ship was being carried to windward by the current faster than she was drifting through the current to leeward. Sail was made to increase her drift, but it was found on taughtening the line that it was still running out at a considerable angle, and before the ship's head could be brought round to steam up to it the line unfortunately parted, and the 'hydra,' two thermometers, and a water-bottle were lost. There was scarcely sufficient strain shown on the accumulators to account for this, and the accident may probably have been caused by a sudden jerk, or by some imperfection in the line. The 'hydra' was again sent down with 3 cwt. of weights on a No. 1 sounding-line as before, and the ship was put before the wind. The line was checked from time to time to insure its being as perpendicular as possible. Each time the line ran out it was carried astern in a bight by the current, and when checked it came forward cutting its way through the water until it was up and down.[1] To keep up to the line the ship was steamed ahead at the rate of three knots an hour, and on the least decrease of speed the line moved ahead, or rather the ship was carried astern. All our usual time-indications of the progress of the sounding were at fault, and in a most perplexing way, for the rate of the current was nearly the same as the ordinary rate of the running out of the sounding-line, so that had the weights reached the bottom the line would have been carried out by the water with the same rapidity as before. When 2,650 fathoms were out there was a lessened strain on the accumulators and the line was hove in, but on the arrival of the instruments at the surface it was found that the weights had not slipped, and had in fact never reached the bottom; although they could not have been far from it, as the thermometers registered a temperature of 1°·8 C.

[1] See Reports of Captain G. S. Nares, R.N., to the Admiralty, published in 1874, p. 6.

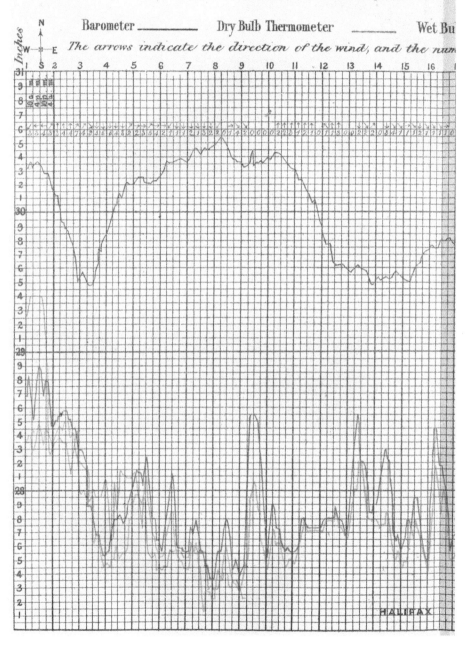

PLATE XIV. *Meteorological Observations*

Barometer ——— Dry Bulb Thermometer ——— Wet Bulb

The arrows indicate the direction of the wind, and the number

HALIFAX

vations for the month of May, 1873.

Bulb Thermometer ——— Temperature of Sea Surface ———

umbers beneath its force according to Beaufort's scale

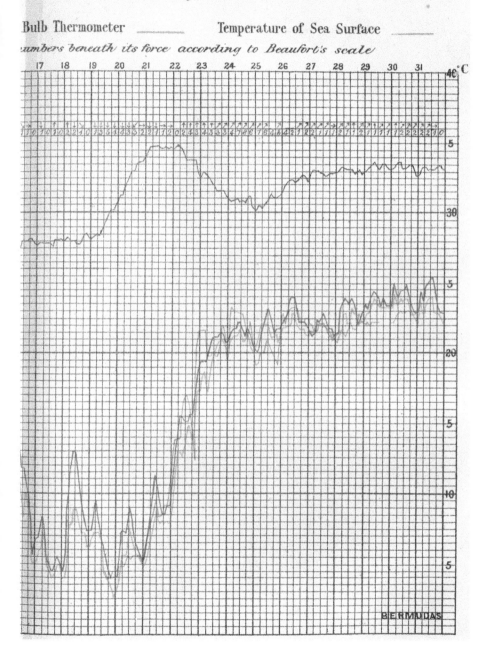

In the course of the afternoon three series of temperature observations were taken which showed that we were on the slope of the so-called 'cold wall.'

Surface	. . . .	23°·9 C.	200 fathoms	. .	11°·1 C.
20 fathoms	. . .	21·8	225 ,,	. . .	10·1
40 ,,	. . .	21·5	250 ,,	. . .	9·2
60 ,,	. . .	21·7	300 ,,	. . .	8·2
80 ,,	. . .	20·0	350 ,,	. . .	7·2
100 ,,	. . .	18·3	400 ,,	. . .	—
125 ,,	. . .	13·8	500 ,,	. . .	5·75
150 ,,	. . .	13·3	550 ,,	. . .	5·3
175 ,,	. . .	12·6	600 ,,	. . .	4·25

The result shows that the Gulf Stream in its restricted sense, that is to say the mass of warm water which issues from the Strait of Florida and courses in a north-easterly direction at a little distance from the coast of North America, was, early in May 1873, at the point where we crossed it and made our observations, about 60 miles in width, 100 fathoms deep, and its rate three knots an hour. I make the statement thus guardedly because descriptions of the stream are somewhat discrepant, and I have no doubt that it varies at a considerable extent both in rapidity and volume, influenced by the season and by different meteorological conditions. It seems evident, as has been already observed by the American coast Surveyors, that the Labrador return current, which is banked up against the American coast within the Gulf Stream, passes also under it to a certain depth. Comparing Fig. 91, a diagram constructed by combining the series of temperatures at Stations 41 and 42 just before entering the Gulf Stream, with Fig. 92 constructed from the soundings near the centre of

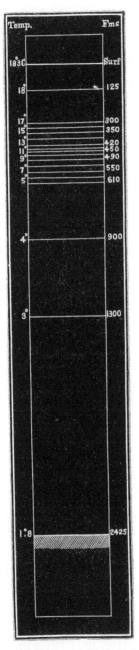 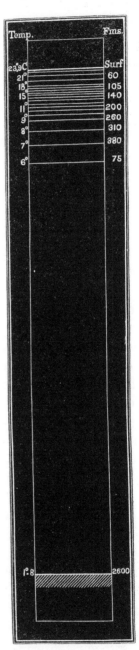

Fig. 91.—Diagram showing the relation between depth and temperature a Stations 41 and 42.

Fig. 92.—Diagram showing the relation between depth and temperature at Station 43.

the stream, we find that at a depth of 100 fathoms the temperature is nearly the same in both; in the former 18°·1 C. and in the latter 18°·3 C.; at 200 fathoms the temperatures are respectively 17°·4 C. and 11°·1 C., and the difference between them 6°·3 C.; at 300 fathoms the temperatures are 17° C. and 8°·2 C., and the difference 8°·8 C.; at 400 they are 13°·7 C. and 6°·6 C., with a difference of 7°·1 C. They now begin once more to approach; at 500 fathoms the temperatures are 8°·7 C. and 5°·75 C., with a difference of 3° C.; and at 600 fathoms they are 5°·3 C. and 4°·25 C., with a difference of about 1° C. The bottom temperature at 2,425 fathoms at Station 42 was 1°·8 C. and at Station 44, just within the Gulf Stream, it was 1°·7, and it is singular that so near one of the sources of cold and directly in the path of one of the most marked polar return-currents the temperature of the bottom water should be higher than that of the deep water of the Middle and South Atlantic.

We sounded on the morning of the 2nd of May at a distance of 209 miles from Sandy Hook in 1,700 fathoms, with a bottom of blueish grey mud, containing a considerable proportion of *globigerinæ*, but not what could be called a true 'globigerina-ooze.' The bottom temperature was 1°·7 C. In the course of the day temperature soundings were taken down to 1,500 fathoms at intervals of 50 and 100 fathoms (Appendix A); and at intervals of 25 to 200.

Surface . . . . 11°· 0 C.	125 fathoms . .	8°· 3 C.
25 fathoms . . . 10 · 5	150 ,, . . .	6 · 6
50 ,, . . . 11 · 4	175 ,, . . .	—
75 ,, . . . 10 · 8	200 ,, . . .	5 . 5
100 ,, . . . 9 · 7		

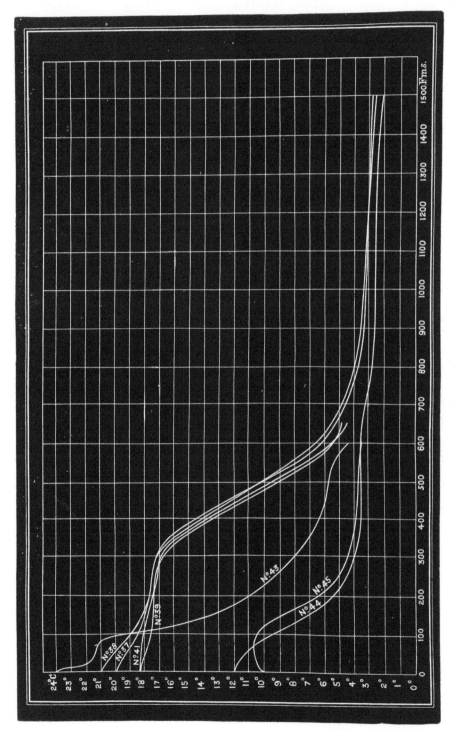

FIG. 93.—Curves constructed from Serial Temperature-Soundings between Bermudas and Sandy Hook

Although on the surface the influence of the Gulf Stream was still felt to a certain extent, the contrast between the observations of this day and those of the day before was most marked; we had crossed the 'cold wall,' and the temperatures registered were almost purely those of the Labrador return current. The dredge was put over shortly after midday, and veered to 2,500 fathoms. It came up in the evening with a considerable quantity of the blueish clay, and the dredge-bag contained many animals of different invertebrate groups, while a large assemblage of larger and more striking forms were on the tangles. The collection as a whole had a decidedly Arctic character, and recalled some of our dredgings on the coasts of Northern Europe, although it seemed that few of the forms were absolutely identical. There were many large foraminifera; most of these were of the arenaceous type, but there were also several calcareous forms, including large examples of *Crystellaria*, *Pulvinulina*, and the delicate *Orbitolites tenuissimus*.

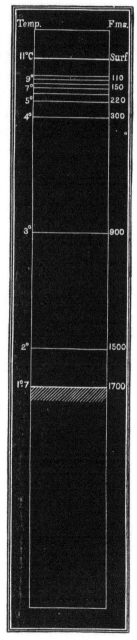

Fig. 94.—Diagram showing the relation between depth and temperature at Station 44.

Star-fishes and ophiurids were abundant chiefly on the tangles; among the former several fine species of *Archaster* apparently undescribed, although approaching very closely forms from the seas of Shetland and Færoe, and among the latter some remarkably large examples of *Ophiomusium lymani.*

We took here a second specimen of an irregular sea-urchin which we had found previously at Station VIII. off Gomera Island in 600 fathoms, for which I propose the name *Aceste bellidifera* (Fig. 95). This

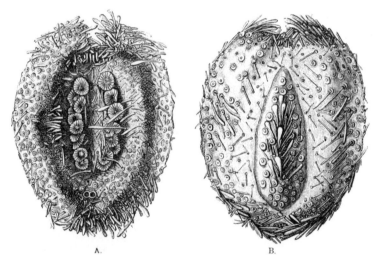

FIG. 95.—*Aceste bellidifera*, WYVILLE THOMSON. A. Upper surface; B. Under surface. Twice the natural size. (No. 44.)

appears to be one of a widespread and characteristic deep-sea family to which *Pourtalesia* belongs. As pointed out by Professor Alexander Agassiz, this family is certainly allied in many respects to the *Ananchytidæ;* but there are important points of divergence, and when we have had an opportunity of comparing them carefully, it may possibly be necessary

THE GULF STREAM.

to define a new family for the reception of a considerable number of kindred forms. In the present species the test is oval and depressed. The apex, with the madreporic tubercle, and two very large ovarial openings, is on the dorsal surface near the posterior extremity; the mouth is rounded or somewhat irregular in form, and is at the bottom of a deep anterior groove; and nearly the whole of the dorsal surface is occupied by a depression beneath which the anterior

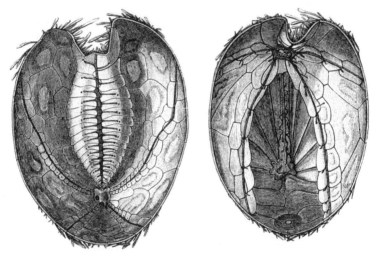

FIG. 96.—*Aceste bellidifera*, WYVILLE THOMSON. Inner surface of the test. Twice the natural size. (No. 44.)

ambulacral canal runs, sending up through a petaloid arrangement of two double rows of pores, two series of tube-feet with very large flower-like terminal disks, the disks supported by an elegant star of calcified tissue. The four paired ambulacra are slightly marked compared with the odd ambulacrum, a slender tube running under the ranges of ambulacral plates and giving off simple processes to single pores.

The excretory opening is at the posterior extremity; only two ovaries are developed, and the eggs appear to be very large at the time of expulsion, corresponding with the great size of the ovarial openings. The apical plates are arranged in the usual way, and not 'disjunct' as in *Pourtalesia*, which might seem to indicate a wide difference between the two genera; but I am inclined to think that the weight of resemblances is in favour of their approximation. In this species there is a singular ingrowth of the plates of the perisom which causes them to overlap, and shows very distinctly, as in *Calveria*, that the ambulacral ranges of plates are essentially *within* the interambulacral (Fig. 96). The specimen examined was slightly distorted.

On Saturday the 3rd we sounded in 1,240 fathoms and lowered the dredge, which again gave us a very full sample of the fauna. Star-fishes allied to the genus *Archaster* were once more most prominent, and among these were several specimens of a very beautiful little sea-star, which I propose to name *Porcellanaster ceruleus*, most nearly allied to *Ctenodiscus*, but presenting many marked differences. The disk in a full-sized example is about 20 mm. in diameter, and length of the arms nearly equals the diameter of the disk. The ad-ambulacral plates are large, and each bears usually two flattened somewhat irregularly-shaped spines. Those plates forming the angles of the mouth are unusually flattened and expanded. The marginal plates are of large size, and arranged in two rows. The surface is finely granular, and each plate of the upper series bears near its inner edge a rounded tubercle. The two terminal marginal plates on each arm are fused together and bear two diverging spines, one on either side, and above these on the dorsal aspect a central spine set on a low tubercle. In the re-entering angles between the arms the two central pairs of marginal plates are closely covered with minute flattened scales inserted on edge and arranged in vertical rows. This is a most characteristic style of ornament; it looks as if there were a little brush between each pair of arms. The

Fig. 97.—*Porcellanaster ceruleus*, WYVILLE THOMSON. Oral surface. Natural size. (No. 45.)

perisom of the dorsal surface is loaded with narrow calcareous plates which run together towards the ends of the arms so as to form an almost continuous calcareous investment; paxillæ are scattered over the disk, and the outer layer of the perisom of the disk has a very delicate colour ranging from a pale to a tolerably strong cobalt-blue. The calcareous plates are clear white with somewhat of a porcellanous lustre, and look harder than the surface plates usually do in star-fishes. The madreporic tubercle is large with sub-parallel grooves and ridges; the excretory opening is very distinct in the centre of the dorsal perisom of the disk. This is a very widely distributed deep-water species. We met with it near Tristan d'Acunha, in the Southern Sea, and in the North Pacific. There is an allied species of the same genus

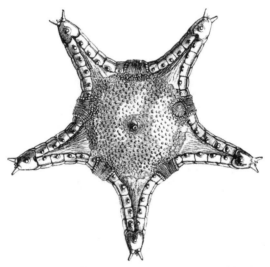

Fig. 98.—*Porcellanaster ceruleus*, Wyville Thomson. Dorsal surface. Natural size. (No. 45.)

somewhat more ornate, and of an orange instead of a blue colour in the China Sea.

The same haul gave us a singular little addition to the family to which *Pourtalesia* and *Aceste* are referred. *Aërope rostrata* (Fig. 99) is a small species, little more than a centimetre in length. Like *Pourtalesia*, it is nearly cylindrical, and the odd interambulacral area is greatly prolonged backwards, and the excretory opening brought up upon the apical aspect.

The apex with a peculiar madreporic tubercle, with

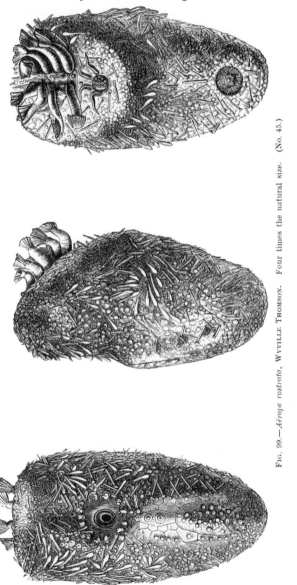

FIG. 99.—*Aërope rostrata*, WYVILLE THOMSON. Four times the natural size. (No. 45.)

large perforations and four ovarial openings with

long exsert-ovarial tubes, is a little in advance of the centre of the dorsal surface and just within a kind of disk surrounded by a semita. The anterior ambulacral canal runs down the centre of this disk, and gives off eight or ten disproportionately large tube-feet with large terminal rosettes as in *Aceste*. The mouth is round and opposite the apex, a little further forward than the centre of the oral surface of the test. The mouth is surrounded by a ring of large oral tentacles. The body is sparsely clothed with fine silky bristles, interspersed with larger paddle-shaped spines.

The temperature both of the air and of the sea was considerably lower than on the previous day. The distance from Sandy Hook was 139 miles at noon. The following day we altered our course towards Halifax and proceeded under steam. On the 6th of May we sounded in 1,350 fathoms, with a bottom temperature of 2°·3 C., the temperature of the surface having now sunk to 4°·4 C. The dredge brought up a small quantity of grey mud, and the tangles were covered with animals of various kinds, particularly star-fishes, including a large number of a fine species of the genus *Zoroaster*; many examples of an *Astropecten* nearly allied to *A. tenuispinus* of which it is probably only a variety, but much more robust and lighter in colour than the European specimens; many gay specimens of *Brisinga endecacnemos*, as usual greatly broken but some retained five or six arms in connection with the disk. During the greater part of the day there was a very brilliant halo round the sun showing prismatic colours. The day was quite calm and slightly hazy.

May 7th.—We sounded and dredged in 1,340

fathoms, a slight current running to the eastward. The bottom was grey mud with many foraminifera. The dredge came up about mid-day with a tremendous strain, and we found that a huge block of syenite weighing 5 cwt. had got jammed between the mouth of the dredge and the arms. We had some difficulty in getting our unwelcome prize safely landed on the bridge. There it remained until we left Halifax, where it excited as much interest as any of our other captures. Dr. Honeyman, the energetic curator of the Halifax Museum, wished to keep it as a souvenir of the expedition, a wish to which we readily acceded; but somehow or other in the hurry of leaving the boulder was forgotten and remained on the bridge; until, on our return voyage, passing not far from the place, we thought it as well to put it back where we found it.

May 8th.—At 5 A.M. we stopped on Le Have bank and sounded in 51 fathoms with a hard bottom. The dredge was put over, and although the dredge-bag was empty, the outside of the dredge and the tangles were crowded with animals, principally echinoderms. The most remarkable of these was *Antedon escrichtii*, the great northern feather-star, which adhered in great numbers by its dorsal cirri; and many of them were infested with their peculiar parasite *Myzostoma*. There were many examples of a large form of *Ophiopholis aculeata*, which seems to be the principal food of the cod on the Nova Scotia and Newfoundland Banks, taking the place of *Ophiothrix fragilis* off Færoe, the north of Scotland, and the Dogger Bank. In the afternoon lines were out all round the ship, and upwards of a hundred cod were taken, the largest about 30 lbs. weight. The stomachs of all of them were full of the *Ophiopholis*. We reached Halifax on the morning of the 9th. The weather was very fine and perfectly still with a light mist, and as we steamed up the bay there was a most extraordinary and bewildering display of mirage. The sea and the land and the sky were hopelessly confused;—all the objects along the shore drawn up out of all proportion, the white cottages standing out like pillars and lighthouses; and all the low rocky islands looking as if they were crowned with battlements and towers. Low hazy islands which had no place on the chart bounded the horizon, and faded away while one was looking at them. The little coasting vessels with their hulls drawn up looked as if they were standing on pedestals, while above them their inverted images on the soft grey mist were more real and definite than they were themselves. None of us had ever seen such an extreme effect before, nor have we seen anything like it since.

Coming directly from the rich vegetation and the lovely sub-tropical spring weather of Bermudas, the first view of the country about Halifax was certainly by no means attractive. The low rolling hills of granite and metamorphic rocks covered with somewhat stunted pine-woods, remind one of some of the tamer parts of Scandinavia. When we arrived the weather was still chilly; on the Saturday before there had been a fall of snow, and the frost was still in the ground. The deciduous trees were leafless, although the large crimson buds swelled fast on the maples during our stay; but the pines and the hemlocks were in great beauty, just beginning to send out their new leaves. Early in the forenoon we were alongside of the coaling wharf. Halifax is not a pretty town. It reminds one greatly of Greenock or a second-rate English seaport, with its dull streets of square houses blackened with coal-smoke. The houses are almost all built of wood, and there is no attempt to lighten the effect by introducing colour. In the centre of the town there are some rather better streets, with good shops and one or two fair public buildings. The Post-office is one of these, and in one of its spare rooms the local Museum has temporary accommodation. The collection is a very miscellaneous one, but it contains many good things, among them some beautifully-stuffed birds, the work of Mr. Downs, an old gentleman who has devoted his life, partly as an amateur and partly as a matter of business, to the preparation of objects of natural history; and, simply by becoming intimately acquainted with them in the field, has acquired a dexterity in reproducing their characteristic attitudes, particularly in repose, which I have never seen surpassed. The collection of specimens illustrating the geology of Nova Scotia, which is under the special superintendence of Dr. Honeyman, the Government geologist, is also very good, and highly instructive to a British naturalist.

At Halifax I had the pleasure of meeting an old Edinburgh friend, Professor Lawson, and he and I had some excursions, as pleasant as the cold damp weather would allow, round the town and into the border of the 'forest primeval,' which stretches away to the westward towards the Bay of Fundy, the Basin of Minas, and the 'beautiful village of Grand-Pré.'

Spring was just breaking, and the spring plants, most of them unfamiliar to Scottish eyes, were beginning to shoot up and show their flower-buds. It was strange to see the *Saracenia*, so prized and so difficult to rear in our conservatories at home, expanding great patches of its ruddy-veined vases in the swamps. I took a quantity of it to Bermudas, and when we left it was still looking fresh and coming into flower; but I should think there was not the least chance of its succeeding there, the conditions are so entirely different. But by far the most charming of the Nova Scotia spring flowers was the 'May-flower' (*Epigæa repens*), which a week or two before had been struggling to form its buds beneath the snow, and was now stretching out its long, trailing sprays covered with trusses of pale flowers, among the grass and moss. Day after day we filled our cabins with wreaths of it, and enjoyed its delicate perfume, which reminded one of that of the fields of *Linnæa borealis* in the pine forests of Dalecarlia; and our thoughts wandered back to the old 'May-flower' and the devoted little band who left all and went so far for the love of God.

We drove a few miles beyond Dartmouth one day to see some gold mines, which they are still working actively. The mines are in a wild, bleak piece of country; from a height you can see the long line of little square sheds with their iron chimneys and their winches, and heaps of rubbish thrown out from the workings, stretching along the strike of the metamorphosed schist, while here and there is a crushing-mill which does the work of several of the mines. The schist is crushed in the usual way by ranges of heavy steel pestles working by steam in steel cylinders, and the gold is then extracted by amalgamation with mercury, being shaken in contact with it by a vibrating motion communicated by the engine, down a long inclined trough. The yield is small but tolerably certain, and now and then they come upon a comparatively rich vein to keep up the gambling stimulus, which makes the direct search for gold so attractive; still it seemed to me that there was less of the gambling spirit among the gold-diggers of Nova Scotia than elsewhere, and they seemed more steady and industrious.

May 16*th*.—We went with the photographer to 'The Point,' a little way out of the town, where there is a very astonishing exhibition of the action of ice. There is a round tower at the top of 'The Point,' with a guard of soldiers and mounting a few cannon, and this tower stands in the middle of an area of one or two acres, where the rock, a highly altered Silurian schist, is perfectly bare and polished. The undulations and contortions in the foliations of the schist are seen in section on the polished surface; and traversing these sinuous markings there is a wonderful system of parallel ruling in grooves of greater or less depth, crushed into the stone by boulders and fragments of rock borne by the ice-cap in its slow progress over it. On our way Dr. Honeyman pointed out to us, and we photographed, a very well marked synclinal axis in the schist, exposed in section in the sea-cliff. This axis seemed to be the key to the structure of the whole district, for the schists dipped down to it, in different directions on either side, for a long distance. Professor Alexander

Agassiz paid us a short visit from Boston during our stay, and we had great pleasure in introducing him to some novelties in his favourite groups, and chatting over our plans with him.

On the morning of the 15th a large number of the members of the Nova Scotia Institute and their friends came on board, under the guidance of Dr. Honeyman, and we took them over the ship and showed them all our appliances; and at 4 o'clock in the afternoon we cast off from the jetty and proceeded on our return voyage to Bermudas.

On the following morning we sounded in 83 fathoms on the eastern extremity of Le Have bank, with a bottom of stones and gravel, and a bottom temperature of $1°·7$ C., that of the surface being $4°·7$ C. Four hauls of the dredge were taken in rapid succession, the dredge remaining on the bottom on each occasion for about an hour. We thus got an extremely good idea of the fauna, which was decidedly sub-arctic in its character, with abundance of the characteristic large amphipod crustaceans and large PYCNOGONIDÆ. Mollusca were more numerous than usual, doubtless on account of the small depth, including species of the genera *Sepiola, Fusus, Buccinum, Trophon, Yoldia, Astarte, Arca,* &c. The annelids were represented by a large form allied to *Aphrodita*, which was in great numbers, *Onuphis, Sabella,* and others; and the echinoderms, which as usual were abundant and prominent, by fine species of the genera *Astrogonium* and *Archaster;* a few urchins, including *Tripylus fragilis;* and many of a small *Psolus,* probably *P. squamatus,* MÜLLER. A series of soundings was taken at every 20 fathoms from the surface,—

Surface	4°·7 C.	60 fathoms	1°·75 C.
20 fathoms	3·2	83 „	1·75
40 „	1·75		

which showed that the minimum temperature of 1°·75 C. was reached at a depth between 20 and 40 fathoms.

We sounded and trawled again on the following day in 1,250 fathoms with a bottom of grey ooze, and a bottom temperature of 2°·7 C. Again echinoderms, including *Antedon, Brisinga, Archaster*, and *Ophiomusium* predominated; but we had in addition some good corals, and among them some specimens of *Caryophyllia borealis* of an unusually large size. Sticking all over the outside of the bag, there were many examples of a small Holothurian, with an outer wall so delicate that in almost every case the intestine, which was loaded with ooze, had broken through it and destroyed the specimen.

The depth on the 22nd was 2,020 fathoms, and the bottom an impure globigerina-ooze. Serial temperatures were taken (Appendix A.), and we essayed to dredge, but the dredge-rope parted at 1,700 fathoms without any apparent cause. There was now a very decided rise in the surface temperature as we approached the northern borders of the Gulf-stream.

Next day we sounded in 2,800 fathoms. We took a series of temperature soundings, but a very heavy swell from the south-west prevented our dredging. At about 1 o'clock A.M. there was a sudden rise of the thermometer at the sea-surface from 12°·2 C. to 18°·2 C., and it rose steadily during the next eight hours up to 22° C., showing that we had entered the Gulf-stream current. The passage from the dull green colour of the Arctic reflux to the deep blue of the Gulf-stream was very perceptible on this occasion. We continued in the Gulf-stream until shortly after midnight on the 25th, when a sudden change in the temperature of the surface of the sea from 22°·2 C. to 18° C. showed that we had crossed its southern limit. In crossing the Gulf-stream in both directions the alternate bands or interdigitations of warm and cold water were very perceptible. Half-hourly temperature observations were taken (Appendix D.), and the diagram Pl. XII. is constructed from the general results.

In a former volume ('The Depths of the Sea,' Chapter VIII.), I have given a general account of the Gulf-stream, and I have entered somewhat fully into the recent controversies regarding its origin and influence. Since that book was written, greater harmony of opinion appears to obtain on these points. It seems to be generally admitted that the Gulf-stream is due to the reflux of the equatorial current, and that it is not in any sense a modified case of a general ocean circulation produced by convection; and most physical geographers seem to be at one as to the very important influence which it exerts in distributing and accumulating tropical warmth in the North Atlantic, and in ameliorating the climatic conditions of the countries which border its eastern shores. We have since had an opportunity of tracing the distribution of temperature in the corresponding region of the North Pacific, and the comparison between the two is very instructive. The differences between them are great, but when carefully considered they are found to be more differences in degree than in kind.

In the Pacific the ocean area is of course vastly greater than in the Atlantic, and the equatorial current is to the full as marked in its phenomena and in its results; but the continuity of the meridional land-belt is broken nearly opposite the point where the current has its greatest force, and a great part of that force is lost, and a considerable portion of the current itself is dissipated, among the passages of the Malayan Archipelago. Perhaps even a larger proportion of the Pacific equatorial current than of the Atlantic current is diverted northwards; for it is guided by the long crescentic broken barrier consisting of the Fiji Islands, the New Hebrides and Papua, and the branch which passes down the east coast of Australia is comparatively insignificant; but the northern division passes at once into the region of the monsoons, where it is baffled for half the year, and one is almost surprised to find the 'Kuro Siwa' a powerful, tolerably permanent warm current sweeping round the south of Japan, and exercising in the North Pacific a thermic influence which is certainly comparable with that of the Gulf-stream. The two diagrams, Fig. 100 and Fig. 101 respectively, are curves constructed from serial soundings taken in the Atlantic and in the Pacific, as near as possible to the parallel 35° N., and they show fairly the correspondence in principle and the divergence in detail in the distribution of heat in the two seas. The abnormal curve No. 44 in Fig. 100 is from the sounding in the Labrador current within the cold-wall of the Gulf-stream; and the abnormal

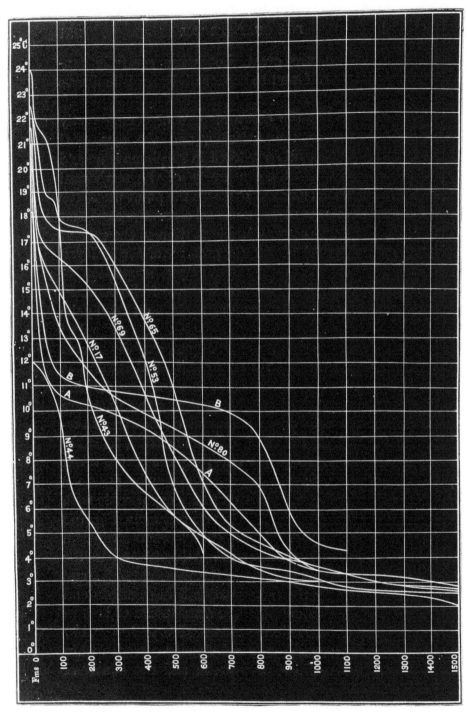

Fig. 100.—Curves constructed from Temperature-Soundings in the North Atlantic.

curve No. 240, Fig. 101, is constructed from a serial sounding in a cold current which passes into the North Pacific from the Sea of Okotsk, probably through Pico or Boussole Channel. This may be only a summer current, and due in a great measure to the melting of the snow over the enormous drainage area of the Amur and the southern Siberian rivers. Curves No. 80 and A. and B. in Fig. 100 are introduced to show what we are inclined to regard as the underlap of the water of the equatorial reflux, steadily cooling but still abnormally warm, against the coast of Europe. The following tables give the positions of the serial soundings:—

In the North Atlantic Ocean. (Fig. 100.)

No. of Station.	Latitude.	Longitude.	Depth in Fathoms.
43	36° 23′ N.	71° 51′ W.	—
44	37 25	71 40	1700
53	36 30	63 40	2650
65	36 33	47 58	2700
69	38 23	37 21	2200
71	38 18	34 48	1675
80	35 3	21 25	2660
A A	Bay of	Biscay.	2090
B B	Coast of	Portugal.	1090

In the North Pacific Ocean. (Fig. 101.)

No. of Station.	Latitude.	Longitude.	Depth in Fathoms.
237	34° 37′ N.	140° 32′ E.	1875
240	35 20	153 39	2900
243	35 24	166 35	2800
245	36 23	174 31	2775
246	36 10	178 0	2050
248	37 41	177° 4′ W.	2900
252	37 52	160 17	2740

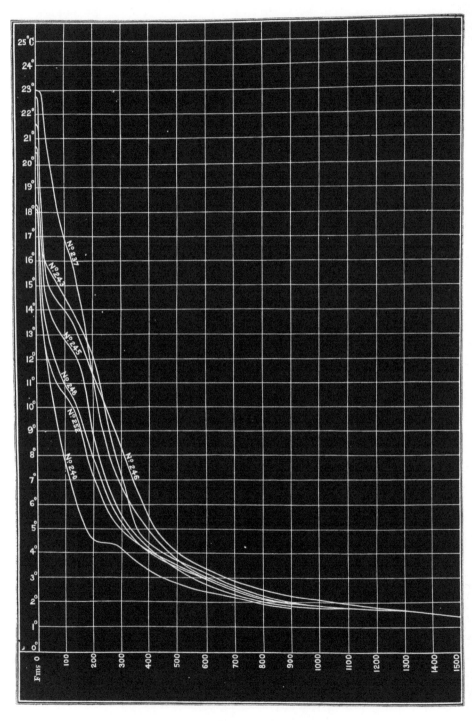

FIG. 101.—Curves constructed from Temperature-Soundings in the North Pacific.

THE GULF-STREAM.

From a depth of 300 fathoms to a depth of 1,500 fathoms, the temperatures in the North Pacific are greatly lower than those in the North Atlantic. In both oceans the temperature gradually falls for every zone of equal depth in passing from west to east, as the influence of the reflux of the equatorial current becomes weakened. The following table, which is constructed from the same serial soundings which are represented graphically in Figures 100 and 101, shows at once the eastward cooling, and the much greater condensation and accumulation of warm water in the basin of the North Atlantic.

Depth in Fathoms.	North Atlantic.		North Pacific.	
	Maximum.	Minimum.	Maximum.	Minimum.
Surface.	24°·0 C.	18°·0 C.	23°·0 C.	18°·2 C.
100	18·0	10·7	17·0	10·5
200	17·5	10·3	12·0	7·4
300	16·0	8·0	8·3	5·0
400	13·7	6·6	5·4	4·0
500	11·0	5·7	4·0	3·4
600	10·2	4·3	3·4	2·7
700	9·9	4·0	2·8	2·4
800	9·0	3·5	2·5	2·1
900	6·0	3·1	2·3	2·0
1000	4·6	2·8	2·1	1·7
1500	2·7	2·0	1·7	1·4

The most marked phenomenon of the Gulf-stream, the condensation and superheating of the water of the equatorial current in the Caribbean Sea and the Gulf of Mexico, and its ejection in a defined hot stream through the strait of Florida, has no parallel in the Pacific, and the 'Kuro Siwa' must be regarded as representing that diffused portion of the reflux of the equatorial current which passes northwards outside the West Indian Islands.

On the 26th we sounded in 2,650 fathoms. Serial temperature-soundings were taken to 800 fathoms; during the operation however we met with rather a serious loss, for the sounding-line with seven thermometers attached fouled the propeller and was carried away. On the 27th, the depth was again 2,650 fathoms, with a bottom of greyish 'red clay.' The trawl was put over in the forenoon, and as this was by far the greatest depth at which we had attempted to employ it, we looked with great interest to the result. In the evening the trawl returned to us in safety, and contained a caridid shrimp, a number of worm-tubes composed chiefly of small foraminifera, two examples of an irregular sea-urchin, and a number of ophiurideans referred to the genera *Amphiura* and *Ophioglypha*. The crustacean may be a pelagic form living at intermediate depths, for such we have reason to believe exist and attain a large size; the annelid we had not an opportunity of determining as the tubes only were present; the urchin is a species new to science and of great interest. *Calymne relicta* (Fig. 102) is at first sight extremely like the normal ANANCHYTIDÆ, indeed it has a close general resemblance to the common chalk form *Ananchytes ovata;* many important characters however separate it from the genus *Ananchytes*, and until we have had an opportunity of comparing the whole series I am not prepared to say that this genus may not find its place in a family as yet undefined, with *Pourtalesia, Aceste,* and *Aërope,* and some wonderful new forms which we have since found in the Southern Sea. The test is 30 mm. in length and 20 mm. in height, and very elegant and symmetrical in form; the outline is oval, slightly truncated posteriorly; a longitudinal ridge from which the sides of the shell slope off with a pleasing curve runs along the apical surface. The oral surface is nearly flat, and a slight keel runs round its edge, defining and limiting it very much, as in *Ananchytes;* a fasciole

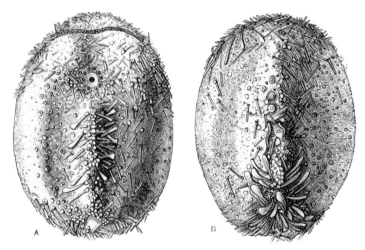

Fig. 102.—*Calymne relicta*, WYVILLE THOMSON. A, oral; B, apical aspect, slightly enlarged. (No. 54.)

follows the keel, only leaving it and appearing for a little part of its course on the oral surface in advance of the mouth. The mouth is oval; its long axis in the direction of the antero-posterior axis of the test. The excretory opening is on the posterior surface above the line of the peripheral ridge. The apical area is, if not disjunct, greatly produced; but it is difficult to make out the exact relations of some of the terminal plates of the ambulacral and interambulacral series. The ambulacra of the trivium meet at an anterior pole on the dorsal surface nearly opposite the mouth, and the two ovarial plates closing the two anterior inter-ambulacral series bear large ovarial openings from which, as in *Aërope*, tubes of considerable length protrude; what appears to be a separate plate immediately behind these bears the madreporic tubercle; only two ovaries are developed, and two plates only are perforated for their ducts. The

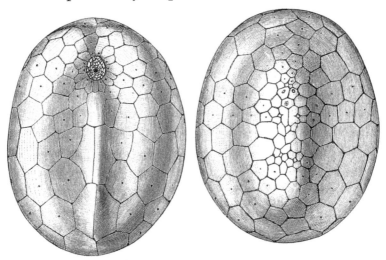

Fig. 103.—*Calymne relicta*, Wyville Thomson. Oral and apical aspects of the denuded test.

two posterior ambulacra end at a secondary pole at a distance of about one-third of the length of the shell from the primary pole near its posterior extremity. The structure of the ambulacra is extremely simple, the ambulacral canal sending a simple diverticulum to a single minute pore near the centre of each ambulacral plate. The mouth is unarmed. The surface of the test bears somewhat sparsely-scattered hair-like spines, and over the central portion of the oral surface and on the apical surface near the posterior pole are groups of delicately striated paddle-shaped spines. The general colour of the test and spines is pale green. Either the same species or one very nearly allied to it was obtained in considerable number near Tristan d'Acunha, but with a test not less than 200 mm. in length. The shell was however so extremely tender and thin that even with the trawl not a single example was got tolerably complete.

Only one specimen occurred of a species of *Amphiura* which has not yet been determined; but there were seven or eight examples of the handsome *Ophioglypha* figured in wood-cuts 104 and 105. *Ophioglypha bullata* belongs to the small subsection of the genus with short, knotted arms, slit-like foot-pores, and marked grooves between the joints of the arms; which already contains two arctic species, *O. nodosa* and *O. stuwitzii*; it is most nearly allied to the former of these, but there are many points of distinction.

The diameter of the disk in a full-grown specimen is 10 mm., and the length of the arm is about twice and a half the diameter of the disk. The dorsal surface of the disk is covered with rather prominent granular scales of different forms and sizes; six of the larger of these form a rosette in the centre of the disk, but they are separated from one another by bands of the scaly perisom, and not in contact as in *O. nodosa*. The radial shields are prominent, slightly longer than broad, and they also are separated throughout their entire length by wedges of perisom. The mouth-shields are egg-shaped, very large and prominent, occupying nearly the whole space between the arms, and running out quite to the edge of the disk. These inflated mouth-shields give a very marked

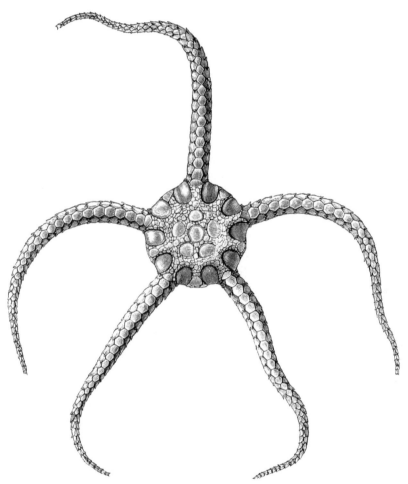

Fig. 104.—*Ophioglypha bullata*, Wyville Thomson. Dorsal aspect. Three times the natural size. (No. 54.)

character to the species. The mouth-papillæ are short and blunt, about twelve on each angle of the mouth; the teeth are six or eight in number on each angle and rather long and pointed. The under arm-plates are square and prominent, apposed to one another in series near the bases of the arms, but rapidly separating and becoming smaller and rounder distally. The upper arm-plates are small and likewise prominent. The lateral arm-plates are large and rounded, and project laterally, which gives its peculiar knotted or beaded appearance to the arm. The arm-spines are very small, in a row of five or six on each lateral plate; on the distal arm-joints they become somewhat longer. The large elliptical foot-pores are fringed all round with closely-set flat papillæ. *Ophioglypha bullata* is a species of great interest on account of its wide distribution at extreme depths. It seems universally distributed in the Atlantic and the Southern Sea. The trawl which we used on this occasion had a beam of eighteen feet long, and as we calculated that it had been dragged for at least a mile over the bottom, it gave us a fair idea of the distribution of the large forms of animal life over a considerable area.

On the following day we sounded in 2,500 fathoms, and on the 29th in 1,075 fathoms in sight of Bermudas, with a bottom of coral mud. The dredge was put over and veered to 1,600 fathoms. It came up at noon with the pasty mortar-like lifeless contents which we find almost constantly on the slopes of coral reefs; the lime sediment was mixed with a large proportion of the shells of Pteropods and Heteropods. Two fine specimens of a Hexactinellid sponge were hanging to the tangles, both unfortunately dead and slightly water-worn. The largest specimen, which seems to be nearly complete, is 120 mm. in height, and

D D

shaped somewhat like an old-fashioned tall champagne-glass. It rests on a very solid hard base of attachment; it then contracts to a kind of stem, and then gradually expands upwards to a width at the top of 40 mm. A deep cavity passes from the upper open end down to the stem-like constriction.

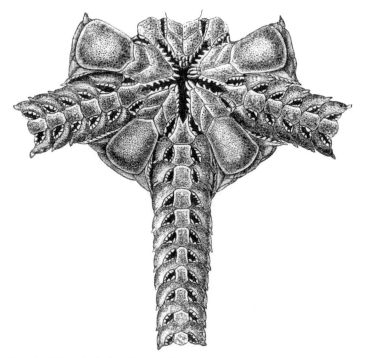

Fig. 105.—*Ophioglypha bullata*, Wyville Thomson. Oral aspect. Five times the natural size. (No. 54.)

The outer surface of the sponge is raised into spiral ridges somewhat as in *Euplectella*, and under the ridges are irregularly spiral lines of large holes. The interior of the cup presents a very remarkable character, which reminds one at once of many of the chalk ventriculites. The inner layer is deeply fluted, thrown into a series of alternating vertical grooves and ridges so that the outline of the cavity in a transverse section is deeply sinuous. The substance of the sponge throughout is composed of a close anastomosing network of siliceous fibres; towards the outside the network much resembles that of *Aphrocallistes*, while on the inner wall the structure is trellis-like, and the form of the meshes square and more regular. The spaces of the network are crowded with small regular hexactinellid spicules, some free, some cemented to the continuous skeleton by attachments of silica. For this beautiful sponge, which I have every reason to believe is undescribed, I propose the name *Lefroyella decora*.

We anchored in the

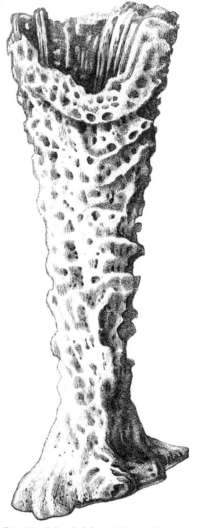

Fig. 106.—*Lefroyella[1] decora*, WYVILLE THOMSON Natural size. (No 56.)

[1] I have associated with this species the name of our kind friend His Excellency Major-General Lefroy, C.B., F.R.S., Governor of Bermudas.

evening on the Chaddock Bank a little to the southwest of Gibbs'-Hill lighthouse, and on the following day we proceeded under steam to sound on the southwest side of the reef. We took one or two hauls of the dredge during the day with but little success, and on the 31st of May we passed through the narrows and returned to Bermudas dockyard.

SAN VICENTE.

APPENDIX A.

Serial Temperature Soundings taken between St. Thomas, Bermudas, and Halifax, during the Months of March, April, and May, 1873.

Depth in Fathoms.	Station No. 25. Lat. 19° 41′ N. Long. 65° 7′ W.	Station No. 27. Lat. 22° 49′ N. Long. 65° 19′ W.	Station No. 28. Lat. 24° 39′ N. Long. 65° 25′ W.	Station No. 29. Lat. 27° 49′ N. Long. 64° 59′ W.	Station No. 37. Lat. 32° 19′ N. Long. 65° 39′ W.	Station No. 38. Lat. 33° 3′ N. Long. 66° 32′ W.	Station No. 39. Lat. 34° 3′ N. Long. 67° 32′ W.
Surface.	24°·5 C	24°·2 C.	23°·9 C.	22°·2 C.	20°·0 C.	21°·1 C.	18°·3 C.
50	...	...	23·3	21·0	...	...	17·9
100	20·3	22·4	19·5	17·7	18·5	18·9	17·6
150	...	...	18·0	17·6	18·2	...	17·6
200	16·7	17·2	17·4	17·0	17·7	17·7	17·8
250	...	...	...	16·1	17·6	...	17·2
300	12·1	13·7	14·5	14·3	17·2	17·1	16·8
350	...	...	...	12·4	15·9	16·1	15·4
400	8·4	9·6	10·8	10·6	14·2	14·5	13·5
450	...	...	...	8·2	...	12·3	11·2
500	6·4	7·0	7·2	6·8	10·0	9·8	8·9
550	...	...	...	...	8·0	7·7	7·1
600	5·0	5·2	5·4	4·6	6·4	6·3	5·3
650	...	...	...	...	...	5·2	4·8
700	...	4·5	4·7	4·7	5·0	4·7	...
800	3·6	4·0	...	4·1	4·0	4·1	...
900	...	3·6	3·8	3·5	3·8	3·9	...
1000	3·1	3·4	3·4	3·4	3·4	3·5	...
1100	2·9	3·1	3·4	3·1	3·3	3·3	...
1200	2·9	3·0	3·1	...	3·1	...	...
1300	2·8	2·9	2·9	2·8	2·9	3·0	...
1400	2·5	2·6	2·1	2·8	2·7	2·8	...
1500	2·4	2·2	2·5	2·4	2·6	2·8	...

Depth in Fathoms.	Station No. 41. Lat. 36° 5' N. Long. 69° 54' W.	Station No. 42. Lat. 35° 58' N. Long. 70° 39' W.	Station No. 43. Lat. 36° 23' N. Long. 71° 51' W.	Station No. 44. Lat. 37° 25' N. Long. 71° 40' W.	Station No. 45. Lat. 38° 34' N. Long. 72° 10' W.	Station No. 50. Lat. 42° 8' N. Long. 63° 39' W.
Surface.	18°·3 C.	°· C.	23°·9 C.	11°·0 C.	10°·0 C.	8°·0 C.
50	18·2	...	...	11·4	...	...
100	18·1	...	18·3	9·7	10·6	7·3
150	17·7	...	13·3	6·6	...	...
200	17·4	...	11·1	5·5	6·2	5·0
250	17·1	...	9·2	4·5	...	...
300	17·0	...	8·2	4·0	4·6	3·3
350	14·9	...	7·2	3·9	...	...
400	13·7	...	...	3·7	...	3·6
450	11·1	..	...	...	...	...
500	8·7	...	5·75	3·6	3·7	4·2
550	7·0	...	5·3	..	...	...
600	5·3	...	4·25	3·4	3·4	3·4
650	4·6	...	...	...	...	...
700	...	...	...	...	3·0	2·3
800	...	...	...	...	...	3·2
900	...	4·2	...	2·7	...	3·1
1000	...	3.5	...	...	...	3·0
1100	...	3.2	...	2·7	...	2·1
1200	...	3·1	...	2·6	...	2·7
1300	...	...	...	2·5	...	...
1400	...	3·1	...	2·3	...	...
1500	...	2·7	...	2·0	...	...

Depth in Fathoms.	Station No. 51. Lat. 41° 19′ N. Long. 63° 12′ W.	Station No. 52. Lat. 39° 44′ N. Long. 63° 22′ W.	Station No. 53. Lat. 36° 30′ N. Long. 63° 40′ W.	Station No. 54. Lat. 34° 51′ N. Long. 63° 59′ W.	Station No. 55. Lat. 33° 20′ N. Long. 64° 37′ W.	Station No. 57. Off Bermudas.
Surface.	15°·0 C.	19°·6 C.	22°·5 C.	21°·4 C.	21°·4 C.	22°·8 C.
50	...	...	19·0	...	...	...
100	12·8	18·2	17·9	18·3	17·8	18·5
150	...	...	17·7	...	...	...
200	8·1	17·6	17·5	17·7	17·0	17·5
250	...	...	...	...	...	...
300	5·4	16·2	15·5	16·4	16·3	15·9
350	...	...	...	15·2	15·2	14·3
400	3·7	11·8	12·2	13·1	13·0	12·1
450	...	...	...	...	...	...
500	3·8	7·9	7·2	...	8·0	8·1
550	...	...	...	...	...	...
600	3·2	5·0	4·1	...	5·0	5·1
650	...	...	...	...	...	...
700	3·1	3·5	4·0	...	3·8	4·2
800	3·0	3·6	3·7	...	3·6	3·8
900	2·8	3·4	...	...	3·4	3·4
1000	2·9	...	...	...	3·2	3·1
1100	3·1	3·9	...	...	...	...
1200	2·4	...	...	...	...	...
1300	2·6	2·2	...	...	...	...
1400	2·5	...	...	...	...	...
1500	2·4	2·6	2·6	...	...	...

APPENDIX B.

Table of the Bottom Temperatures between St. Thomas and Halifax, observed during the Months of March, April, and May, 1873.

Station No.	Date, 1873.	Latitude N.		Longitude W.		Depth in Fathoms.	Bottom Temperature.	
27	March 28	22°	49′	65°	19′	2960	1°·	5 C.
28	,, 29	24	39	65	25	2850	1 ·	7
29	,, 31	27	49	64	59	2700	1 ·	6
30	April 1	29	5	65	1	2600	1 ·	8
31	,, 3	31	24	64	0	2475	1 ·	7
32	,, 3	31	49	64	55	2250	1 ·	8
34	,, 21	...		...		1375	3 ·	0
35a	,, 22	32	39	65	6	2450	1 ·	6
35b	,, 22	32	26	65	9	2100	1 ·	6
37	,, 24	32	19	65	39	2650	1 ·	6
38	,, 25	33	3	66	32	2600	1 ·	8
39	,, 27	34	3	67	32	2850	1 ·	8
42	,, 30	35	58	70	39	2425	1 ·	8
44	May 2	37	25	71	40	1700	1 ·	7
45	,, 3	38	34	72	10	1240	2 ·	4
46	,, 6	40	17	66	48	1350	2 ·	3
49	,, 20	43	3	63	39	83	1 ·	75
50	,, 21	42	8	63	39	1250	2 ·	7
51	,, 22	41	19	63	12	2020	1 ·	5
52	,, 23	39	44	63	22	2800	1 ·	5
53	,, 26	36	30	63	40	2650	1 ··	8

APPENDIX C.

Specific Gravity Observations taken between St. Thomas, Bermudas, and Halifax, during the Months of March, April, and May, 1873.

Date, 1873	Latitude N.	Longitude W.	Depth of the Sea.	Depth (d.) at which Water was taken.	Temperature (t.) at d.	Temperature (t'.) during Observation.	Specific Gravity at (t'.) Water at 4° = 1 C.	Specific Gravity at 15°·5 Water at 4° = 1 C.	Specific Gravity at (t.) Water at 4° = 1 C.
			Fms.						
March 25	St Thos	...		Surface.	25°·0C	25°·7C.	1·02432	1·02714	1·02467
26	19° 41'	65° 7'	3870	Bottom.	...	23 ·7	1·02418	1·02647	1·02845
	...	...		Surface.	24 ·4	25 ·2	1·02435	1·02702	1·02456
27	21 21	65 12	2800	Surface.	25 ·2	25 ·4	1·02441	1·02713	1·02434
	...	...		Bottom.	...	24 ·3	1·02363	1·02600	1·02798
28	22 49	65 19	2960	Bottom.	1 ·5	24 ·7	1·02358	1·02607	1·02805
	...	...		Surface.	25 ·1	25 ·2	1·02453	1·02720	1·02455
29	24 39	65 25	2850	Bottom.	1 ·67	25 ·2	1·02351	1·02616	1·02814
	...	...		Surface.	24 ·3	25 ·0	1·02459	1·02719	1·02478
30	26 26	65 18		Surface.	23 ·8	24 ·2	1·02504	1·02737	1·02507
31	27 49	64 59	2700	Surface.	22 ·8	23 ·4	1·02534	1·02746	1·02536
,,				Bottom.	1 ·67	22 ·6	1·02427	1·02615	1·02814
,,				100	17 ·72	23 ·9	1·02563	1·02788	1·02733
,,				200	17 ·00	28 ·6	1·02498	1·02716	1·02679
,,				300	14 ·33	24 ·1	1·02447	1·02679	1·02714
,,				400	10 ·62	23 ·4	1·02435	1·02647	1·02744
,,				500	6 ·84	22 ·6	1·02432	1·0,620	1·02773
April 1	29 5	65 1	2600	Surface.	22 ·2	24 ·1	1·02510	1·02741	1·02564
	...	...		Bottom.	1 ·8	23 ·1	1·02606	1·02780	1·02980
2	29 42	65 7		Surface.	21 ·0	22 ·1	1·02564	1·02737	1·02595
3	31 49	64 55	2475	Bottom.	1 ·7	20 ·5	1·02527	1·02653	1·02352
,,	...	...	2250	Bottom.	1 ·8	21 ·2	1·02464	1·02608	1·02807
	...	...		Surface.	21 ·0	21 ·7	1·02571	1·02731	1·02588
22	32 26	65 9	2450	Surface.	20 ·3	20 ·7	1·02585	1·02714	1·02592
25	33 4	66 33	2600	Surface.	21 ·1	21 ·7	1·02566	1·02729	1·02584
26	34 11	67 37		Surface.	18 ·6	18 ·4	1·02635	1·02707	1·02627
27	34 3	67 32	2850	Surface.	18 ·3	18 ·0	1·02642	1·02704	1·02638
28	34 51	68 30	2675	Surface.	20 ·8	18 ·1	1·02637	1·02702	1·02563
29	36 5	69 54		Surface.	19 ·2	19 ·5	1·02605	1·02709	1·02615
30	35 58	70 39	2425	Surface.	18 ·3	17 ·6	1·02647	1·02697	1·02626
	...	...		Bottom.	1 ·8	18 ·0	1·02609	1·02671	1·02869
May 1	36 23	71 51		Surface.	23 ·9	22 ·8	1·02486	1·02675	1·02445
2	37 25	71 40	1700	Surface.	13 ·3	14 ·4	1·02566	1·02538	1·02584
3	38 84	72 10	1250	Surface.	9 ·7	10 ·6	1·02603	1·02504	1·02613
5	39 50	69 14		Surface.	7 ·8	11 ·1	1·02652	1·02561	1·02701
6	40 17	66 48	1250	Surface.	5 ·6	6 ·1	1·02568	1·02407	1·02576
7	41 15	65 45	1340	Surface.	6 ·0	6 ·2	1·02552	1·02422	1·02586
20	43 3	63 39	83	Bottom.	1 ·7	5 ·6	1·02570	1·02406	1·02576
	...	...		Surface.	4 ·7	6 ·0	1·02520	1·02357	1·02536
21	42 10	63 39	1250	Surface.	8 ·1	8 ·6	1·02582	1·02453	1·02588
	...	...		Bottom	2 ·7	10 ·6	1·02645	1·02538	1·02735
22	41 19	63 11	2020	Surface.	1 ·5	11 ·8	1·02588	1·02588	1·02786
,,	...	...		Surface.	15 ·1	15 ·1	1·02635	1·02625	1·02633
,,	...	...		250	6 ·5	12 ·5	1·02662	1·02598	1·02755
,,	...	...		500	3 ·75	12 ·7	1·02652	1·02591	1·02780
23	39 44	63 22	2800	Surface.	19 ·7	20 ·1	1·02601	1·02719	1·02610
,,	...	...		100	18 ·15	19 ·9	1·02602	1·02712	1·02642
,,	...	...		300	16 ·2	19 ·8	1·02538	1·02695	1·02678
	...	...		Bottom.	1 ·5	19 ·8	1·02596	1·02702	1·02900
26	36 30	63 40	2650	Bottom.	1 ·8	22 ·3	1·02526	1·02700	1·02898
,,	...	...		250	16 ·8	23 ·7	1·02497	1·02718	1·02684
	...	...		Surface.	22 ·8	23 ·9	1·02489	1·02714	1·02518
27	34 51	63 59	2650	Surface.	21 ·7	22 ·8	1·02527	1·02721	1·02558
28	33 20	64 37	2500	Surface.	21 ·4	21 ·4	1·02529	1·02716	1·02563

APPENDIX D.

Table of Meteorological Observations made in Crossing and Recrossing the Gulf-Stream.

Date and Position.	Hours.	Barometer.	Temperature of Air.		Temperature of Sea Surface.
			Dry Bulb.	Wet Bulb.	
		Inches.	Deg. Cent.	Deg. Cent.	Deg. Cent.
April 23rd, 1873 . .	Noon.	29·93	20·6	17·8	...
	1	...	...	...	...
Off Bermudas . . .	2	29·95	22·2	20·3	...
	3	...	...	...	...
	4	29·89	21·1	19·8	...
	5	...	...	...	...
	6	29·84	21·1	20·0	19·7
	7	...	...	...	19·4
	8	29·84	20·8	20·0	19·4
	9	...	...	...	19·7
	10	29·82	20·6	19·4	19·4
	11	...	...	...	19·7
	Midn.	29·80	20·0	19·4	19·4
April 24th .	1	...	...	...	19·4
	2	29·85	20·0	19·4	19·4
	3	...	...	...	19·4
	4	28·85	19·4	18·3	19·4
	5	...	...	...	19·7
	6	29·90	20·0	18·3	19·4
	7	...	...	...	19·4
	8	29·94	20·3	18·6	19·4
	9	...	...	...	19·7
	10	29·97	20·3	18·9	20·0
	11	...	...	...	20·0
Lat. 32° 19′ N. . . Long. 65° 39′ W. . .	Noon.	29·98	21·2	19·4	20·0
	1	...	...	...	20·0
	2	29·97	21·1	20·0	19·7
	3	...	...	...	19·7
	4	29·96	21·1	20·0	19·9

THE GULF-STREAM.

Date and Position.	Hours.	Barometer.	Temperature of Air.		Temperature of Sea Surface.
			Dry Bulb.	Wet Bulb.	
		Inches.	Deg. Cent.	Deg. Cent.	Deg. Cent.
	5	...	...	...	20·0
	6	29·98	21·1	19·7	20·0
	7	...	...	...	20·2
	8	30·02	21·7	19·4	20·2
	9	...	...	...	20·2
	10	30·02	21·1	19·2	20·2
	11	...	...	...	20·2
	Midn.	30·00	21·1	19·7	20·2
April 25th	1	...	...	...	20·0
	2	29·87	20·6	17·8	20·0
	3	...	...	...	20·0
	4	29·84	20·0	17·8	20·0
	5	...	...	...	20·6
	6	29·97	20·6	19·4	20·9
	7	...	...	...	21·1
	8	30·00	21·5	20·6	21·1
	9	...	...	...	21·1
	10	29·99	22·8	20·8	21·1
	11	...	...	...	21·1
Lat. 32° 4′ N. Long. 66° 33′ W.	Noon.	29·99	23·1	20·8	21·1
	1	...	...	...	21·1
	2	29·94	23·3	20·6	21·1
	3	...	...	...	21·1
	4	29·92	22·2	20·6	21·3
	5	...	...	...	21·3
	6	29·86	21·7	20·0	21·1
	7	...	...	...	20·6
	8	29·80	21·1	20·0	20·6
	9	...	...	...	19·4
	10	29·73	20·6	19·7	19·7
	11	...	...	...	19·7
	Midn.	29·71	20·6	19·2	19·7
April 26th	1	...	...	...	19·4
	2	29·74	18·9	16·7	18·9
	3	...	...	...	18·9
	4	29·75	17·8	15·0	18·3
	5	...	...	...	18·9
	6	29·77	16·1	13·3	18·9
	7	...	...	...	18·9
	8	29·80	16·1	13·3	18·9
	9	...	...	...	18·9
	10	29·89	15·3	11·7	18·9
	11	...	...	...	18·6
Lat. 34° 11′ N. Long. 67° 37′ W.	Noon.	29·92	16·1	12·2	18·6

Date and Position.	Hours.	Barometer.	Temperature of Air.		Temperature of Sea Surface.
			Dry Bulb.	Wet Bulb.	
		Inches.	Deg. Cent.	Deg. Cent.	Deg. Cent.
April 27th	1	...	...	...	18·5
	2	29·94	16·1	12·5	18·9
	3	...	...	...	18·3
	4	29·96	16·4	12·2	18·0
	5	...	...	...	18·3
	6	30·01	14·4	11·7	18·3
	7	...	...	...	18·3
	8	30·02	14·4	11·7	18·3
	9	...	...	...	18·3
	10	30·04	14·3	11·1	18·3
	11	...	...	...	18·3
	Midn.	30·05	14·7	11·1	18·3
	1	...	...	...	18·6
	2	30·07	15·0	11·6	18·3
	3	...	...	...	18·3
	4	30·07	14·4	11·7	18·3
	5	...	...	...	18·3
	6	30·12	14·7	11·0	18·3
	7	...	...	...	18·3
	8	30·19	15·3	10·6	18·3
	9	...	...	...	18·3
	10	30·19	15·0	10·6	18·3
	11	...	...	...	18·3
Lat. 34° 3′ N. Long. 67° 32′ W.	Noon.	30·20	15·0	10·6	18·3
	1	...	...	...	18·3
	2	30·07	16·1	12·2	18·7
	3	...	...	...	18·6
	4	30·07	16·1	13·0	18·6
	5	...	...	...	18·9
	6	30·12	16·7	12·6	18·9
	7	...	...	...	18·6
	8	30·19	15·7	12·2	18·9
	9	...	...	...	18·0
	10	30·19	15·5	12·2	18·0
	11	...	...	...	18·0
	Midn.	30·20	15·5	12·2	18·0
April 28th	1	...	...	...	18·9
	2	30·25	16·1	11·4	18·9
	3	...	...	...	19·0
	4	30.25	16·1	12·2	19·4
	5	...	...	...	19·4
	6	30·26	18·3	12·2	19·6
	7	...	...	...	20·0
	8	30·27	17·8	13·0	20·3
	9	...	...	...	20·6

THE GULF-STREAM.

Date and Position.	Hours.	Barometer.	Temperature of Air.		Temperature of Sea Surface.
			Dry Bulb.	Wet Bulb.	
		Inches.	Deg. Cent.	Deg. Cent.	Deg. Cent.
Lat. 34° 51′ N. Long. 68° 30′ W.	10	30·30	19·4	14·3	20·8
	11	...	...	...	20·8
	Noon.	30·32	18·9	13·3	...
	1	...	...	...	...
	2	...	...	...	...
	3	...	...	...	...
	4	...	...	...	...
	5	...	...	...	20·8
	6	30·22	17·8	12·5	20·8
	7	...	...	...	20·0
	8	30·23	17·2	12·7	20·0
	9	...	...	...	19·7
	10	30·21	17·2	13·3	19·4
	11	...	...	...	19·4
	Midn.	30·14	16·7	13·9	19·4
April 29th	1	...	...	...	18·3
	2	30·00	16·1	16·1	18·3
	3	...	...	...	18·3
	4	29·90	15·5	15·5	18·3
	5	...	...	...	17·8
	6	29·81	15·8	15·6	17·8
	7	...	...	...	17·8
	8	29·68	19·4	18·6	18·9
	9	...	...	...	18·9
	10	29·66	19·4	18·3	18·9
	11	...	...	...	19·2
Lat. 36° 5′ N. Long. 69° 54′ W.	Noon.	29·70	19·7	18·0	19·2
	1	...	...	...	18·9
	2	29·83	18·6	16·7	18·9
	3	...	...	...	18·9
	4	29·88	18·6	16·2	18·3
	5	...	...	...	18·3
	6	29·92	17·8	16·1	18·3
	7	...	...	...	18·3
	8	29·99	17·2	16·1	18·3
	9	...	...	...	18·3
	10	30·00	17·5	15·6	18·3
	11	...	...	...	18·3
	Midn.	30·01	17·5	15·8	18·3
April 30th	1	...	...	...	18·3
	2	29·95	17·8	16·1	18·3
	3	...	...	...	18·3
	4	29·94	16·6	16·1	18·3
	5	...	...	...	18·3

Date and Position.	Hours.	Barometer.	Temperature of Air.		Temperature of Sea Surface.
			Dry Bulb.	Wet Bulb.	
		Inches.	Deg. Cent.	Deg. Cent.	Deg. Cent.
	6	29·93	16·6	15·6	18·3
	7	...	...	...	...
	8	30·00	16·1	15·3	...
	9	...	...	...	18·3
	10	30·07	16·1	13·9	18·3
	11	...	...	...	18·3
Lat. 35° 58′ N. Long. 70° 39′ W.	Noon.	30·07	16·5	13·9	18·3
	1	...	...	...	18·6
	2	30·12	16·7	13·9	22·0
	3	...	...	...	22·0
	4	30·16	16·7	13·9	21·9
	5	...	...	...	22·8
	6	30·18	16·7	13·9	22·5
	7	...	...	...	22·8
	8	30·23	16·7	13·9	22·8
	9	...	...	...	22·2
	10	30·30	16·7	12·8	22·2
	11	...	...	...	22·0
	Midn.	30·32	16·7	13·9	21·7
May 1st.	1	...	...	...	22·2
	2	30·28	16·7	13·9	22·2
	3	...	...	...	22·2
	4	30·31	16·7	13·6	22·5
	5	...	...	...	22·5
	6	30·31	16·7	14·0	22·9
	7	...	...	...	23·9
	8	30·37	18·3	14·0	23·9
	9	...	...	...	23·7
	10	30·33	15·0	12·8	23·9
	11	...	...	...	23·9
Lat. 36° 23′ N. Long. 71° 51′ W.	Noon.	30·34	15·6	12·8	23·9
	1	...	...	...	23·9
	2	30·37	18·9	13·9	23·9
	3	...	...	...	23·9
	4	30·35	18·9	14·7	23·9
	5	...	...	...	23·9
	6	30·33	16·7	13·3	23·9
	7	...	...	...	23·3
	8	30·29	17·2	13·6	23·0
	9	...	...	...	20·6
	10	30·29	17·8	14·4	19·4
	11	...	...	...	19·4
	Midn.	30·25	17·2	14·4	13·3
May 2nd	1	...	...	...	12·2

THE GULF-STREAM.

Date and Position.	Hours.	Barometer.	Temperature of Air.		Temperature of Sea Surface.
			Dry Bulb.	Wet Bulb.	
		Inches.	Deg. Cent.	Deg. Cent.	Deg. Cent.
	2	30·20	14·7	13·3	12·5
	3	...	...	...	12·5
	4	30·15	14·7	13·2	12·8
	5	...	...	...	12·8
	6	30·11	15·0	13·7	13·3
	7	...	...	...	13·5
	8	30·11	15·0	14·4	13·5
	9	...	...	...	13·7
	10	30·04	15·3	14·6	13·7
	11	...	...	...	...
Lat. 37° 25′ N. Long. 71° 40′ W.	Noon.	29·92	15·3	14·4	...
	1	...	...	...	13·3
	2	29·91	15·6	15·0	13·3
	3	...	...	...	13·3
	4	29·86	15·6	15·0	13·3
	5	...	...	...	12·8
	6	29·85	15·0	15·0	12·2
	7	...	...	...	12·8
	8	29·77	15·0	13·9	11·1
	9	...	...	...	12·2
	10	29·73	14·4	13·9	12·2
	11	...	...	...	...
	Midn	29·66	14·4	14·3	13·3
May 3rd	1	...	...	...	11·1
	2	29·50	13·3	12·8	11·4
	3	...	...	...	10·0
	4	29·53	12·8	12·7	10·0
	5	...	...	...	10·0
	6	29·56	12·8	12·5	9·7
	7	...	...	...	10·0
	8	29·53	12·2	12·0	10·0
	9	...	...	...	9·4
	10	29·52	11·7	11·7	9·4
	11	...	...	...	9·7
Lat. 38° 34′ N. Long. 72° 10′ W.	Noon.	29·48	8·9	8·9	9·7
	1	...	...	...	9·7
	2	29·47	8·9	8·9	9·7
	3	...	...	...	9·4
	4	29·49	8·9	8·6	10·0
	5	...	...	...	9·4
	6	29·59	6·6	6·3	9·4
	7	...	...	...	9·4
	8	29·74	6·6	7·1	9·4
	9	...	...	...	10·4
	10	29·73	6·6	5·3	9·4

Date and Position.	Hours.	Barometer.	Temperature of Air.		Temperature of Sea Surface.
			Dry Bulb.	Wet Bulb.	
		Inches.	Deg. Cent.	Deg. Cent.	Deg. Cent.
	11	...	...	...	9·4
	Midn.	29·77	5·8	5·0	9·4
May 4th	1	...	...	...	9·4
	2	29·80	5·2	4·3	10·0
	3	...	...	...	10·6
	4	29·83	5·2	4·6	10·6
	5	...	...	...	10·0
	6	29·89	5·5	4·7	10·0
	7	...	...	...	9·4
	8	29·94	5·5	4·4	6·6
	9	...	...	...	6·6
	10	30·00	6·5	5·0	6·4
	11	...	...	...	6·6
Lat. 39° 13′ N. Long. 71° 20′ W.	Noon.	30·05	6·9	5·0	6·6
	1	...	...	...	7·0
	2	30·05	7·8	6·1	10·5
	3	...	...	...	7·8
	4	30·19	7·8	5·0	8·6
	5	...	...	...	11·1
	6	30·10	8·1	5·3	11·4
	7	...	...	...	11·1
	8	30·15	7·5	5·3	11·1
	9	...	...	...	11·1
	10	30·21	7·8	5·6	11·1
	11	...	...	...	11·1
	Midn.	30·21	8·9	6·7	10·8
May 5th	1	...	...	...	10·8
	2	30·20	9·4	8·0	10·8
	3	...	...	...	10·8
	4	30·20	10·0	8·0	10·8
	5	...	...	...	11·1
	6	30·22	10·6	9·4	11·1
	7	...	...	...	10·0
	8	30·26	11·4	9·7	9·4
	9	...	...	...	9·4
	10	30·26	11·1	9·4	8·3
	11	...	...	...	7·8
Lat. 39° 50′ N. Long. 69° 14′ W.	Noon.	30·26	11·1	9·2	7·2
	1	...	...	...	7·8
	2	30·22	10·4	9·4	9·4
	3	...	...	...	9·4
	4	30·21	12·2	10·6	9·4
	5	...	...	...	5·5
	6	30·21	10·4	8·9	5·5
	7	...	...	...	5·5

THE GULF-STREAM.

Date and Position.	Hours.	Barometer.	Temperature of Air.		Temperature of Sea Surface.
			Dry Bulb.	Wet Bulb.	
		Inches.	Deg. Cent.	Deg. Cent.	Deg. Cent.
May 6th	8	30·21	8·9	7·8	5·2
	9	...	...	...	4·4
	10	30·24	7·8	6·7	5·0
	11	...	...	...	5·3
	Midn.	30·24	7·8	6·7	5·3
	1	...	...	...	4·4
	2	30·25	6·1	5·6	4·1
	3	...	...	...	4·3
	4	30·27	6·1	5·4	4·4
	5	...	...	...	4·1
	6	30·28	5·5	5·3	4·1
	7	...	...	...	4·4
	8	30·31	6·7	5·4	4·4
	9	...	...	...	4·4
	10	30·37	7·8	6·7	4·4
	11	...	...	...	5·6
Lat. 40° 17′ N. Long. 66° 48′ W.	Noon.	30·36	8·3	7·2	5·6
	1	...	...	...	5·5
	2	30·37	10·0	8·3	5·5
	3	...	...	...	5·5
	4	30·37	11·1	8·3	5·5
	5	...	...	...	5·5
	6	30·38	8·9	7·0	5·5
	7	...	...	...	5·5
	8	30·38	5·8	5·3	5·5
	9	...	...	...	5·5
	10	30·39	5·6	5·0	5·5
	11	...	··	...	5·0
	Midn.	30·38	5·6	5·0	5·0
May 7th	1	...	...	...	5·0
	2	30·37	5·5	4·4	5·0
	3	...	...	...	5·0
	4	30·37	5·5	4·4	5·0
	5	...	...	...	5·0
	6	30·39	5·5	4·4	5·0
	7	...	...	...	5·5
	8	30·42	7·2	6·1	5·5
	9	...	...	...	5·8
	10	30·46	6·9	5·6	5·8
	11	...	...	...	5·8
Lat. 41° 15′ N. Long. 65° 45′ W.	Noon.	30·46	7·8	6·1	6·0
	1	...	...	...	6·1
	2	30·45	6·7	4·4	5·5
	3	...	...	...	5·5

F E

Date and Position.	Hours.	Barometer.	Temperature of Air.		Temperature of Sea Surface.
			Dry Bulb.	Wet Bulb.	
		Inches.	Deg. Cent.	Deg. Cent.	Deg. Cent.
	4	30 43	5·5	4·3	5·5
	5	...	...	...	5·5
	6	30·45	5·5	2·8	5·5
	7	...	...	...	5·5
	8	30 46	4 4	1·2	5·5
	9	...	...	...	5·5
	10	30·45	4 4	2·9	5·5
	11	...	...	...	3·9
	Midn.	30·46	3·3	2·8	2·8
May 8th	1	...	...	...	3·3
	2	30·47	3·3	2·8	3·0
	3	...	...	...	2·8
	4	30·48	3·3	2·2	2·8
	5	...	...	...	2·8
	6	30·49	3·3	2·2	2·8
	7	...	...	...	2·8
	8	30·53	5·5	4·4	2·8
	9	...	...	...	2·8
	10	30·55	5·5	3·6	3·3
	11	...	...	...	3·3
Lat. 43° 2′ N. Long. 64° 2′ W.	Noon.	30·53	6·7	4·8	3·3
	1	...	...	...	3·9
	2	30·50	7·8	5·6	3·9
	3	...	...	...	3·9
	4	30·48	7·5	5·8	3·9
	5	...	...	...	3 3
	6	30·42	6·1	4·4	3·0
	7	. .	...	...	3·0
	8	30·40	5·5	3 3	2·8
	9	...	...	...	3·3
	10	30 38	4·4	3·3	3·3
	11	...	...	...	3·3
	Midn.	30·38	5·0	3·3	3·3
May 9th	1	...	...	...	3·3
	2	30·36	4·2	3·3	3 3
	3	...	...	...	3·0
	4	30·36	3·5	2·7	3·0
	5	...	...	...	1·7
	6	30·33	4·4	3·6	2·2
	7	...	...	...	2·2
	8	30·33	4·7	3·9	2·2
	9	...	...	...	...
	10	30·34	8·3	6·1	...
	11	...	...	...	...
At Halifax	Noon.	30·35	15·0	9·4	...

Date and Position.	Hours.	Barometer.	Temperature of Air.		Temperature of Sea Surface.
			Dry Bulb.	Wet Bulb.	
		Inches.	Deg. Cent.	Deg. Cent.	Deg. Cent.
May 19th .... At Halifax	Noon.	29·84	8·9	7·2	...
	1	...	...	...	...
	2	29·88	6·7	6·1	...
	3	...	...	...	...
	4	...	...	...	...
	5	...	...	...	...
	6	29·92	5·5	5·0	...
	7	...	...	...	...
	8	29·99	4·7	4·2	4·4
	9	...	...	...	4·2
	10	30·01	4·1	3·6	3·9
	11	...	...	...	3·9
	Midn.	30·02	3·9	3·3	3·9
May 20th	1	...	...	...	4·4
	2	30·03	3·9	2·5	4·4
	3	...	...	...	4·4
	4	30·07	3·9	3·0	4·4
	5	...	...	...	4·7
	6	30·14	4·4	3·9	4·7
	7	...	...	...	...
	8	30·13	7·0	4·4	4·7
	9	...	...	...	...
	10	30·20	7·2	5·6	4·7
	11	...	...	...	...
	Noon.	30·22	7·0	5·6	4·7
	1	...	...	...	4·7
	2	30·24	7·2	6·2	4·9
	3	...	...	...	5·0
	4	30·28	8·9	7·5	5·5
	5	...	...	...	5·3
	6	30·30	7·2	6·7	5·3
	7	...	...	...	5·3
	8	30·32	6·4	5·8	5·5
	9	...	...	...	5·5
	10	30·36	5·8	5·6	5·5
	11	...	...	...	5·0
	Midn.	30·36	5·6	5·6	5·5
May 21st	1	...	...	...	5·3
	2	30·37	5·5	5·3	5·3
	3	...	...	...	5·0
	4	30·38	5·0	5·0	5·0
	5	...	...	...	5·0
	6	30·43	6·1	5·8	...
	7	...	...	...	...
	8	30·49	7·2	5·7	7·2
	9	...	...	...	7·2

Date and Position.	Hours.	Barometer.	Temperature of Air.		Temperature of Sea Surface.
			Dry Bulb.	Wet Bulb.	
		Inches.	Deg. Cent.	Deg. Cent.	Deg. Cent.
	10	30·48	8·9	7·8	7·5
	11	...	...	...	8·0
Lat. 42° 19′ N. Long. 63° 30′ W.	Noon.	30·48	9·4	8·3	8·0
	1	...	...	...	...
	2	30·48	11·1	8·9	8·9
	3	...	...	...	...
	4	30·48	10·5	8·9	9·2
	5	...	...	...	8·3
	6	30·47	9·4	7·8	8·3
	7	...	...	...	8·3
	8	30·47	8·9	8·0	8·3
	9	...	...	...	8·4
	10	30·47	8·3	7·8	8·6
	11	...	...	...	8·6
	Midn.	30·46	8·9	7·9	8·9
May 22nd	1	...	...	...	8·9
	2	30·47	8·9	7·8	8·9
	3	...	...	...	10·0
	4	30·45	9·4	8·9	10·0
	5	...	...	...	11·1
	6	30·46	10·5	8·9	12·5
	7	...	...	...	14·2
	8	30·49	12·8	10·8	13·9
	9	...	...	...	13·9
	10	30·48	13·9	11·7	13·9
	11	...	...	...	14·4
Lat. 41° 19′ N. Long. 63° 11′ W.	Noon.	30·47	15·5	12·8	15·2
	1	...	...	...	15·3
	2	30·42	15·3	12·8	15·8
	3	...	...	...	16·4
	4	30·39	15·0	12·5	16·7
	5	...	...	...	17·0
	6	30·39	15·0	13·3	17·0
	7	...	...	...	17·0
	8	30·38	15·5	14·2	16·4
	9	...	...	...	14·4
	10	30·38	15·4	14·8	15·3
	11	...	...	...	14·4
	Midn.	30·38	15·5	13·9	12·2
May 23rd	1	...	...	...	13·2
	2	30·29	18·3	16·4	20·8
	3	...	...	...	21·0
	4	30·26	18·9	17·2	21·6
	5	...	...	...	21·6

THE GULF-STREAM.

Date and Position.	Hours.	Barometer.	Temperature of Air.		Temperature of Sea Surface.
			Dry Bulb.	Wet Bulb.	
		Inches.	Deg. Cent.	Deg. Cent.	Deg. Cent.
	6	30·30	19·4	18·0	21·6
	7	...	...	...	21·4
	8	30·28	19·4	18·6	21·6
	9	...	...	...	19·3
	10	30·27	19·4	17·2	19·6
	11	...	...	...	19·6
Lat. 39° 44' N. Long. 63° 22' W.	Noon.	30·25	20·0	18·9	19·7
	1	...	...	...	19·4
	2	30·21	20·8	19·7	19·7
	3	...	...	...	20·0
	4	30·20	21·1	20·0	20·0
	5	...	...	...	20·0
	6	30·15	21·1	20·3	19·7
	7	...	...	...	19·4
	8	30·15	21·1	20·6	20·0
	9	...	...	...	20·6
	10	30·15	21·4	20·8	20·3
	11	...	...	...	20·0
	Midn.	30·15	21·4	20·8	21·4
May 24th	1	...	...	...	22·0
	2	30·14	21·4	20·6	21·8
	3	...	...	...	21·8
	4	30·12	20·7	20·1	19·4
	5	...	...	...	19·4
	6	30·10	21·1	20·6	21·7
	7	...	...	...	22·8
	8	30·09	...	...	23·1
	9	...	...	...	22·8
	10	30·09	21·7	21·1	22·8
	11	...	...	...	22·8
Lat. 38° 22' N. Long. 63° 36' W.	Noon.	30·10	21·7	21·1	22·7
	1	...	...	...	22·5
	2	30·06	22·2	21·6	22·7
	3	...	...	...	22·8
	4	30·08	22·0	21·4	22·8
	5	...	...	...	22·8
	6	30·08	21·4	20·8	22·8
	7	...	...	...	22·5
	8	30·06	21·1	20·8	21·1
	9	...	...	...	19·4
	10	30·11	21·7	21·1	21·7
	11	...	...	...	22·2
	Midn.	30·11	21·1	21·1	22·2
May 25th	1	...	...	...	22·2

Date and Position.	Hours.	Barometer.	Temperature of Air.		Temperature of Sea Surface.
			Dry Bulb.	Wet Bulb.	
		Inches.	Deg. Cent.	Deg. Cent.	Deg. Cent.
	2	30·04	20·3	20·3	19·4
	3	...	...	...	19·2
	4	30·04	20·2	20·1	19·2
	5	...	...	...	19·2
	6	30·02	20·0	20·0	19·0
	7	...	...	...	18·0
	8	30·05	20·2	20·2	19·2
	9	...	...	...	19·4
	10	30·03	20·6	20·6	19·8
	11	...	...	...	19·8
Lat. 37° 9′ N. Long. 62° 30′ W.	Noon.	30·06	21·7	21·4	19·8
	1	...	...	...	19·8
	2	30·07	22·8	22·2	20·0
	3	...	...	...	20·6
	4	30·08	23·1	22·2	20·9
	5	...	...	...	20·9
	6	30·14	22·8	21·7	20·9
	7	...	...	...	20·2
	8	30·10	21·7	21·1	20·1
	9	...	...	...	20·1
	10	30·10	21·4	21·1	19·7
	11	...	...	...	20·0
	Midn.	30·11	21·7	21·4	19·7
May 26th	1	...	...	...	19·2
	2	30·13	21·7	20·6	19·2
	3	...	...	...	19·8
	4	30·13	22·0	21·7	21·4
	5	...	...	...	21·4
	6	30·18	22·2	21·7	22·8
	7	...	...	...	21·7
	8	30·22	22·5	21·8	22·2
	9	...	...	...	22·5
	10	30·23	22·8	22·0	22·5
	11	...	...	...	22·5
Lat. 36° 30′ N. Long. 63° 40′ W.	Noon.	30·24	23·9	22·0	22·8
	1	...	...	...	22·2
	2	30·23	23·9	22·8	23·1
	3	...	...	...	23·1
	4	30·22	23·9	22·8	23·1
	5	...	...	...	22·9
	6	30·24	23·3	22·2	23·1
	7	...	...	...	22·8
	8	30·25	22·2	21·7	22·8
	9	...	...	...	22·3

THE GULF-STREAM.

Date and Position.	Hours.	Barometer.	Temperature of Air.		Temperature of Sea Surface.
			Dry Bulb.	Wet Bulb.	
		Inches.	Deg. Cent.	Deg. Cent.	Deg. Cent.
	10	30·28	22·2	21·7	22·2
	11	...	...	...	22·2
	Midn.	30·26	22·2	21·7	22·2
May 27th	1	...	...	...	22·5
	2	30·23	22·0	21·4	22·5
	3	...	...	...	22·5
	4	30·24	21·4	21·1	22·5
	5	...	...	...	21·1
	6	30·28	21·1	21·1	22·0
	7	...	...	...	22·0
	8	30·32	21·4	21·4	22·0
	9	...	...	...	21·7
	10	30·30	21·4	21·4	21·4
	11	...	...	...	21·7
Lat. 34° 51′ N. Long. 63° 59′ W.	Noon.	30·30	22·3	21·6	21·7
	1	...	...	...	22·0
	2	30·28	22·1	21·7	22·2
	3	...	...	...	22·2
	4	30·28	22·3	22·0	22·2
	5	...	...	...	22·2
	6	30·27	22·8	22·2	22·2
	7	...	...	...	22·2
	8	30·27	22·0	21·7	21·7
	9	...	...	...	21·1
	10	30·28	21·7	21·6	21·5
	11	...	...	...	21·2
	Midn.	30·29	21·7	21·2	21·2
May 28th	1	...	...	...	21·1
	2	30·29	21·1	20·8	21·0
	3	...	...	...	20·8
	4	30·29	21·4	21·1	20·6
	5	...	...	...	...
	6	30·30	21·4	21·4	20·5
	7	...	...	...	20·8
	8	30·33	21·5	21·4	21·2
	9	...	...	...	21·4
	10	30·33	23·3	22·8	21·4
	11	...	...	...	21·4
Lat. 33° 20′ N. Long. 64° 37′ W.	Noon.	30·32	23·9	22·8	21·4
	1	...	...	...	...
	2	30·31	23·6	22·2	21·2
	3	...	...	...	...
	4	30·31	23·3	22·2	21·7
	5	...	...	...	...

Date and Position.	Hours.	Barometer.	Temperature of Air.		Temperature of Sea Surface.
			Dry Bulb.	Wet Bulb.	
		Inches.	Deg. Cent.	Deg. Cent.	Deg. Cent.
May 29th	6	30·30	23·9	23·2	22·0
	7	...	...	...	...
	8	30·28	23·0	22·8	21·7
	9	...	...	...	...
	10	30·30	22·8	22·8	21·7
	11	...	...	...	...
	Midn.	30·32	22·2	22·0	22·2
	1	...	...	...	...
	2	30·29	22·7	22·5	22·7
	3	...	...	...	...
	4	30·30	22·8	22·2	22·2
	5	...	...	...	...
	6	30·32	23·3	22·8	22·0
	7	...	...	...	...
	8	30·33	23·0	22·9	22·2
	9	...	...	...	...
	10	...	...	...	22·0
	11	...	...	...	...
Off Bermudas . . .	Noon.	30·36	24·1	23·4	22·5
	1	...	...	...	...
	2	30·34	24·4	23·2	22·2
	3	...	...	...	...
	4	...	...	...	...
	5	...	...	...	...
	6	30·33	23·9	23·5	22·2
	7	...	...	...	...
	8	30·35	23·4	23·4	...
	9	...	...	...	...
	10	30·35	23·9	23·4	...
	11	...	...	...	...
	Midn.	30·35	23·3	23·3	...

END OF VOL. I.